The Numerical Universe

By Anthony Morris

The grand book of the universe... was written in the language of mathematics, and its characters are triangles, circles and other geometrical figures, without which it is impossible to understand a single word of it. - GALILEO GALILEI, 1623.

We don't have to believe, we can know.

Copyright © 2017 Anthony Morris

All rights reserved.

ISBN-13: 978-1522930860

Anthony Morris

DEDICATION

To my family, and for everyone else's.

ACKNOWLEDGMENTS

I acknowledge the great work and support of a great many others in producing this book, including, in no particular order:

John Michell, Walter Russell, James Heyworth, Bonnie Gaunt, Dr. Mark Leaning, Dr. Jordi Soler, Vernon Jenkins, Vic Showell, Stuart Mitchell, Randy Powell, Marcus Weston, Dr. Jean Claude Perez, Professor Ken Ono, Dr. Rupert Sheldrake, Plato, Pythagoras, Aldous Huxley, Arthur Young, Aryeh Kaplan, Michael Schneider, Johannes Kepler, Karen Elkins, Dr. Talal Ghannam, Tim & Liz Farrell, Graham Hancock, Dr. Carl Calleman, Marko Rodin, John Stuart Reid, Richard McGough, Dale Pond, Scott Onstott, Clay Taylor, Professor Jay Kappraff, Professor Leon Crickmore, Martin Doutre, Richard Heath, Robin Heath, Julian Shelbourne, Rich Merrick, Mark Rossi, Dr. Peter Plichta, Michael Joyce, Neil Shotton, Stephen Wells, Malcolm Rutherford, Christine Rhone, Eamonn Loughran, and Alex Syed.

My thanks also to the various Publishers who have allowed me to reproduce the work of others:- Inner Traditions, W M Norton, Thames & Hudson and The University of Science and Philosophy.

CONTENTS

Dedication		i
Acknowledgements		pg 1
Contents		pg 2
Authors Note		pg 3
Introduction	In Quest of the Canon by John Michell.	pg 4

Part 1 - The Numerical Structure.

Chapter 1	Modular Arithmetic	pg 7
Chapter 2	Number Pairs and Number Groups	pg 9
Chapter 3	Important Number Sequences	pg 11
Chapter 4	Music	pg 20
Chapter 5	The Re-Classification of Prime and Composite Numbers	pg 30
Chapter 6	Numbers and the Numerical Structure of Reality	pg 39
Chapter 7	The Vortex Theory of Atoms	pg 44
Chapter 8	Canonical Numbers and Mod 99	pg 48
Chapter 9	Recurring Decimals	pg 52

Part 2 – DNA & Amino Acids – The Game of Life.

Chapter 1	Introduction to DNA	pg 55
Chapter 2	DNA Analysis	pg 58
Chapter 3	Side Chain Analysis	pg 59
Chapter 4	Standard Block Analysis	pg 62
Chapter 5	Combined Analysis	pg 64
Chapter 6	Caveat and First Proof	pg 75

Part 3 – Integer Partition Theory.

Chapter 1	Introduction	pg 78
Chapter 2	792 Partitions of Number 21	pg 81
Chapter 3	Origins - Integer Partition Theory	pg 83
Chapter 4	Locked Potentials for the Number 21	pg 86
Chapter 5	Key Numbers - 12 and 21	pg 97
Chapter 6	Key Numbers - 27 and 37	pg 101
Chapter 7	Key Numbers - 55	pg 110
Chapter 8	Key Numbers - 37 and 73	pg 118
Chapter 9	Key Numbers - 19 and 91	pg 123
Chapter 10	Key Numbers - 27 and 72	pg 125
Chapter 11	Preceding Fulcra Numbers	pg 129
Chapter 12	Number 11 - 19 v 37 - Unlocked v Locked Potentials	pg 131
Chapter 13	Inside the Integer Partition Table	pg 134
Chapter 14	The 11 Octaves of Physical Reality	pg 141
Chapter 15	Remaining Hologram Projector Numbers for Number 21	pg 147
Chapter 16	Remaining Hologram Projector Numbers for 12th Stage Value	pg 152
Chapter 17	Conway's Game of Life	pg 156
Chapter 18	The Numerical World Soul	pg 168
Chapter 19	Insights and Predictions About the Universe	pg 229
Chapter 20	A Holographic Universe	pg 231
Chapter 21	Conclusion	pg 240
	About the Author	pg 243

Author's Note

The Numerical Universe sets out to show that there exists a primordial, numerical, geometric and musical structure to the Universe. The proposal is simply that there is only order in the universe; that there is no chaos and that we are not all here by virtue of some incredible stroke of luck. The universe is instead shown to be the effect produced by a perfectly balanced, always in equilibrium, numerical order, inherent to the decimal system of numbers 0 to 9.

A complete understanding of this system of Number is the gateway to both a perfect understanding of everything in the universe and a personal enlightenment. We do not have to believe in a creation, we can know, absolutely.

Schumpeter's idea of 'Creative Destruction' which describes the process of an inherent and incessant revolution in the economic structure, forever destroying the old one, constantly creating the new one, is again ushering in the perennial philosophy advanced by Plato and Pythagoras which will once more lead us out of the darkness. The perennial philosophy tends to reassert itself in time of political chaos as something that is immutable and enduring. When the true nature of Nature is revealed so will it in turn lead to a return to living in harmony with Nature and the Universe instead of obviating natural cycles with Man's technology. It is ignorance that causes war and divisiveness among Men, nothing more.

The book starts with a look at the numerical structure and how the decimal system of numbers work in specific pairs and groups, making use of modular arithmetic - the reader won't need any formal mathematical training to follow the arguments and analysis. I then move on to analyse the group of 20 amino acids which are common to all life and which reveals the canonical, numerical, geometric and ultimately musical biological template that pervades the created Universe.

In the final section I demonstrate how the Integer Partition Table of Numbers might be the numerical algorithm used to create and structure the whole of the universe. Numbers are seen to organise themselves in such a way that they unfold as an elegant geometric integration to produce a holographic universe.

Hard evidence of this ancient knowledge may be found in analysis of the Bible, the Great Pyramid, Stonehenge and all the megalithic sites, that are shouting this numerical canon that it may never be lost.

The theory offers extraordinary new insights into the central question of natural philosophy, the origin of the Fine Structure Constant, the force that essentially holds us and the universe together and that stops us from flying apart, and the famous number 137, that has so obsessed some of the greatest thinkers of the 20th Century.

Enduring mysteries concerning Prime Numbers, Photosynthesis, Plato, Dante's 55 Stelle, the 153 Fish in the Bible, and the SATOR Square recovered from the ruins of Pompeii - all can be readily explained and understood by application of this number theory which can potentially project our learning and advancement like no theory yet conceived.

From a religious point of view I can honestly say that I have no axe to grind. I was brought up Catholic and seriously disillusioned in formative years to the point of rejection of Catholic doctrine but not the rejection of the God that resides in me (and that resides in us All), and which connection had always been very strongly felt in me as a child and now in adulthood.

I don't think of myself as a scientist or as an academic. I am just very curious, with a wide range of interests and have always been attracted to Numbers, but not necessarily the mathematics I was taught. It is unfortunate that I was never exposed to Pythagoras and Plato during my education and thus had no appreciation of the quality of numbers until 2010.

In this effort I have pieced together the work of many great men and women and it is by standing upon their shoulders that this theory of everything can come to you at all. Otherwise, I have simply let intuition be my guide and have been rewarded by a flow of incredible synchronicity that has allowed me to progress the theory to this stage.

I do not claim to have all the answers by any means but my hope is that my work will provide the inspiration for the research of many others much cleverer than I. To me it feels like I have opened a crack in a door that has been shut for a very long time.

To set the scene for the whole of this work I can think of no better introduction than the one written by John Michell in his brilliant book, The Dimensions of Paradise, with which all of my work is strongly aligned.

Introduction.

In Quest of the Canon by John Michell.

'Ancient science was based, like that of today, on number, but whereas number is now used in the quantitative sense for secular purposes, the ancients regarded numbers as symbols of the universe, finding parallels between the inherent structure of number and all types of form and motion. Theirs was a very different view of the world from that which now obtains. They inhabited a living universe, a creature of divine fabrication, designed in accordance with reason and thus to some extent comprehensible by the human mind.

The special regard paid to mathematical studies in the ancient world arose from the understanding that number is the mean term in the progression from divine reason to its imperfect reflection in humanity. At some very early period, by a process quite beyond explanation, certain groups of numbers were brought together and codified. Thus was created that numerical standard, or canon of proportion, which was at the root of all ancient cultures and was everywhere attributed to some form of miraculous revelation. It was taken to be the nucleus and activating principle of number generally, a summary of all the types of progressions and relationships that occur within the field of number and thus a faithful image of the numerically created universe.

In the known civilizations of antiquity, such as China, Babylon, and Egypt, the canon of number was venerated as the source of all knowledge and a guide to rightful conduct. Its influence extended from art and music to affairs of state. Every branch of science expressed its theories and observations in terms of that same small group of numbers which are investigated in this book. One numerical code has fashioned the whole of ancient mathematics, music, astronomy, chronology, metrology, and every variety of craft. It has left its mark on every relic and tradition of ancient cultures. There is nothing artificial about it, for the conclusion to these researches is that the various orders of natural phenomena do indeed conform to certain similar patterns of number, which also provide the framework of number itself. This allows the eventual reconstitution of that scientific standard which supported the fabric of ancient societies over periods of time that, by modern reckoning, seem remarkably long.

The author is frankly partial to the traditional order of philosophy associated in the West with Plato but also expressed in every different culture. It is called idealistic because it is concerned with causes rather than effects and with ideal forms rather than appearances. It is also called perennial, meaning that it grows naturally in the human mind and blossoms at certain seasons. The reason for its constant, universal recurrence is that it is mathematically based. Thus it provides a most realistic view of the world, balanced and made fully human by its transcendental aspect, the traditional doctrine of the soul's immortality.

The number series that is demonstrated throughout this book not only was the source of all traditional arts and sciences but also gave birth to the system of philosophy adopted by Plato. That philosophy occurs of its own accord in the mind of whoever studies these numbers, their relationships, and their applications in different branches of nature. Through such studies the Pythagorean dictum All Is Number comes to life, and thus is opened a new outlook on the world in general. This new outlook, which is not in fact new but traditional, has certain consequences, discussed in the final chapter. It opens possibilities for the future, when the lack of a common, humanistic, scientific standard in affairs has become even more glaring than it is now, and necessity compels the search for some universal guiding principle.

To that future this book is dedicated. Its contents may seem in part dense and obscure. That is mainly due to the author's lack of ability to do justice to this worthiest of all subjects. But it is due also to the modern eclipse of the traditional worldview that gives significance to these studies. The science here unfolded is of no obvious relevance to the modern world, and the type of philosophy that goes with it has been supplanted by other, temporary orthodoxies. To give sense and context to the following studies in the ancient canon of number, the traditional sciences relating to it are briefly described in separate chapters, with reference to the grand alchemical science to which they all contributed and the cosmological outlook that engendered them. This last is the most important, for the purpose and methods of the old sciences are apparent only in the light of traditional cosmology. That light is still hidden below the horizon of modern consciousness, but it can never be extinguished. And when next it arises, demanding the forms of science appropriate to it, the subjects raised in this book will be once again of paramount importance.

This book is the outcome of its author's quest, pursued over many years, for the legendary key to universal knowledge alluded to in esoteric traditions and early texts. Its point of departure was Plato's statement in the Laws that the Egyptian priests possessed a canon of lawful proportions and harmonies, by means of which their civilized standards had been preserved uncorrupted for literally thousands of years. The discovery and maintenance of true cultural standards was the main theme of Plato's own writings. His scheme for a well-governed city, described in his Laws, was based on a certain numerical formula, often referred to but specified by only one of its components, the number 5,040. From this and his other mathematical allusions, the inference is that Plato himself had studied the laws of harmony he attributed to the Egyptians. In pure mathematical form, those laws were made the cornerstone of his proposed reforms in education and politics. The following chapters on Platonic number show how the laws of harmony were expressed numerically, as the dimensions of a city, a scale of music, or intervals in astronomy. In all his cosmological demonstrations, Plato used the same set of numbers and similar geometrical diagrams, applying them to such apparently different things as music and the order of the planets, and thus illustrating his belief that number is the "natural bond" that holds together the entire universe.

One of the conclusions from this study is that Plato's symbolic arithmetic was not a contemporary discovery but a heritage from the distant past. The ground plan of his imaginary City consists of the same combined shapes and numbers as the Stonehenge plan laid down some 1,500 years earlier. Their common units of measure were derived from the same archetype, the numerical image of the cosmos. In chapter 3, the ancient units are analyzed and given their exact values, from which it appears that their lengths represent subdivisions of certain basic standards. The standards referred to are not in the first instance physical: they are indeed reflected in the actual dimensions of the earth and the solar system, but in essence they are purely numerical. And the numbers that express ancient units of length are the same as those that denote the scales of traditional music. The forms of music and measure known to Plato were defined and codified thousands of years before his time. Their common source was the canon of number, which Plato either learned wholly from certain teachers or partly reconstructed. His own concern as a would-be reformer and cultural revivalist was to renew the influence of the canon and make it once more effective as an instrument for universal harmony and stability.

It was probably through the Mystery schools, the select institutions of scholarship and mystical inquiry in classical and early Christian times, that the esoteric tradition which Plato drew on was passed down to the founders of Christianity. They were the cause of its last flourishing, soon to be blighted. St. John's New Jerusalem, a visionary form of Plato's ideal City and numerically identical to it, was a token of the "new heaven and new earth" that the prophet foresaw as the issue of renaissance through Christianity and the restoration of the true cosmic standard. Early Christian traditions are particularly useful through indicating the relative meanings and importance attributed to those numbers that occur in earlier sacred contexts. Thus, for instance, we learn of the supreme significance to Christian mystics of the number 3,168, which was also the paramount number in the ancient canon. Those Christian sects who practiced the numerical theology claimed that, through assimilating the pagan science, Christianity had become the legitimate heir to the ancient religious tradition.

The significance of this present subject can be summed up in many different ways. To artists, architects, and musicians, the study of number and proportion has been of traditional interest, and when the current vogue for novelty and individualism has run its course, it will become so again. With scientists of all disciplines, the case is similar. Bereft of guidance by any common philosophy, their researches and products are determined by the whims of commerce, militarism, national pride, and similar vanities; and the world of scholarship is likewise dominated by faddish intellectualism. Thus are created weird, aberrant thought-forms and monstrous manifestations. At such periods of philosophical anarchy, says Plato in the Republic, when there is no common means of distinguishing between beneficial and destructive products, popular demand arises for a standard of judgment. This is usually answered by some tyrant with his own prescription for standards. The demand, of course, is for an objective standard, one that is rooted in nature and reflects no particular theory or ideology. With this consideration begins the quest for the venerable cosmic standard or unified world image that is numerically structured to represent, in essence, the entire universe.'

The Dimensions of Paradise by John Michell published by Inner Traditions International and Bear & Company, ©2008. All rights reserved. http://www.Innertraditions.com Reprinted with permission of publisher.

I had originally intended to try to avoid bringing the overtly emotional subjects of Religion, God, Gematria, and Kabbalah to any of this work as I felt that it might instantly turn a great many people off. However, as the research has progressed it has become obvious that what I am working on will instantly clarify other philosophical research which is correct and true, including these areas which are so clearly in line with what I have uncovered throughout the research.

This has allowed me to reveal the hidden meanings in mysteries that are millennia old - these are covered in the final section of my work in more depth. Now, while the numerical work might bear out key aspects of Kabbalah say, this does not mean necessarily that the scholars who have interpreted it have done a good job. I am no expert and not the person to judge. However, what we can say is that the basic structure is correct.

We don't have to believe. We can know. This is a numerical universe and everything can be decoded with a full understanding of the decimal system and how numbers operate as a clear and congruent system to produce an innate order in the Universe, even if this order is beyond our own egoic perception which has been conditioned to believe in chaos.

The Number system is the glue that underpins the operation of all things, the true nature of Nature, the divine intelligence.

Part 1 – The Numerical Structure

"Simplicity is the ultimate sophistication." – Leonardo Da Vinci

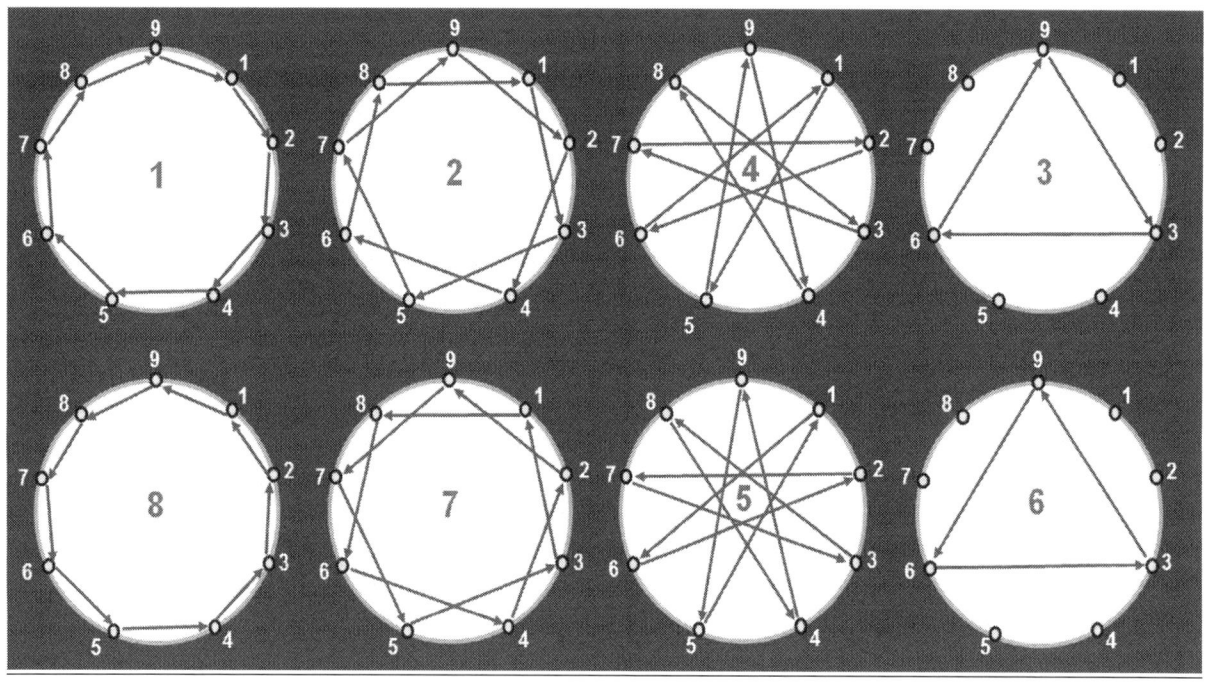

Chapter 1: Modular Arithmetic

My story begins on Wednesday September 1st 2010, a day I will never forget.

After a sabbatical following a long and arduous contract with a leading hedge fund during the financial crisis, I was back to my desk and considering the best way forward for my career. I had been sat down for just 15 minutes when I happened across a quote attributed to Nikola Tesla that changed my life:

'If you only knew the power of the 3 6 and 9 you would have the power of the Universe at your disposal.'

I sat there perplexed as I took the quote in. It didn't make any sense. I knew a little about Tesla, enough to know he was thought to be an exceptional genius, one of the greatest to walk the Earth in modern times.

I googled the quote and for 3 weeks solid, in a zone like trance, for 18 hours a day, I became transfixed by this enigma, quickly finding my way to a rather obscure and esoteric, fringe, branch of mathematics named Vortex Based Mathematics (VBM) pioneered by the eccentric Marko Rodin and brought forward by Randy Powell who was a particular rock of support throughout the very early stages of this work. Vortex Base Mathematics is certainly well outside mainstream academia and yet I knew they were on to something very deep and very important.

Very briefly, VBM can be defined as the study of resonance and ratio that attempts to identify the natural flow of energy, modelled on what is called a torus, a geometric object that looks like a doughnut.

The Torus

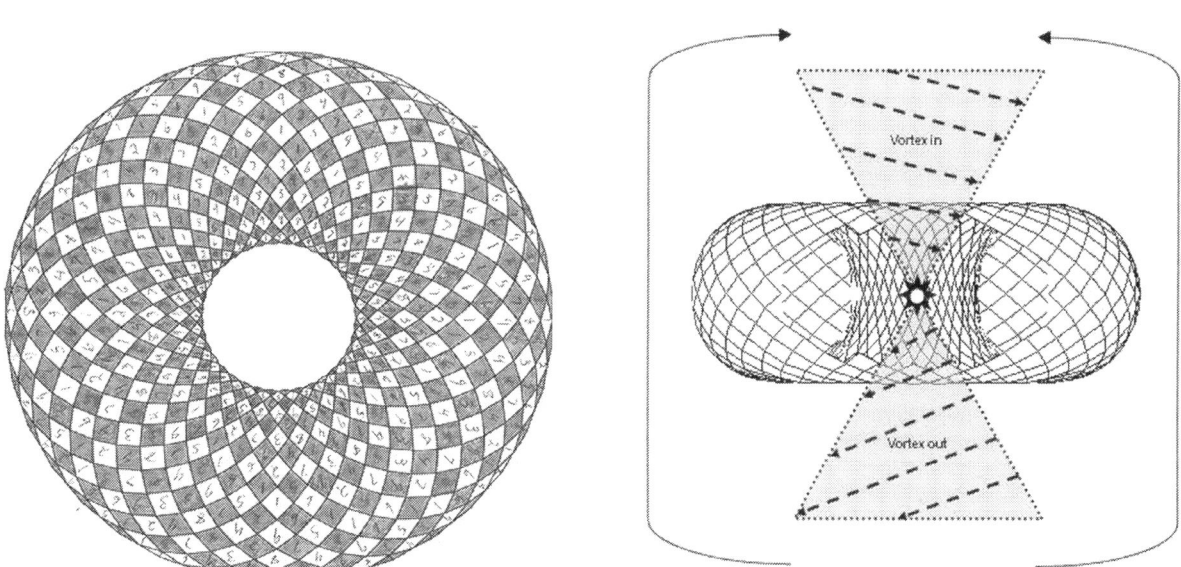

I have no scientific background whatsoever, and being totally ignorant about how electricity and magnetism work, I found it hard to take in what they were talking about. However, what I really zeroed in, and what really caught my attention, was their use of Modular Arithmetic, also known as clock arithmetic or as I prefer, the arithmetic of remainders. Of particular interest was Mod 9 which is central to VBM and of course the only Modular base to include all the numbers we have - 0 1 2 3 4 5 6 7 8 9.

It was this that has led me to more of an understanding of what Tesla may have meant by that quotation above, and this research is dedicated to that end.

His quote unequivocally infers that Numbers are real in some causal way and it is in understanding the archetypal properties or 'idiosyncrasies' of each Number 0 to 9 that we can potentially know, and not guess at, how everything works, everywhere in the Universe.

It will hopefully become clear that Mod 9 can and must be used to extract useful information from data available to us that would not otherwise be readily identifiable and that reveals, clearly, the causal explanation for the effects we see and measure.

Included here as evidence for my assertions, I start by showing some of the basic VBM work pioneered by Marko Rodin and progressed by Randy Powell and others, which covers Cellular Mitosis and part of the section on the Fibonacci Sequence.

This will then lead on to my original applications of Mod 9 to other Integer Sequences, to Musical tuning scales, to Prime Numbers, to Music and ultimately to Particle Physics and the numerical and geometric basis of Light.

It has been quite a journey.

Throughout this section I want to show the patient reader how the numbers we have, 0 to 9 behave specifically as a system that pervades all of our natural world. In order to do this I am going to use modular arithmetic which is much simpler than it sounds.

In mathematics, modular arithmetic is a system of arithmetic for integers (the term used for whole numbers), where numbers "wrap around" after reaching a certain value - the modulus (plural moduli).

The modern approach to modular arithmetic was developed by Carl Friedrich Gauss in his book Disquisitiones Arithmeticae, published in 1801.

One use of modular arithmetic that everyone will be familiar with is in the 12-hour clock, where the day is divided into two 12-hour periods.

If the time is 7 o'clock now, then 8 hours later it will be 3 o'clock. Usual addition would suggest that the later time should be 7 + 8 = 15, but this is not the answer because clock time "wraps around" every 12 hours. Because the hour number starts over after it reaches 12, this is arithmetic, modulus 12.

I like to think of modular arithmetic as the arithmetic of remainders. To illustrate modular arithmetic very simply, let us take the number 12 as an example:

In Mod 9 (Mod being short for Modulus), the number 12 is 3 because 9 goes into 12 once, leaving a remainder of 3.

In mathematical terms we write 12 = 3 Mod 9.

35 = 8 Mod 9 because 9 goes into 35 three times, leaving a remainder of 8.

When working in other Moduli the answer would be achieved in the same way.

For example, take that number 35 again, but this time in Mod 8.

8 goes into 35 four times and leaves a remainder of 3 so we would write 35 = 3 Mod 8.

There is a neat trick for quickly computing the Mod 9 answer for any number: - By simply adding up the digits of the number we can ascertain the answer. So for 35 we would simply go 3 + 5 = 8 Mod 9 and the answer is correct. If we take the number 1231 for example, the answer would just be 1 + 2 + 3 + 1 = 7 Mod 9. A number greater than 9 as a sum of its digits would be treated in the same way so that 3459 for example would be 3 + 4 + 5 + 9 = 21 = 2 + 1 = 3 Mod 9.

While the mathematician might argue that the patterns I show in the following sections can be found whatever modulus one might use, for the purposes of my research I am only interested in Mod 9 because of its exclusive properties where all of our numbers are included and utilised. A lower modulus, for example Mod 6 can only make use of numbers 0 to 6 which obviously leaves out the numbers 7, 8 and 9.

As we will see, Mod 9 shows how the decimal system of numbers behaves as a very specific system from which we can infer deep insights into the nature of Nature.

Chapter 2: Number Pairs and Number Groups.

Multiplication Tables – Mod 9.

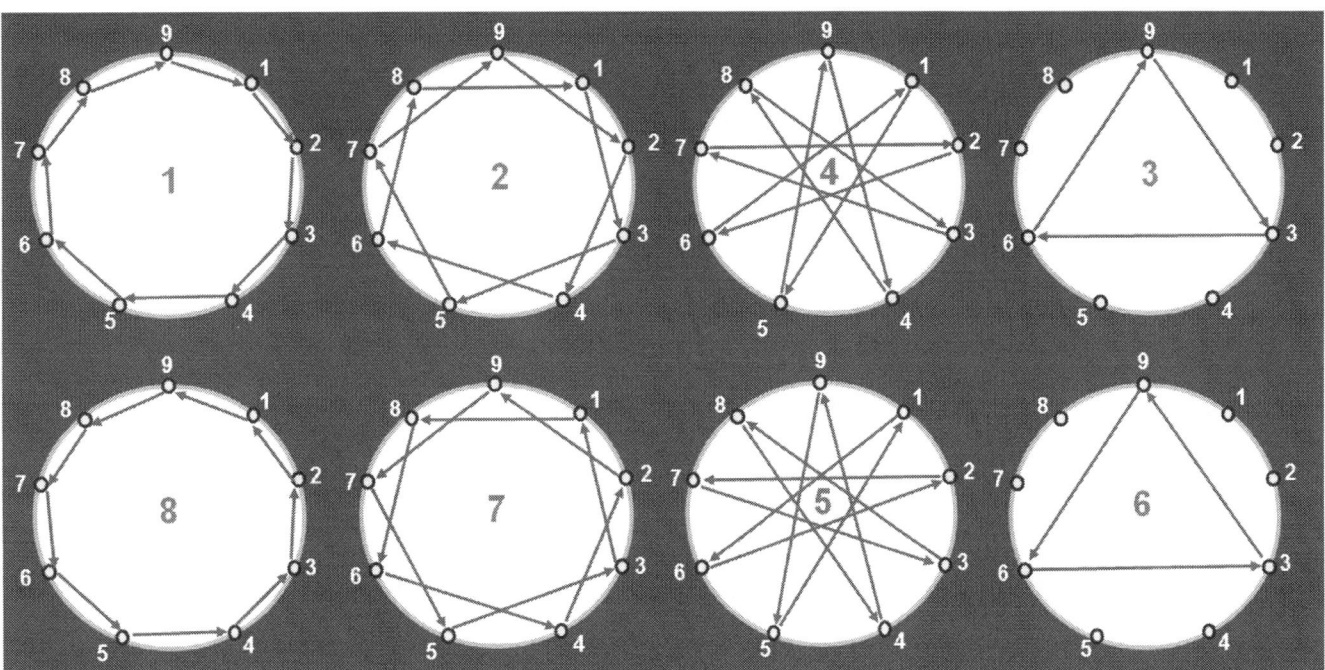

In the graphic above I have used a circle as a control for the 9 natural numbers of our decimal system and then used the multiplication tables for each number. When we get a result, for example, 4 x 4, it gives the natural result 16. Mod 9 is then applied to the result to get the answer. So, in this case, 4 x 4 = 16 = 7 Mod 9 because 9 goes into 16 leaving a remainder of 7.

To create the geometries you see above, we simply go from result to result for each number's multiplication table and join the lines up to form geometric shapes. Above we can clearly see the geometric qualities exhibited by each of the first 8 numbers when we apply Mod 9.

For those who are new to modular arithmetic and were wondering what happened to the 9 and 0? Well, whatever Modular base you work in will necessarily result back to itself or zero as there is no remainder. In Mod 9, the Number 9 is paired with the Number 0.

To clarify absolutely how we get to the above geometries, let's start with the 1 times table: 1 x 1 = 1, 2 x 1 = 2, 3 x 1 = 3 etc. You can see that the one times table creates a geometric shape called a nonagon and you can see that the direction of the flow is clockwise.

Now the 2 times table goes 2, 4, 6, 8, 10 is 1 (Mod 9), 12 is 3 (Mod 9), 14 is 5 (Mod 9) etc. Here we have a different geometry. Again, we see the flow is in a clockwise direction.

The 4 times table goes 4, 8, 12 is 3 (Mod 9), 16 is (7 Mod 9), 20 is 2 (Mod 9) etc. It produces a 9 pointed star and again flowing in a clockwise direction.

The 3 times table goes 3, 6, 9, 12 is 3 (Mod 9) 15 is 6 (Mod 9), 18 is 9 (Mod 9) and so on, creating the triangle and again the flow is clockwise.

The 8 times table goes 8, 16 is 7 (Mod 9) 24 is 6 (Mod 9), 32 is 5 (Mod 9), 40 is 4 (Mod 9) etc. Notice the 8 shares the same geometry as the 1, except it is flowing anticlockwise.

Similarly for the 7, 5 and 6 which produce the same geometry as the 2, 4 and 3 respectively.

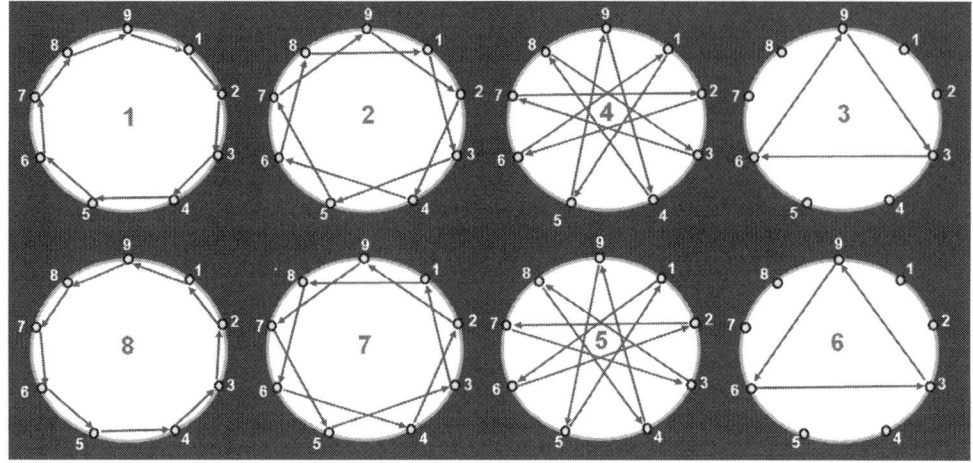

We can see clearly that the numbers are paired geometrically and have a Positive / Negative, Clockwise / Anti-clockwise, relationship. These form what are called the Number Pairs. Mod 9 is the only modular base system that displays this kind of symmetry where **ALL** the numbers we have 0 to 9 are paired off geometrically to form 5 pairs.

$$\text{1 and 8} \quad \text{2 and 7} \quad \text{4 and 5} \quad \text{3 and 6} \quad \text{0 and 9}$$

Two Distinct 'Circuits'

You can see from the graphic that there are 2 distinct 'circuits' at work here by noticing that the geometries of the numbers 1 2 4 5 7 and 8 all touch each of the 9 numbers around the circle, whereas the 3 6 and 9 keep to themselves. Seeing numbers pictorially in this way was very new and interesting to me as I am sure it will be to many readers.

Number Groups - Numerical Analysis.

Here I looked at the permutations of the group of 3 digit numbers that make up each Number Group to see

3 6 9 (0)

$$369 + 396 + 963 + 936 + 639 + 693 = 3996 = 9 \text{ or } 0 \text{ Mod9}$$

Note $3996 = 108 \times 37$

$$360 + 306 + 630 + 036 + 603 + 063 = 1998 = 9 \text{ or } 0 \text{ Mod9}$$

Note $1998 = 54 \times 37$

2 5 8

$$258 + 285 + 528 + 582 + 825 + 852 = 3330 = 9 \text{ or } 0 \text{ Mod9}$$

Note $3330 = 90 \times 37$

1 4 7

$$147 + 174 + 741 + 714 + 471 + 417 = 2664 = 9 \text{ or } 0 \text{ Mod9}$$

Note $2664 = 72 \times 37$

Chapter 3: Important Number Sequences.

Cellular Mitosis.

Essential to life, cellular mitosis is essentially the process of cell division that allows humans to grow and for cells to repair themselves. It gives the following sequence: 1, 2, 4, 8, 16, 32, 64, 128, 256, 512, etc. – i.e. the cells simply double and double and double.

When we apply Mod 9 to this basic sequence we see a regular repeating sequence of 6 numbers emerge.

The Mod 9 results are in brackets.

$$1\ 2\ 4\ 8\ 16\ (7)\ 32\ (5)$$
$$1\ 2\ 4\ 8\ 7\ 5 \quad \text{Mod 9}$$
$$64\ (1)\ 128\ (2)\ 256\ (4)\ 512\ (8)\ 1024\ (7)\ 2048\ (5)$$
$$1\ 2\ 4\ 8\ 7\ 5 \quad \text{Mod 9}$$
$$4096\ (1)\ 8192\ (2)\ 16384\ (4)\ 32768\ (8)\ 65536\ (7)\ 131072\ (5)$$
$$1\ 2\ 4\ 8\ 7\ 5 \quad \text{Mod 9}$$

When we arrange this 6 number sequence in two rows of 3 numbers with the first three above the second three and then read vertically, we see the Number Pairs that were produced by the multiplication tables for each number in each column. I call this two dimensional numerical symmetry.

$$1\ 2\ 4$$
$$8\ 7\ 5$$

Further, when we arrange this sequence in two rows of 3 numbers and read vertically we see what are called the Number Groups. I call this three dimensional numerical symmetry. We will see more of this going forward in other Integer sequences and Prime Numbers.

$$1\ 2$$
$$4\ 8$$
$$7\ 5$$

As we saw from the multiplication table results Mod 9, one of the circuits 1 2 4 5 7 8 is represented by this particular sequence. This circuit breaks down into two of the three distinct number groups 1 4 and 7, and 2 5 and 8.

The other circuit in the multiplication table results Mod 9, represented by the 3 and 6 is not represented in this sequence but will appear further along in other important sequences. These two numbers are joined by the interoperable 0 and 9 to form the third Number Group.

To recap I have shown the numbers behaving in Pairs and Groups.

5 Number Pairs

1 and 8   2 and 7   4 and 5   3 and 6   0 and 9

3 Number Groups

1 4 7   2 5 8   3 6 9 (0)

The Fibonacci Sequence - Origins.

The Fibonacci sequence appears in Indian mathematics where the clearest exposition of the sequence arises in the work of Virahanka (c. 700 AD), whose own work is lost, but is available in a quotation by Gopala (c. 1135):

From Wikipedia – 'Outside India, the Fibonacci sequence first appears in the book *Liber Abaci* (1202) by Fibonacci after whom the sequence is named. Fibonacci considered the growth of an idealised, though biologically unrealistic rabbit population, assuming that: a newly born pair of rabbits, one male, one female, are put in a field; rabbits are able to mate at the age of one month so that at the end of its second month a female can produce another pair of rabbits; rabbits never die and a mating pair always produces one new pair (one male, one female) every month from the second month on. The puzzle that Fibonacci posed was: how many pairs will there be in one year?

- At the end of the first month, they mate, but there is still only 1 pair.
- At the end of the second month the female produces a new pair, so now there are 2 pairs of rabbits in the field.
- At the end of the third month, the original female produces a second pair, making 3 pairs in all in the field.
- At the end of the fourth month, the original female has produced yet another new pair, and the female born two months ago also produces her first pair, making 5 pairs.

At the end of the nth month, the number of pairs of rabbits is equal to the number of new pairs (that is, the number of pairs in month $n - 2$) plus the number of pairs alive last month (that is, $n - 1$). This is the nth Fibonacci number.'

The Fibonacci sequence is found everywhere in Nature. Below we see it in the hand, the sequence displayed as a spiral and the sunflower.

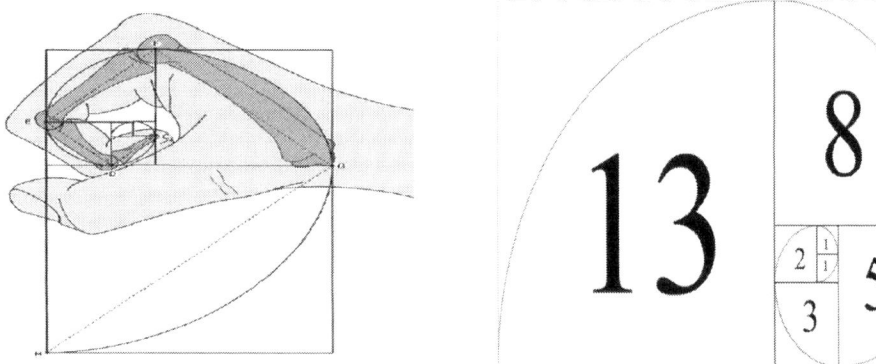

Source: www.creationresearches.com

https://www.flickr.com/photos/lucapost/694780262

THE FIBONACCI SEQUENCE

Term	Fibonacci Sequence	Mod 9
1	1	1
2	1	1
3	2	2
4	3	3
5	5	5
6	8	8
7	13	4
8	21	3
9	34	7
10	55	1
11	89	8
12	144	9
13	233	8
14	377	8
15	610	7
16	987	6
17	1597	4
18	2584	1
19	4181	5
20	6765	6
21	10946	2
22	17711	8
23	28657	1
24	46368	9
25	75025	1
26	121393	1
27	196418	2
28	317811	3
29	514229	5
30	832040	8
31	1346269	4
32	2178309	3
33	3524578	7
34	5702887	1
35	9227465	8
36	14930352	9
37	24157817	8
38	39088169	8
39	63245986	7
40	102334155	6
41	165580141	4
42	267914296	1
43	433494437	5
44	701408733	6
45	1134903170	2
46	1836311903	8
47	2971215073	1
48	4807526976	9

Having more than a passing acquaintance with Fibonacci and the Golden Mean through my work on financial markets, I immediately looked at the sequence in a new light having thrown the Mod 9 lens over it.

For those that do not know, the Fibonacci sequence is formed by adding the previous two terms of the sequence to get the next term. It begins:-

0, 1, 1, 2, 3, 5, 8, 13, 21, 34, 55, 89, 144, 233, 377, 610, 987 etc.

It is the organising structure found throughout nature and evident in every living thing and all physical phenomena. As an example it is found throughout the human body, even in our DNA. The DNA molecule, the program for all life, measures 34 angstroms long by 21 angstroms wide for each full cycle of its double helix spiral.

When we divide any term in the sequence by the previous term we approximate to what is called the Phi or Golden Mean or Ratio relationship of 1.618.

When we divide any term in the sequence by the next term we approximate to 0.618, also called phi but with a little p. Further, when we divide any term in the sequence by the term two before it we find that they are in a relationship that approximates to 2.618.

Using Mod 9 analysis of this sequence I discovered to my amazement that if you go far enough out you soon realise that there is an infinitely repeating 24 number sequence:

1,1,2,3,5,8,4,3,7,1,8,9,8,8,7,6,4,1,5,6,2,8,1,9

Arranged as 2 lines of 12 digits we can see the Number Pairs

1 1 2 3 5 8 4 3 7 1 8 9
8 8 7 6 4 1 5 6 2 8 1 9

Arranged as 3 lines of 8 digits we can see the Number Groups

1 1 2 3 5 8 4 3
7 1 8 9 8 8 7 6
4 1 5 6 2 8 1 9

Arranged as 6 lines of 4 digits we can see the Doubling Circuit that we see in the mitosis squence in the First Column, the Number Pair 1 and 8 in the second, another Doubling Circuit, and finally the 3 6 9 Number Group.

1 1 2 3
5 8 4 3
7 1 8 9
8 8 7 6
4 1 5 6
2 8 1 9

Other Natural Integer Sequences.

Below I have applied the same technique to all the other important number sequences, using Mod 9, to find repeating cycles of numbers, or moduli, underpinning them all and all either showing the Number Pairs or Number Groups in two or three dimensional numerical symmetry.

Term	Lucas	Mod 9	Pell	Mod 9	Jacobsthal	Mod 9	Padovan	Mod 9
1	2	2	1	1	1	1	1	1
2	1	1	2	2	1	1	0	9
3	3	3	5	5	3	3	0	9
4	4	4	12	3	5	5	1	1
5	7	7	29	2	11	2	0	9
6	11	2	70	7	21	3	1	1
7	18	0	169	7	43	7	1	1
8	29	2	408	3	85	4	1	1
9	47	2	985	4	171	9	2	2
10	76	4	2378	2	341	8	2	2
11	123	6	5741	8	683	8	3	3
12	199	1	13860	9	1365	6	4	4
13	322	7	33461	8	2731	4	5	5
14	521	8	80782	7	5461	7	7	7
15	843	6	195025	4	10923	6	9	9
16	1364	5	470832	6	21845	2	12	3
17	2207	2	1136689	7	43691	5	16	7
18	3571	7	2744210	2	87381	9	21	3
19	5778	0	6625109	2	174763	1	28	1
20	9349	7	15994428	6	349525	1	37	1
21	15127	7	38613965	5	699051	3	49	4
22	24476	5	93222358	7	1398101	5	65	2
23	39603	3	225058681	1	2796203	2	86	5
24	64079	8	543339720	9	5592405	3	114	6
25	103682	2	1311738121	1	11184811	7	151	7
26	167761	1	3166815962	2	22369621	4	200	2
27	271443	3	7645370045	5	44739243	9	265	4
28	439204	4	18457556052	3	89478485	8	351	9
29	710647	7	44560482149	2	178956971	8	465	6
30	1149851	2	107578520350	7	357913941	6	616	4
31	1860498	0	259717522849	7	715827883	4	816	6
32							1081	1
33							1432	1
34							1897	7
35							2513	2
36	MOD 9 REPEATING SEQUENCES						3329	8
37	HIGHLIGHTED IN TAN AND YELLOW						4410	9
38							5842	1
39							7739	8
40							10252	1
41							13581	9
42							17991	9
43							23833	1
44							31572	9
45							41824	1

Lucas Sequence - Mod 9
24 digit recurring sequence Mod 9 showing the Number Pairs.

2 1 3 4 7 2 0 2 2 4 6 1 7 8 6 5 2 7 0 7 7 5 3 8

2 1 3 4 7 2 0 2 2 4 6 1

7 8 6 5 2 7 0 7 7 5 3 8

Pell Numbers - Mod 9
24 digit recurring sequence Mod 9 showing the Number Pairs.

1 2 5 3 2 7 7 3 4 2 8 9 8 7 4 6 7 2 2 6 5 7 1 9

1 2 5 3 2 7 7 3 4 2 8 9

8 7 4 6 7 2 2 6 5 7 1 9

Jacobsthal Numbers - Mod 9
18 digit recurring sequence Mod 9 showing the Number Pairs.

1 1 3 5 2 3 7 4 9 8 8 6 4 7 6 2 5 9

1 1 3 5 2 3 7 4 9

8 8 6 4 7 6 2 5 9

Padovan Sequence - Mod 9
39 digit recurring sequence Mod 9 showing the Number Groups.

1 9 9 1 9 1 1 1 2 2 3 4 5 7 9 3 7 3 1 1 4 2 5 6 7 2 4 9 6 4 6 1 1 7 2 8 9 1 8

1 9 9 1 9 1 1 1 2 2 3 4 5

7 9 3 7 3 1 1 4 2 5 6 7 2

4 9 6 4 6 1 1 7 2 8 9 1 8

The Lucas Sequence is the same as the Fibonacci series except it begins 2 1 instead of 0 1 that often occurs when working with the Fibonacci series.

The Pell numbers are an infinite sequence of integers that have been known since ancient times, the denominators of the closest rational approximations to the square root of 2. This sequence of approximations begins 1/1, 3/2, 7/5, 17/12, and 41/29, so the sequence of Pell numbers begins with 1, 2, 5, 12, and 29. Pell numbers grow exponentially, proportionally to powers of the silver ratio 1 + sq root 2.

The Jacobsthal sequence starts with 0 and 1, then each following number is found by adding the number before it to twice the number before that.

The Numerical Universe

The Padovan sequence produces the side lengths of a spiral of equilateral triangles.

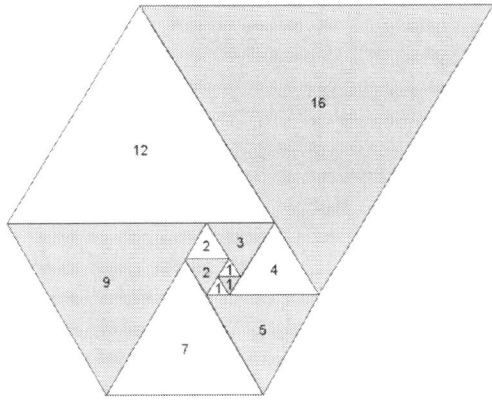

https://commons.wikimedia.org/w/index.php?curid=24501622

Term	Lines	Mod 9	Term	Ellipses	Mod 9	Term	Circles	Mod 9
1	1	1	1	2	2	1	2	2
2	2	2	2	6	6	2	4	4
3	4	4	3	14	5	3	8	8
4	7	7	4	26	8	4	14	5
5	11	2	5	42	6	5	22	4
6	16	7	6	62	8	6	32	5
7	22	4	7	86	5	7	44	8
8	29	2	8	114	6	8	58	4
9	37	1	9	146	2	9	74	2
10	46	1	10	182	2	10	92	2
11	56	2	11	222	6	11	112	4
12	67	4	12	266	5	12	134	8
13	79	7	13	314	8	13	158	5
14	92	2	14	366	6	14	184	4
15	106	7	15	422	8	15	212	5
16	121	4	16	482	5	16	242	8
17	137	2	17	546	6	17	274	4
18	154	1	18	614	2	18	308	2
19	172	1	19	686	2	19	344	2
20	191	2	20	762	6	20	382	4
21	211	4	21	842	5	21	422	8
22	232	7	22	926	8	22	464	5
23	254	2	23	1014	6	23	508	4
24	277	7	24	1106	8	24	554	5
25	301	4	25	1202	5	25	602	8
26	326	2	26	1302	6	26	652	4
27	352	1	27	1406	2	27	704	2
28	379	1	28	1514	2	28	758	2
29	407	2	29	1626	6	29	814	4
30	436	4	30	1742	5	30	872	8
31	466	7	31	1862	8	31	932	5
32	497	2	32	1986	6	32	994	4
33	529	7	33	2114	8	33	1058	5
34	562	4	34	2246	5	34	1124	8
35	596	2	35	2382	6	35	1192	4
36	631	1	36	2522	2	36	1262	2

DIVIDING A PLANE

Maximal Number of Regions into which Lines Divide a Plane.

This sequence returns a repeating 9 digit sequence showing a Palindrome and the 147 Number Group. Palindromic numbers or sequences can be read backwards or forwards.

1 2 4 7   2   7 4 2 1

1 2 4
7 2 7
4 2 1

Maximal Number of Regions into which Ellipses Divide a Plane.

This sequence returns a repeating 9 digit sequence showing a Palindrome and the 258 Number Group.

2 6 5 8   6   8 5 6 2

2 6 5
8 6 8
5 6 2

Maximal Number of Regions into which Circles Divide a Plane.

This sequence returns a repeating 9 digit sequence showing a Palindrome and the 258 Number Group.

2 4 8 5   4   5 8 4 2

2 4 8
5 4 5
8 4 2

Term	Spheres	Mod 9	Term	Planes	Mod 9
1	0	0	1	1	1
2	2	2	2	2	2
3	4	4	3	4	4
4	8	8	4	8	8
5	16	7	5	15	6
6	30	3	6	26	8
7	52	7	7	42	6
8	84	3	8	64	1
9	128	2	9	93	3
10	186	6	10	130	4
11	260	8	11	176	5
12	352	1	12	232	7
13	464	5	13	299	2
14	598	4	14	378	0
15	756	0	15	470	2
16	940	4	16	576	0
17	1152	0	17	697	4
18	1394	8	18	834	6
19	1668	3	19	988	7
20	1976	5	20	1160	8
21	2320	7	21	1351	1
22	2702	2	22	1562	5
23	3124	1	23	1794	3
24	3588	6	24	2048	5
25	4096	1	25	2325	3
26	4650	6	26	2626	7
27	5252	5	27	2952	0
28	5904	0	28	3304	1
29	6608	2	29	3683	2
30	7366	4	30	4090	4
31	8180	8	31	4526	8
32	9052	7	32	4992	6
33	9984	3	33	5489	8
34	10978	7	34	6018	6
35	12036	3	35	6580	1
36	13160	2	36	7176	3
37	14352	6	37	7807	4
38	15614	8	38	8474	5
39	16948	1	39	9178	7
40	18356	5	40	9920	2
41	19840	4	41	10701	0
42	21402	0	42	11522	2
43	23044	4	43	12384	0

DIVIDING SPACE

Maximal Number of Regions into which Spheres Divide Space

This sequence returns a repeating 27 digit Mod 9 sequence.

0 2 4 8 7 3 7 3 2 6 8 1 5 4 0 4 0 8 3 5 7 2 1 6 1 6 5

0 2 4 8 7 3 7 3 2

6 8 1 5 4 0 4 0 8

3 5 7 2 1 6 1 6 5

Maximal Number of Regions into which Planes Divide Space

This sequence returns a repeating 27 digit Mod 9 sequence.

1 2 4 8 6 8 6 1 3 4 5 7 2 0 2 0 4 6 7 8 1 5 3 5 3 7 0

1 2 4 8 6 8 6 1 3

4 5 7 2 0 2 0 4 6

7 8 1 5 3 5 3 7 0

Notice the Number Groups displaying this three dimensional numerical symmetry.

Geometric Sequences & Figurate Numbers.

The mathematical study of figurate numbers is said to have originated with Pythagoras, possibly based on Babylonian or Egyptian precursors. Figuracy of number is something all children should be taught about early on, so that they form a relationship with numbers through the visual nature of geometry as opposed to a meaningless symbol good only for use at the shops.

A number can be described as figurate if a regular geometric shape may be made up from the available quantity of building blocks or counters. For example, 3 blocks or counters may be arranged in a triangle, four in a square, etc. This gives rise to sequences of numbers, triangular, square, pentagonal etc.

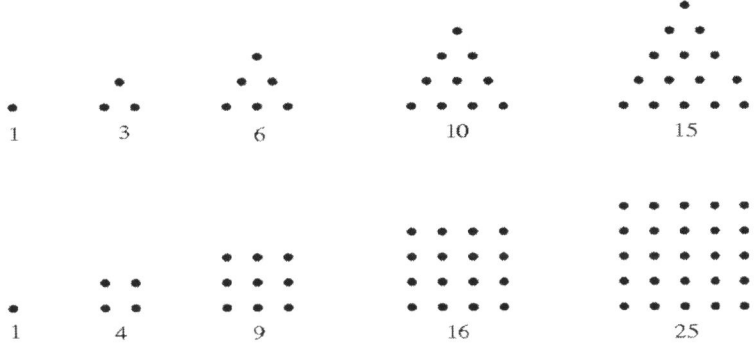

Some numbers can be represented by two or more regular shapes. These are known as poly-figurate numbers. If two shapes can be created, the number is called Bi-figurate. There are however only two Tri-Figurate Numbers, numbers which can make three regular geometric shapes.

37 as Centred Hexagon, Hexagram and Centred Octagon
91 as Triangle, Centred Hexagon and 5 layer Square Pyramid

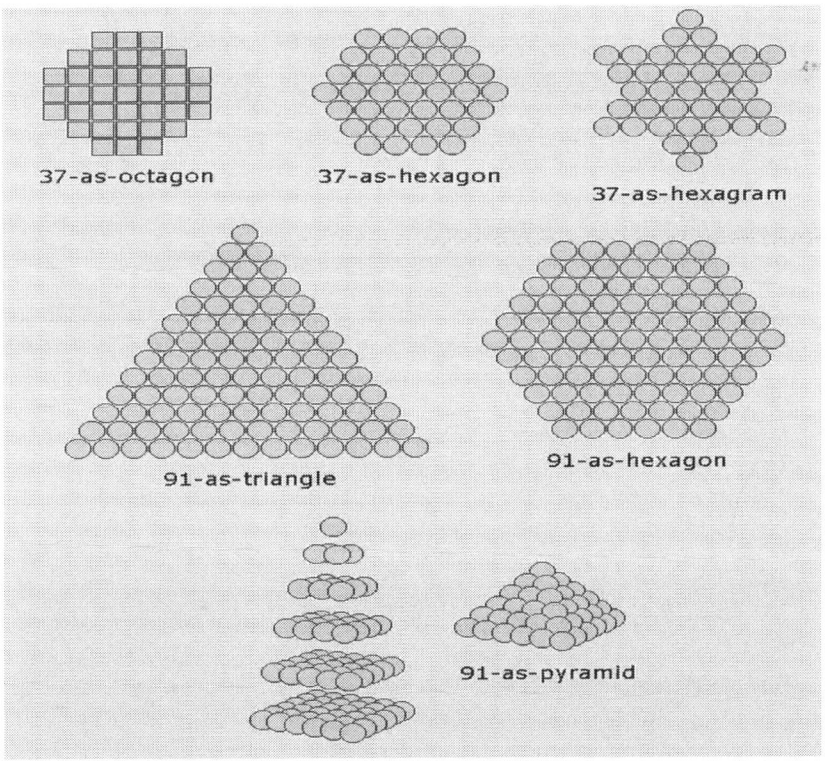

http://homepage.virgin.net/vernon.jenkins/Symb.htm

Below we can see that all figurate numbers display a 9 number sequence that repeats Mod 9. Notice again that each of these sequences possess the three dimensional numerical symmetry that display the Number Groups.

Term	Triangle	Mod 9	Square	Mod 9	Pentagon	Mod 9	Hexagon	Mod 9	Heptagon	Mod 9	
1	1	1	1	1	1	1	1	1	1	1	**Triangle Numbers**
2	3	3	4	4	5	5	6	6	7	7	1 3 6 1 6 3 1 0 0
3	6	6	9	0	12	3	15	6	18	0	
4	10	1	16	7	22	4	28	1	34	7	1 3 6
5	15	6	25	7	35	8	45	0	55	1	1 6 3
6	21	3	36	0	51	6	66	3	81	0	1 0 0
7	28	1	49	4	70	7	91	1	112	4	**Square Numbers**
8	36	0	64	1	92	2	120	3	148	4	1 4 0 7 7 0 4 1 0
9	45	0	81	0	117	0	153	0	189	0	
10	55	1	100	1	145	1	190	1	235	1	1 4 0
11	66	3	121	4	176	5	231	6	286	7	7 7 0
12	78	6	144	0	210	3	276	6	342	0	4 1 0
13	91	1	169	7	247	4	325	1	403	7	**Pentagonal Numbers**
14	105	6	196	7	287	8	378	0	469	1	1 5 3 4 8 6 7 2 0
15	120	3	225	0	330	6	435	3	540	0	
16	136	1	256	4	376	7	496	1	616	4	1 5 3
17	153	0	289	1	425	2	561	3	697	4	4 8 6
18	171	0	324	0	477	0	630	0	783	0	7 2 0
19	190	1	361	1	532	1	703	1	874	1	**Hexagonal Numbers**
20	210	3	400	4	590	5	780	6	970	7	1 6 6 1 0 3 1 3 0
21	231	6	441	0	651	3	861	6	1071	0	
22	253	1	484	7	715	4	946	1	1177	7	1 6 6
23	276	6	529	7	782	8	1035	0	1288	1	1 0 3
24	300	3	576	0	852	6	1128	3	1404	0	1 3 0
25	325	1	625	4	925	7	1225	1	1525	4	**Heptagonal Numbers**
26	351	0	676	1	1001	2	1326	3	1651	4	1 7 0 7 1 0 4 4 0
27	378	0	729	0	1080	0	1431	0	1782	0	
											1 7 0
											7 1 0
											4 4 0

Term	Octagon	Mod 9	Nonagon	Mod 9	Decagon	Mod 9	Hendecagonal	Mod 9	Dodecagon	Mod 9	
1	1	1	1	1	1	1	1	1	1	1	**Octagonal Numbers**
2	8	8	9	0	10	1	11	2	12	3	1 8 3 4 2 6 7 5 0
3	21	3	24	6	27	0	30	3	33	6	
4	40	4	46	1	52	7	58	4	64	1	1 8 3
5	65	2	75	3	85	4	95	5	105	6	4 2 6
6	96	6	111	3	126	0	141	6	156	3	7 5 0
7	133	7	154	1	175	4	196	7	217	1	**Nonagonal Numbers**
8	176	5	204	6	232	7	260	8	288	0	1 0 6 1 3 3 1 6 0
9	225	0	261	0	297	0	333	0	369	0	
10	280	1	325	1	370	1	415	1	460	1	1 0 6
11	341	8	396	0	451	1	506	2	561	3	1 3 3
12	408	3	474	6	540	0	606	3	672	6	1 6 0
13	481	4	559	1	637	7	715	4	793	1	**Decagonal Numbers**
14	560	2	651	3	742	4	833	5	924	6	1 1 0 7 4 0 4 7 0
15	645	6	750	3	855	0	960	6	1065	3	
16	736	7	856	1	976	4	1096	7	1216	1	1 1 0
17	833	5	969	6	1105	7	1241	8	1377	0	7 4 0
18	936	0	1089	0	1242	0	1395	0	1548	0	4 7 0
19	1045	1	1216	1	1387	1	1558	1	1729	1	**Hendecagonal Numbers**
20	1160	8	1350	0	1540	1	1730	2	1920	3	1 2 3 4 5 6 7 8 0
21	1281	3	1491	6	1701	0	1911	3	2121	6	
22	1408	4	1639	1	1870	7	2101	4	2332	1	1 2 3
23	1541	2	1794	3	2047	4	2300	5	2553	6	4 5 6
24	1680	6	1956	3	2232	0	2508	6	2784	3	7 8 0
25	1825	7	2125	1	2425	4	2725	7	3025	1	**Dodecagonal Numbers**
26	1976	5	2301	6	2626	7	2951	8	3276	0	1 3 6 1 6 3 1 0 0
27	2133	0	2484	0	2835	0	3186	0	3537	0	
											1 3 6
											1 6 3
											1 0 0

We can see that when we apply Mod 9 analysis to figurate numbers there are 3 distinct groups that are separated by 3.

Triangle Hexagonal Nonagonal Dodecagonal - 3 6 9 12 and so on. The 3 6 9 Grouping.
Square Heptagonal Decagonal - 4 7 10 and so on. The 1 4 7 Grouping.
Pentagonal Octagonal Hendecagonal - 5 8 11 and so on. The 2 5 8 Grouping

Platonic Solids.

The Greek philosopher Plato, born around 430 B.C., wrote about these five solids in a work called Timaeus. Historical accounts vary a bit, but it is usually agreed that the solids were discovered by the early Pythagoreans; perhaps by 450 B.C. Plato associated each of the four classical elements (earth, air, water, and fire) with a regular solid as shown below. Of the fifth Platonic solid, the dodecahedron, Plato obscurely remarks, "...the god used [it] for arranging the constellations on the whole heaven".

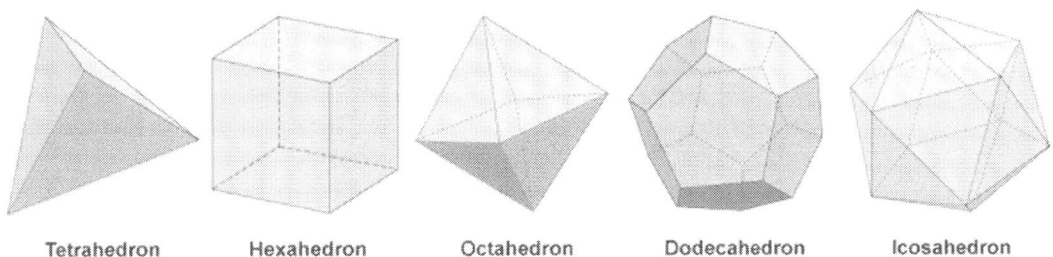

Tetrahedron   Hexahedron   Octahedron   Dodecahedron   Icosahedron

https://www.technologyuk.net/mathematics/geometry/platonic-solids.shtml

The Platonic Solids each show a repeating sequence Mod 9 again illustrating the Number Groups 1 4 7, 2 5 8 and 3 6 9 (0).

Tetrahedral	Mod 9	Octahedral	Mod 9	Cube	Mod 9	Dodecahedral	Mod 9	Icosahedral	Mod 9
1	1	1	1	1	1	1	1	1	1
4	4	6	6	8	8	12	3	20	2
10	1	19	1	27	0	48	3	84	3
20	2	44	8	64	1	124	7	220	4
35	8	85	4	125	8	255	3	455	5
56	2	146	2	216	0	456	6	816	6
84	3	231	6	343	1	742	4	1330	7
120	3	344	2	512	8	1128	3	2024	8
165	3	489	3	729	0	1629	0	2925	0
220	4	670	4	1000	1	2260	1	4060	1
286	7	891	0	1331	8	3036	3	5456	2
364	4	1156	4	1728	0	3972	3	7140	3
455	5	1469	2	2197	1	5083	7	9139	4
560	2	1834	7	2744	8	6384	3	11480	5
680	5	2255	5	3375	0	7890	6	14190	6
816	6	2736	0	4096	1	9616	4	17296	7
969	6	3281	5	4913	8	11577	3	20825	8
1140	6	3894	6	5832	0	13788	0	24804	0
1330	7	4579	7	6859	1	16264	1	29260	1
1540	1	5340	3	8000	8	19020	3	34220	2
1771	7	6181	7	9261	0	22071	3	39711	3
2024	8	7106	5	10648	1	25432	7	45760	4
2300	5	8119	1	12167	8	29118	3	52394	5
2600	8	9224	8	13824	0	33144	6	59640	6
2925	0	10425	3	15625	1	37525	4	67525	7
3276	0	11726	8	17576	8	42276	3	76076	8
3654	0	13131	0	19683	0	47412	0	85320	0
4060	1	14644	1	21952	1	52948	1	95284	1
4495	4	16269	6	24389	8	58899	3	105995	2
4960	1	18010	1	27000	0	65280	3	117480	3
5456	2	19871	8	29791	1	72106	7	129766	4
5984	8	21856	4	32768	8	79392	3	142880	5
6545	2	23969	2	35937	0	87153	6	156849	6
7140	3	26214	6	39304	1	95404	4	171700	7
7770	3	28595	2	42875	8	104160	3	187460	8
8436	3	31116	3	46656	0	113436	0	204156	0

Platonic Solid Moduli

Tetrahedral Numbers

1 4 1 2 8 2 3 3 3 4 7 4 5 2 5 6 6 6 7 1 7 8 5 8 0 0 0

1 4 1 2 8 2 3 3 3
4 7 4 5 2 5 6 6 6
7 1 7 8 5 8 0 0 0

Octahedral Numbers

1 6 1 8 4 2 6 2 3 4 0 4 2 7 5 0 5 6 7 3 7 5 1 8 3 8 0

1 6 1 8 4 2 6 2 3
4 0 4 2 7 5 0 5 6
7 3 7 5 1 8 3 8 0

Cube Numbers

1 8 0

Icosahedral Numbers

1 3 3 7 3 6 4 3 0

1 3 3
7 3 6
4 3 0

Dodecahedral Numbers

1 2 3 4 5 6 7 8 0

1 2 3
4 5 6
7 8 0

Anthony Morris

Chapter 4: Music.

I preface the following by the admission that I have developed little to no musical aptitude as yet and have never studied music theory. However, that does not seem to have stopped me from uncovering what looks to be some very interesting observations having applied Mod 9 to the frequencies generated by the black and white keys of the musical scale.

1955 saw the introduction of the International Standard Tuning of 440 Hz on the A of Middle C Octave. This standardised all instruments around the world. There are 4 tuning systems but the one I looked at was the natural tuning, called Just. Just tuning is deeply connected to the numbers that pervade all of my work.

Standard Tuning - 440 Hz

OCTAVE / NOTE (hz)	C	INT	D	INT	E	F	INT	G	INT	A	INT	B
6 ABOVE	4.125	0.5156	4.6406	0.4219	5.0625	5.5	0.6875	6.1875	0.6875	6.875	0.8594	7.7344
5 ABOVE	8.25	1.0313	9.2813	0.8438	10.125	11	1.375	12.375	1.375	13.75	1.7188	15.469
4 ABOVE	16.5	2.0625	18.563	1.6875	20.25	22	2.75	24.75	2.75	27.5	3.4375	30.938
3 ABOVE	33	4.125	37.125	3.375	40.5	44	5.5	49.5	5.5	55	6.875	61.875
2 ABOVE	66	8.25	74.25	6.75	81	88	11	99	11	110	13.75	123.75
1 ABOVE	132	16.5	148.5	16.5	165	176	22	198	22	220	27.5	247.5
MIDDLE	264	33	297	33	330	352	44	396	44	440	55	495
1 BELOW	528	66	594	66	660	704	88	792	88	880	110	990
2 BELOW	1056	132	1188	132	1320	1408	176	1584	176	1760	220	1980
3 BELOW	2112	264	2376	264	2640	2816	352	3168	352	3520	440	3960
4 BELOW	4224	528	4752	528	5280	5632	704	6336	704	7040	880	7920
5 BELOW	8448	1056	9504	1056	10560	11264	1408	12672	1408	14080	1760	15840
6 BELOW	16896	2112	19008	2112	21120	22528	2816	25344	2816	28160	3520	31680

Mod 9 Analysis

OCTAVE / NOTE	C	INT	D	INT	E	F	INT	G	INT	A	INT	B
6 ABOVE	3	6	9	6	6	1	8	9	8	8	1	9
5 ABOVE	6	3	9	3	3	5	4	9	4	4	5	9
4 ABOVE	3	6	9	6	6	7	2	9	2	2	7	9
3 ABOVE	6	3	9	3	3	8	1	9	1	1	8	9
2 ABOVE	3	6	9	6	6	4	5	9	5	5	4	9
1 ABOVE	6	3	9	3	3	2	7	9	7	7	2	9
MIDDLE	3	6	9	6	6	1	8	9	8	8	1	9
1 BELOW	6	3	9	3	3	5	4	9	4	4	5	9
2 BELOW	3	6	9	6	6	7	2	9	2	2	7	9
3 BELOW	6	3	9	3	3	8	1	9	1	1	8	9
4 BELOW	3	6	9	6	6	4	5	9	5	5	4	9
5 BELOW	6	3	9	3	3	2	7	9	7	7	2	9
6 BELOW	3	6	9	6	6	1	8	9	8	8	1	9

The C's were all 3 & 6, same for C sharp, then D's are all 9's and on until I came to F which revealed the 1 2 4 8 7 5 sequence, in order. This was very exciting yet obviously made sense as the one thing I did know was that frequencies double octave to octave.

You will notice that using this tuning at 440 Hz we see that:

5 sections of the octave are 1 2 4 8 7 5

4 sections are 3 & 6

3 sections are 9

Immediately the Pythagorean 3 4 5 triangle springs to mind.

It also became quickly apparent that I had no need to show as many octaves as I have shown opposite because the Mod 9 sequence repeats every 6 octaves or 6 x 12 = 72 notes. 72 is a very important number in my overall theory, as we shall see. For example, the Mod 9 sequence is the same for the Middle octave, the octave 6 Above and the octave 6 Below.

72 NUMBER SEQUENCE

OCTAVE / NOTE	C	INT	D	INT	E	F	VOID	INT	G	INT	A	INT	B
2 ABOVE	3	6	9	6	6	4	9	5	9	5	5	4	9
1 ABOVE	6	3	9	3	3	2	9	7	9	7	7	2	9
MIDDLE	3	6	9	6	6	1	9	8	9	8	8	1	9
1 BELOW	6	3	9	3	3	5	9	4	9	4	4	5	9
2 BELOW	3	6	9	6	6	7	9	2	9	2	2	7	9
3 BELOW	6	3	9	3	3	8	9	1	9	1	1	8	9

As I did with the multiplication tables for the numbers 1 to 8, I decided to draw each of the octaves, again, using the circle as the control to see what geometry they produced and if any information might be gleaned.

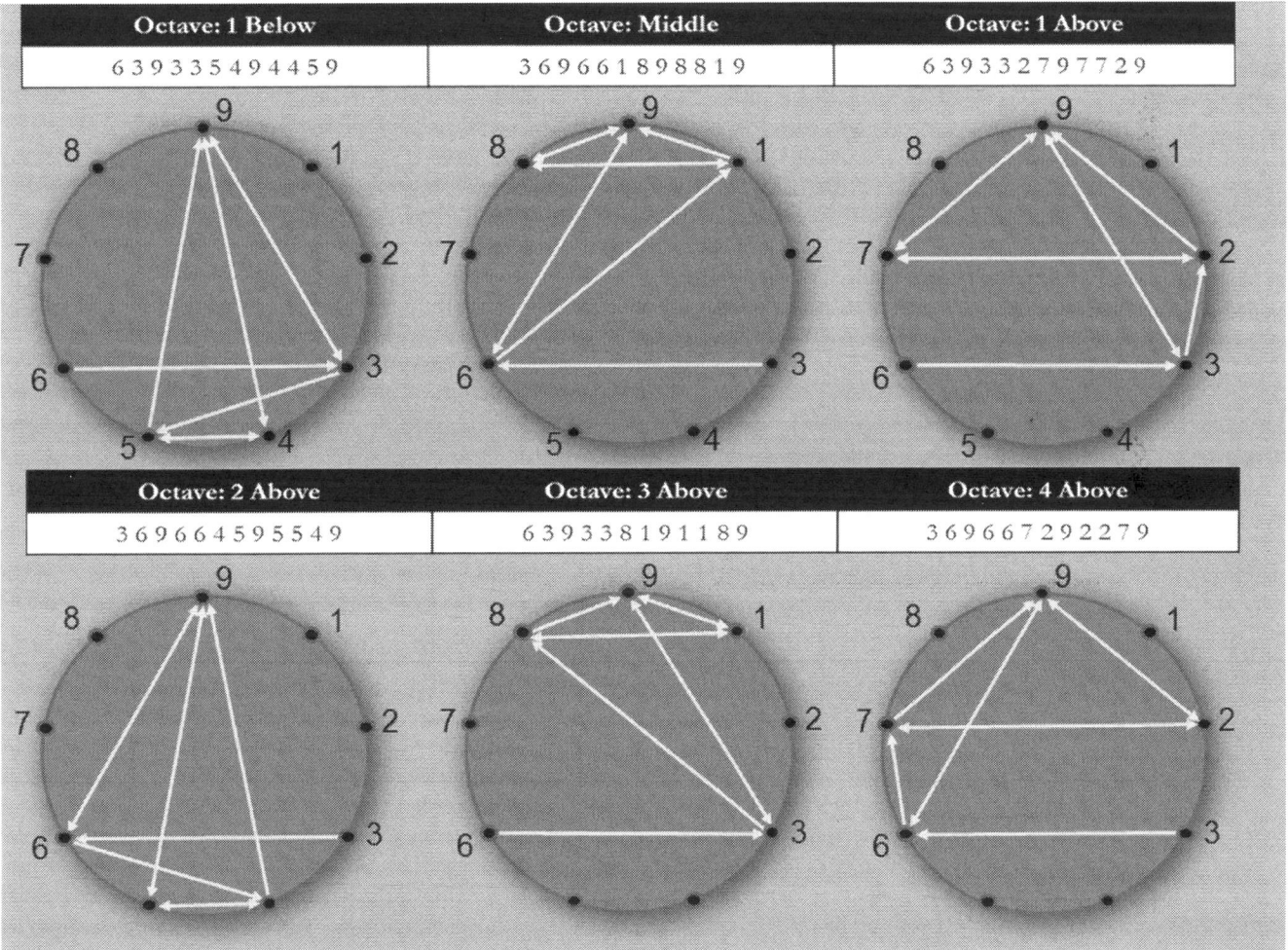

Above, we can clearly see that, as with the numbers, the octaves are paired up symmetrically and reflected, separated by 3 octaves.

For example the octave 1 Below is qualitatively or 'archetypically' the same as 2 Above and the Middle octave is the 'same' as 3 Above but they are mirrors, or reflections, of each other.

Just before the end of 2010 I discovered, in my ignorance, and let me assure you I have plenty of that, that the current International Standard tuning of 440 Hz on Middle A had only been around since 1955. Given what I had discovered so far, I was very interested to investigate what came before and what that would look like under the same microscope.

Interestingly, I discovered that the first proponent of the current standard was none other than Nazi Propaganda chief, Josef Goebbels, in 1939, which was incidentally the same year that televisions began to be mass produced and widely available. The introduction of a major Nazi player aroused my interest as those boys knew exactly what they were doing and had all the brainpower in the world at their disposal. Germans won a ridiculous percentage of all Nobel Prizes prior to 1939.

I naturally wondered what the Nazi motivation was to get it changed and of course now I had to know what had been in use before, that they wanted to get away from. Very quickly, I discovered the Pythagorean Musical Scale, the tuning favoured by Mozart & Bach. This tuning is based on 432Hz on the A of the Middle octave instead of the current standard which is at 440 Hz that we analysed above.

Pythagorean Musical Scale - 432 Hz

OCTAVE / NOTE (hz)	C	INT	D	INT	E	F	INT	G	INT	A	INT	B
6 ABOVE	4	0.5	4.5	0.5625	5.0625	5.333	0.666	6	0.75	6.75	0.8438	7.5938
5 ABOVE	8	1	9	1.125	10.125	10.666	1.333	12	1.5	13.5	1.6875	15.188
4 ABOVE	16	2	18	2.25	20.25	21.333	2.666	24	3	27	3.375	30.375
3 ABOVE	32	4	36	4.5	40.5	42.666	5.333	48	6	54	6.75	60.75
2 ABOVE	64	8	72	9	81	85.333	10.666	96	12	108	13.5	121.5
1 ABOVE	128	16	144	18	162	170.67	21.333	192	24	216	27	243
MIDDLE	256	32	288	36	324	341.33	42.666	384	48	432	54	486
1 BELOW	512	64	576	72	648	682.67	85.333	768	96	864	108	972
2 BELOW	1024	128	1152	144	1296	1365.3	170.67	1536	192	1728	216	1944
3 BELOW	2048	256	2304	288	2692	2730.7	341.33	3072	384	3456	432	3888
4 BELOW	4096	512	4608	576	5184	5461.3	682.67	6144	768	6912	864	7776
5 BELOW	8192	1024	9216	1152	10368	10923	1365.3	12288	1536	13824	1728	15552
6 BELOW	16384	2048	18432	2304	20736	21845	2730.7	24576	3072	27648	3456	31104

Mod 9 Analysis

OCTAVE / NOTE	C	INT	D	INT	E	F	INT	G	INT	A	INT	B
6 ABOVE	4	5	9	9	9	9 (3)	9 (6)	6	3	9	9	9
5 ABOVE	8	1	9	9	9	9 (6)	9 (3)	3	6	9	9	9
4 ABOVE	7	2	9	9	9	9 (3)	9 (6)	6	3	9	9	9
3 ABOVE	5	4	9	9	9	9 (6)	9 (3)	3	6	9	9	9
2 ABOVE	1	8	9	9	9	9 (3)	9 (6)	6	3	9	9	9
1 ABOVE	2	7	9	9	9	9 (6)	9 (3)	3	6	9	9	9
MIDDLE	4	5	9	9	9	9 (3)	9 (6)	6	3	9	9	9
1 BELOW	8	1	9	9	9	9 (6)	9 (3)	3	6	9	9	9
2 BELOW	7	2	9	9	9	9 (3)	9 (6)	6	3	9	9	9
3 BELOW	5	4	9	9	9	9 (6)	9 (3)	3	6	9	9	9
4 BELOW	1	8	9	9	9	9 (3)	9 (6)	6	3	9	9	9
5 BELOW	2	7	9	9	9	9 (6)	9 (3)	3	6	9	9	9
6 BELOW	4	5	9	9	9	9 (3)	9 (6)	6	3	9	9	9

Look how the numbers turn out when we again apply Mod 9 to the frequencies and see the difference in the sections of the octave occur between the two tunings. We see that:-

2 sections that are 1 2 4 8 7 5, versus 5, using the 440 Hz tuning.

2 sections that are 3 6, versus 4, using the 440 Hz tuning.

8 sections that are 9 versus only 3, using the 440 Hz tuning.

Here too I saw there was a 72 number sequence.

72 NUMBER SEQUENCE

OCTAVE / NOTE	C	INT	D	INT	E	F	VOID	INT	G	INT	A	INT	B
2 ABOVE	1	8	9	9	9	9 (3)	9	9 (6)	6	3	9	9	9
1 ABOVE	2	7	9	9	9	9 (6)	9	9 (3)	3	6	9	9	9
MIDDLE	4	5	9	9	9	9 (3)	9	9 (6)	6	3	9	9	9
1 BELOW	8	1	9	9	9	9 (6)	9	9 (3)	3	6	9	9	9
2 BELOW	7	2	9	9	9	9 (3)	9	9 (6)	6	3	9	9	9
3 BELOW	5	4	9	9	9	9 (6)	9	9 (3)	3	6	9	9	9

Again we see that the octaves are paired and mirrored as they were in the 440 Hz tuning scale.

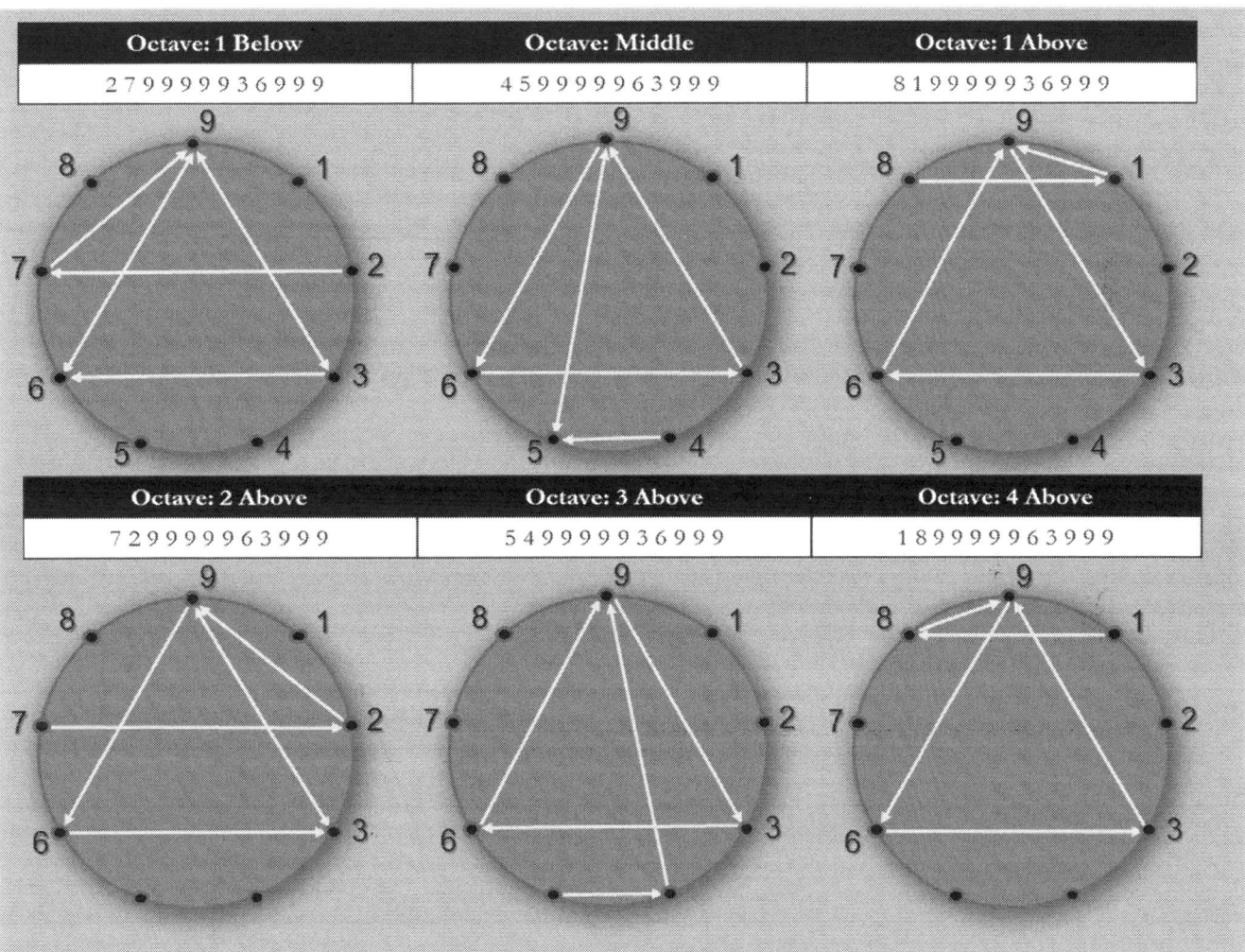

Looking at the respective tuning scales, we can see clearly that there is a lot more 9 and a lot less 1 2 4 8 7 5 and 3 6 in the 432 Hz scale when you compare it to the 440 Hz tuning. My conclusion, using Mod 9, is that this 432 Hz tuning scale is unique in that, it, alone, produces the least amount of electro-magnetic interference versus the largest amount of Source or pure energy. In other words this is the purest direct experience a human being can have of Source and why the works of Bach and Mozart are so imbued with the sublimity that they are.

Now, turning our attention back to the 440 Hz tuning we operate on now. It clearly has too much electro-magnetic interference, with likely, highly negative, disruptive, physical and non-physical side effects.

When we put the octaves together in their respective 72 number sequences we see something interesting.

Here below, we see the geometry created by the 72 number sequence created by the 440 Hz scale.

OCTAVE	C	INT	D	INT	E	F	INT	G	INT	A	INT	B
1 Below	6	3	9	3	3	5	4	9	4	4	5	9
Middle	3	6	9	6	6	1	8	9	8	8	1	9
1 Above	6	3	9	3	3	2	7	9	7	7	2	9
2 Above	3	6	9	6	6	4	5	9	5	5	4	9
3 Above	6	3	9	3	3	8	1	9	1	1	8	9
4 Above	3	6	9	6	6	7	2	9	2	2	7	9

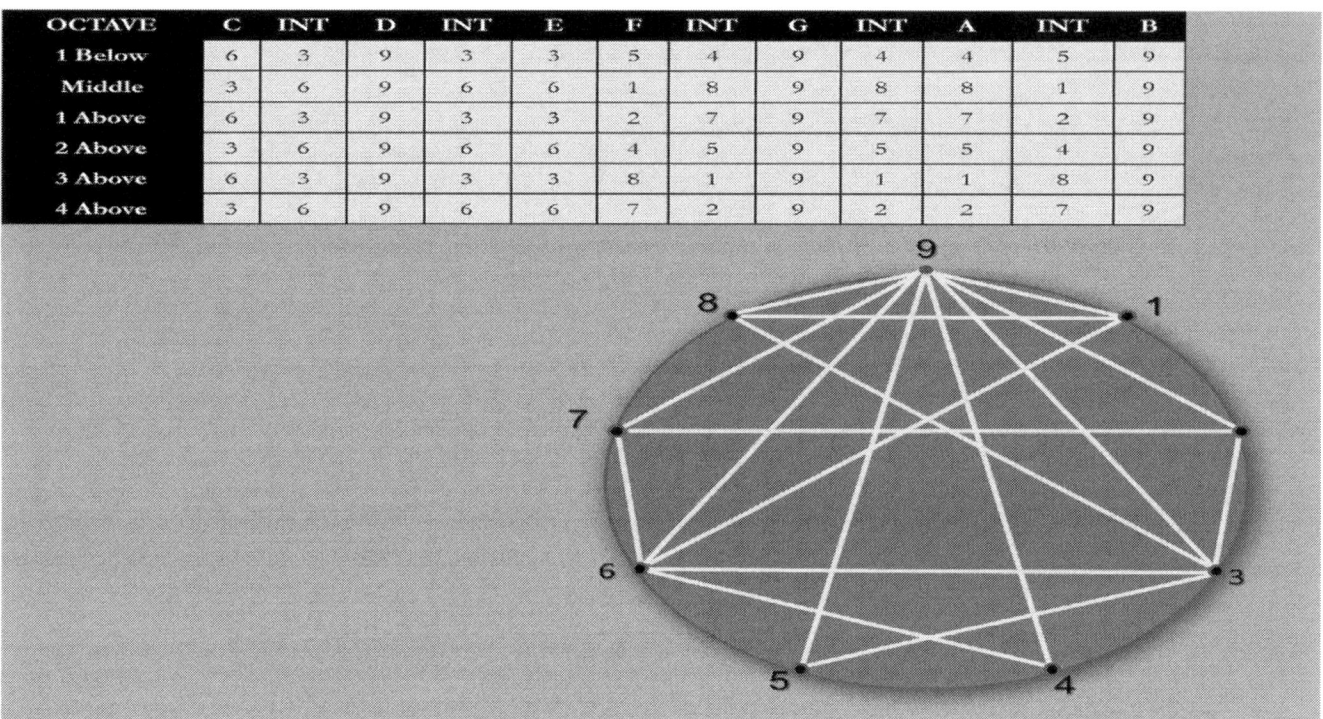

The electro-magnetic interference effects can be seen where we have the perverted 1 6, 8 3, 4 6, 3 5 connections that are removed by the 432 Hz scale.

Below is the geometry created by the 432 Hz scale. Beautiful isn't it?

OCTAVE	C	INT	D	INT	E	F	INT	G	INT	A	INT	B
1 Below	2	7	9	9	9	9	9	3	6	9	9	9
Middle	4	5	9	9	9	9	9	6	3	9	9	9
1 Above	8	1	9	9	9	9	9	3	6	9	9	9
2 Above	7	2	9	9	9	9	9	6	3	9	9	9
3 Above	5	4	9	9	9	9	9	3	6	9	9	9
4 Above	1	8	9	9	9	9	9	6	3	9	9	9

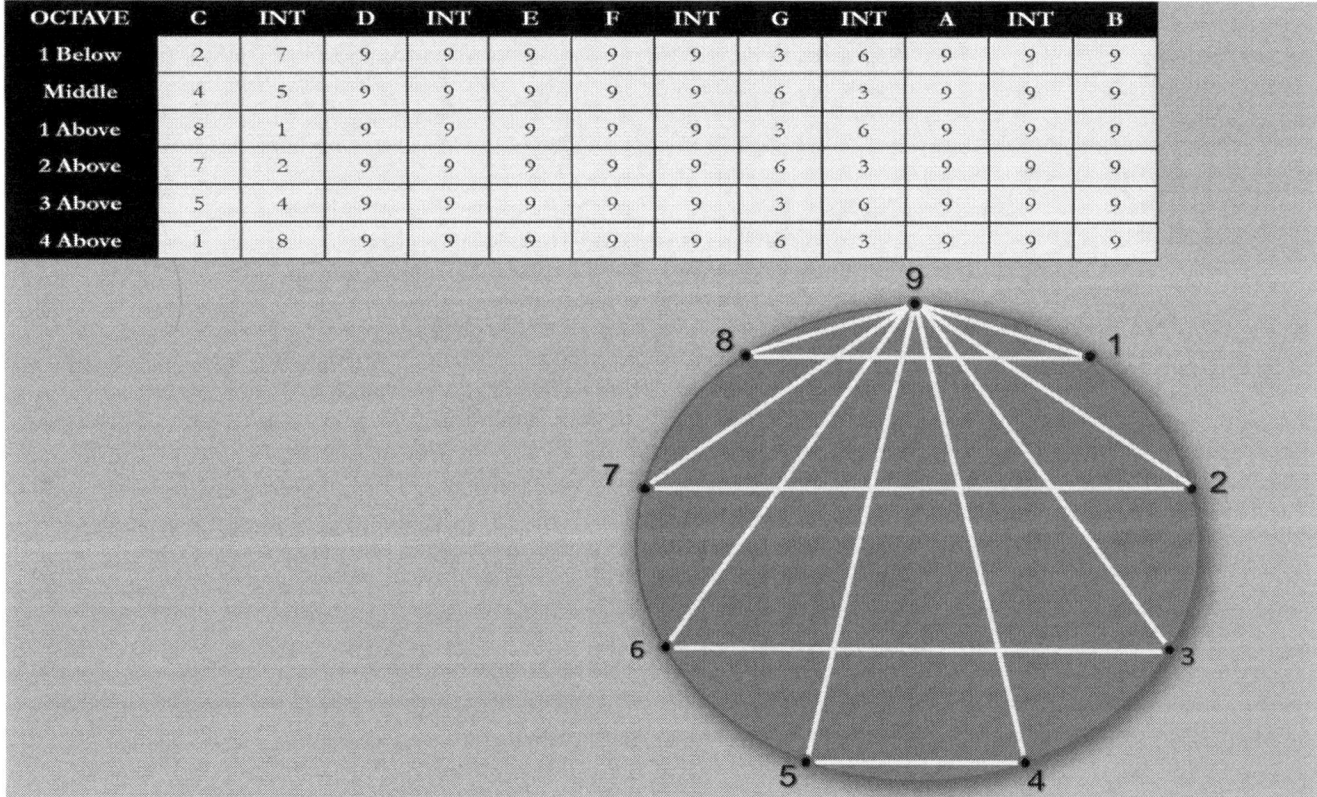

Music of the Decimal System.

Everyone can agree that the laws of Music are not Man made and they were instead discovered by Man. It is my contention that these musical laws represent the underlying harmony responsible for the formation of the Cosmos and that Numbers, and the Decimal System within which they operate, also, are not Man Made, but were discovered in just the same way as Music, except metaphysically.

When we consider these important number pairs, that I showed earlier, we can see that the Decimal System is nothing but Musical and here is how.

(00) 09 18 27 36 45 54 63 72 81 90 (99)

I simply looked at the differences between each of the Number Pairs, or rather how one may travel from one to the next, to see all the Music of the Universe there in the Number Pairs of the Decimal System.

09 to 18 is 09 x 2 which is the Octave

18 to 27 is 18 x 3/2 is the Perfect Fifth

27 to 36 is 27 x 4/3 is the Perfect Fourth

36 to 45 is 36 x 5/4 is the Major Third

45 to 54 is 45 x 6/5 is the Minor Third

54 to 63 is 54 x 7/6 is the Septimal Minor Third

63 to 72 is 63 x 8/7 is the Septimal Whole Tone

72 to 81 is 72 x 9/8 is the Major Tone

81 to 90 is 81 x 10/9 is the Minor Tone

$$9/8 : 10/9 = 81 : 80$$

Their size differs by exactly one Syntonic comma 81:80

where it is important to note that:-

$$80/81 = 0.987654321 \text{ and } 1/81 = 0.123456789$$

More simply, visually, we see the rather elegant:-

2   1 1/2   1 1/3   1 1/4   1 1/5   1 1/6   1 1/7   1 1/8   1 1/9   1 1/10   etc.

These number pairs, and therefore the decimal system are musical and are therefore inherent and universal.

Anthony Morris

Seeking a Vibrational Foundation to Music.

Collaborator Julian Shelbourne has kindly allowed me to reproduce some of his excellent work.

'The foundation of music is harmonics: Just touching a vibrating string at a whole number division of its length yields another tone.

For example, touching a vibrating guitar string at precisely 1/3 of its length yields a musical note called "the fifth" (the fifth note in the western scale - do-re-mi-fa-**so**) which has a vibrational frequency 3 times higher than the original. You can continue touching the vibrating string at whole-number divisions of its length to yield five tones of the western harmonic scale: ½ (octave), ¼ (octave), 1/5 (major "third" – do-re-**mi**), 1/6 (an octave of a fifth), 1/7 (minor seventh), 1/8 (octave), 1/9 (ninth – do-**re**). Doing this from a single string doesn't provide all 7 of the notes necessary to construct the western scale; but doing this on two strings - one tuned to the loudest harmonic of the first string (its "fifth") - does.

Each note generated in this way is an exact, mathematically-aligned reflection of the starting vibration - but it begs the question: which vibrational frequency *should* we start with?

Anthony has explored the differences between the modern standard where A=440 Hertz, and the naturally resonant standard of A=432 Hertz - which was documented by Zarlino in the 16th century and in other sources.

I wanted to add my own personal experience which bolsters the notion of A being correct at 432 Hertz, but also indicates that A is just part of a universal musical scale for which the foundational vibration is B-flat, at 7.2 Hertz - or some sub-audible octave of this.

This insight came as a result of exploring sub-audible frequencies on a tone generator app on my iPhone using Bluetooth headphones. I discovered that at around 11 Hertz there would be a slow, swooshing sound that would slow to a stop at precisely 10.8 Hertz. The swooshing sound would speed up and then stop again as I lowered the frequency to 7.2 Hertz, and again at 5.4 Hertz.

In music, the term "beating" is used to describe the oscillation heard when two tones are just slightly out of tune with each other. I believe the swooshing sound I was witnessing was the beating between the frequency I was generating on my tone-generator and the overtones of some kind of "background radiation" of sub-audio sound.

The frequencies themselves are interesting. 108, 72 and 54 are all numbers that appear in sacred geometry, show Anthony's Number Pairs and all return 0 or 9 Mod 9.

If you take 7.2 Hertz as your starting frequency, and generate whole number harmonics from it, you realize that 10.8 is exactly 3/2 (or a musical <u>fifth</u>), and 5.4 Hertz is exactly an octave below that. So, this apparently random standing-wave phenomenon is musically harmonic.

If you construct a harmonic series based on these frequencies, you get the following five frequencies. (Here I have started 32 octaves above the frequency I found: 7.2 Hertz x 32 = 230.4 Hz):

Multiplier	Interval	First Harmonic Series		
		Note	Harmonic Frequency	Mod 9
1	Unison	B-flat	230.4	2+3+4=9
3	Fifth	F	345.6	3+4+5+6=18=9
5	Third	D	288	2+8+8=18=9
7	Seventh	G#/Ab	201.6	2+1+6=9
9	Ninth	C	518.4	5+1+8+4=18=9

Opposite we can see the second harmonic series - based on the second frequency I discovered, raised by 32 octaves to give us more familiar frequencies. (10.8 Hertz x 32 = 345.6). Note that A comes out to be 432 Hertz!

Multiplier	Interval	Second Harmonic Series		
		Note	Harmonic Frequency	Mod 9
1	Unison	F	345.6	3+4+5+6=18=9
3	Fifth	C	518.4	5+1+8+4=18=9
5	Third	A	432	4+3+2=9
7	Seventh	D#/Eb	302.4	3+2+4=9
9	Ninth	G	388.8	3+8+8+8=27=9

Finding A = 432 Hz in this way was not my intent. It just so happens that A = 432 Hertz is the major-third harmonic of 10.8 Hertz standing wave frequency that I had discovered. But this discovery bolsters the notion that A should be 432 Hertz rather than 440 Hertz, while also indicating that it is B-flat which is the fundamental resonance of the universe from which all other vibrations are generated - rather than A. Putting these two harmonic series together yields us a musical scale of 7 notes which I believe is resonant with the core of the universe:-

Multiplier	Interval	Harmonic Musical Scale		
		Note	Harmonic Frequency	Mod 9
1	Unison	B-flat	230.4	2+3+4=9
9	2nd/9th	C	259.2	2+5+9+2=18=9
5	3rd	D	288	2+8+8=18=9
3 x 7	4th	D#/Eb	302.4	3+2+4=9
3	5th	F	345.6	3+4+5+6=18=9
3 x 9	6th	G	388.8	3+8+8+8=27=9
7	7th	G#/Ab	403.2	4+3+2=9
3 x 5	Maj-7th	A	432	4+3+2=9

Another interesting thing I discovered is that 8 octaves below my "magic frequency" of 7.2 Hertz, B-flat is 0.9 Hertz. Working from the bottom of this table upwards, if you multiply 0.9 Hertz by whole numbers according to the harmonic series, the frequencies you get each have the number on the left of the decimal point rising by one, while the number on the right of the decimal point goes down by one. They're all harmonics of our magic frequency 7.2 hertz and they're all exhibit 0 or 9 Mod 9.

Multiplier	Note	Harmonic Frequency	Mod 9	Interval
10	D	9.0	9+0=9	3rd
9	C	8.1	8+1=9	9th
8	B-flat	7.2	7+2=9	Octave
7	A-flat	6.3	6+3=9	7th
6	F	5.4	5+4=9	5th
5	D	4.5	4+5=9	3rd
4	B-flat	3.6	3+6=9	Octave
3	F	2.7	2+7=9	5th
2	B-flat	1.8	1+8=9	Octave
1	B-flat	0.9	0+9=9	Unison

In a universe constructed based on 9s, is it really so surprising that the fundamental frequency that drives the sympathetic vibrations of all others might itself be 0.9 Hertz?! I have identified other phenomena which correspond to these "magic frequencies" on my web-site at HarmonicsOfNature.com - including NASA's recordings of B-flat being generated by black-holes, ancient musical instruments tuned to these frequencies and cymatics which first form at these frequencies.' - Julian Shelbourne.

Cymatics.

Cymatics basically reveals the geometry of sound through vibration. This was originally researched and published by Dr. Hans Jenny in 1967 in his book The Study of Wave Phenomena. Ever since I first saw this information I knew it was fantastic and a major piece of the puzzle. I just hope my work brings those involved in this field more attention. I need to quickly cover this area as it is consistent with the insights gleaned from the work on the Pythagorean Musical Scale.

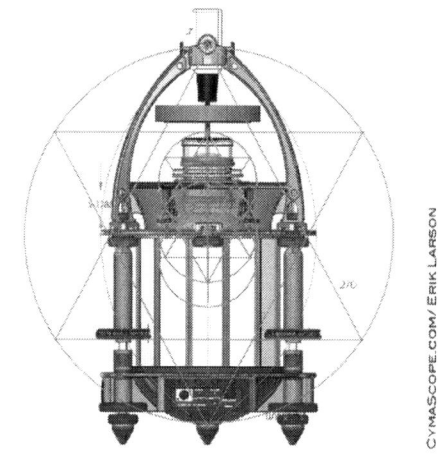

The Cymascope is a new type of scientific instrument that makes sound visible. Its development by John Stuart Reid and Erik Larson began in 2002 with a prototype that featured a thin, circular, P.V.C. membrane; later they used latex. Fine particulate matter was used as the revealing media. However, it was soon discovered that far greater detail could be obtained by imprinting sonic vibrations on the surface of ultra-pure water. The surface tension of water has high flexibility and fast response to imposed vibrations, with transients even as short-lived as a few milliseconds.'

When they looked at the geometry of 432 Hz under the Cymascope they found that it pops out as a triangle, every time they imaged it. They initially thought there was something wrong with the CymaScope but after trying for more than an hour they concluded that the number 3 was somehow universally connected to 432 Hz.

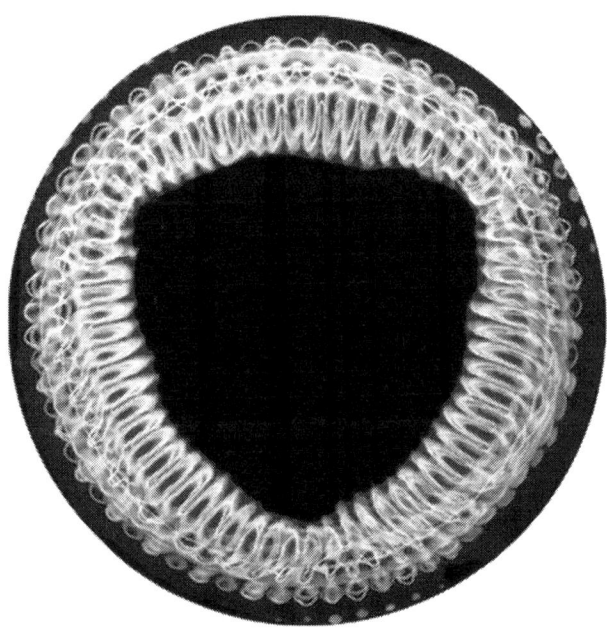

I realised that this was in fact the same shape as one can see when one draws 3 intersecting circles as we can see below.

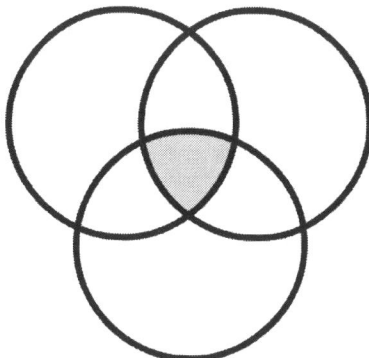

Reid researched the reason why it takes up this geometry and it turns out to be an interesting case:-

'When A is tuned to 432Hz the frequencies of the other A's shift (within a decimal point) to 27 Hz, 54, 108, 216, 864, 1728 in other octaves. D becomes 576 Hz which becomes 9 Hz, 18, 36, 72, 144, 288, 1152 in other octaves. E becomes 324 Hz which becomes 81 Hz, 162, 648, 1296 in other octaves. All of these frequencies are divisible by 3.' - John Stuart Reid.

Brian Collins, who wrote to John about taking a look at 432Hz under the Cymascope, said John had confirmed 'what seems to be a scientific interpretation of what I believe in my personal opinion, what Rudolph Steiner in his esoteric knowledge of the Mysteries of Golgotha and cosmic principals stated; that a prime centre of tone for living enlightenment in mankind is a C = 128 Hz note with the octave at C = 256 Hz (within 432 Hz Concert pitch) or Michael Sun tone A = 432 Hz which ties the spiritual tetra or trinity of spherical scalar wave patterns and evolution of man ascendant together in sound and geometry of the super hologram of consciousness. The medium of 432 Hz is certain evidence of divine intelligent design of the universe and light.

A 53 note temperament scale based on A = 432 Hz pitch goes less than 1 Hz difference out of harmonic synchronicity over the entire 53 note scale, most other pitches are out of sync after only two or three notes up the octave, an excited engineer confirmed that scales tuned to A=432 Hz are surely the fingerprint evidence of an intelligent divine creator.'

Please do your own research into this exceptional, transformative new Science. I believe it has the ability to create a whole new way of healing disease in the body by using vibration and in fact this has been borne out by recent research in this field and is being used to treat tumours - http://www.cancerresearchuk.org/about-cancer/cancer-in-general/treatment/other/high-intensity-focused-ultrasound-hifu

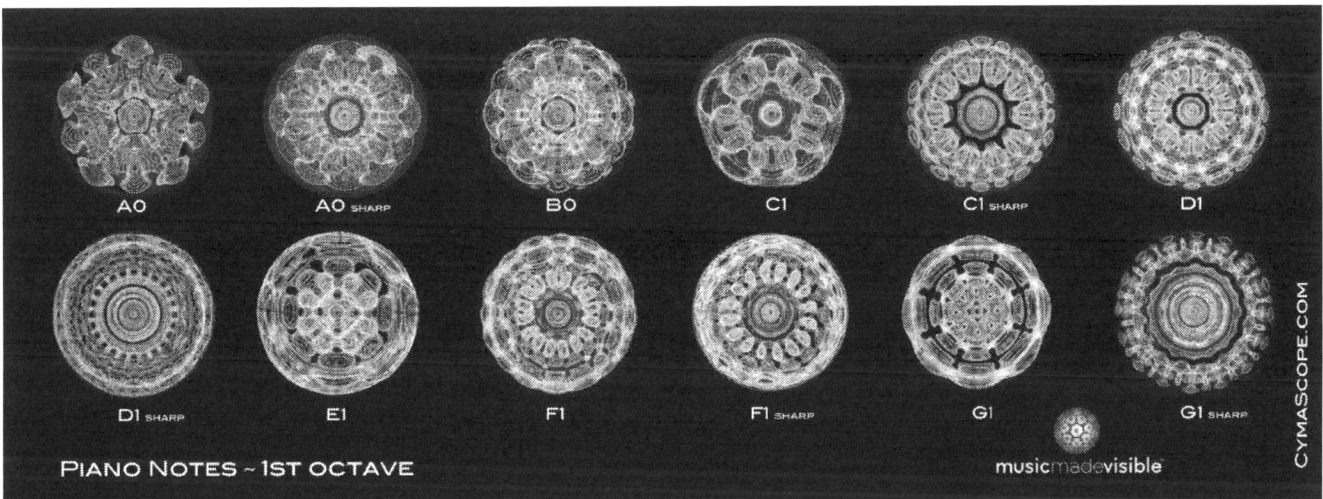

As the body has a high water content and the organisation of water responds by creating beautiful patterns as seen above, might it explain why the classical music of Mozart and Bach, tuned at 432Hz, has such a profound effect upon people who listen to it?

Chapter 5: The Re-Classification of Prime and Composite Numbers.

Term	Prime Numbers	Mod 9
1	2	2
2	3	3
3	5	5
4	7	7
5	11	2
6	13	4
7	17	8
8	19	1
9	23	5
10	29	2
11	31	4
12	37	1
13	41	5
14	43	7
15	47	2
16	53	8
17	59	5
18	61	7
19	67	4
20	71	8
21	73	1
22	79	7
23	83	2
24	89	8
25	97	7
26	101	2
27	103	4
28	107	8
29	109	1
30	113	5
31	127	1
32	131	5
33	137	2

PRIME NUMBERS.

In order to properly understand the system of Numbers it is crucial that each number be classified correctly. I am asserting that this is not currently the case.

It is thought that Numbers may easily be divided into two groups, those that are called Prime and those that are called Composite.

A Prime Number is recognised as a Number that has factors of itself and 1 only.

A Composite Number is recognised as a Number that has more than these 2 factors.

As we shall see, this is not the whole story or the right way of looking at the picture.

The first thing I did was to simply apply Mod 9 to the first 65000 Prime numbers.

This revealed the surprise that results exclusively returned 1 2 4 5 7 or 8 for all Prime numbers greater than or equal to 5.

However, as with all the previous number sequences, there was no regularity to these Mod 9 results, that is there was no repeating sequence that was immediately discernible.

1	2	3	4	5	6	C	1	2	3	4	5	6
7	8	9	10	11	12	O	7	8	9	1	2	3
13	14	15	16	17	18	N	4	5	6	7	8	9
19	20	21	22	23	24	V	1	2	3	4	5	6
25	26	27	28	29	30	E	7	8	9	1	2	3
31	32	33	34	35	36	R	4	5	6	7	8	9
37	38	39	40	41	42	T	1	2	3	4	5	6
43	44	45	46	47	48	S	7	8	9	1	2	3
49	50	51	52	53	54		4	5	6	7	8	9
55	56	57	58	59	60	T	1	2	3	4	5	6
61	62	63	64	65	66	O	7	8	9	1	2	3
67	68	69	70	71	72		4	5	6	7	8	9

Prime Numbers in Red   Composite Numbers is Black

The 72 Number Sequence that I had discovered in Music, 6 octaves of 12, led me to arrange the Numbers 1 to 72 into 6 columns as laid out above.

On the left we have the numbers and on the right we can see the Mod 9 results.

Laying the numbers out like this quickly shows us why Prime numbers will exclusively return these remainder values infinitely, because the Prime numbers only occur in Column 1 and Column 5 which simply repeat 1 7 4 and 5 2 8 respectively, showing that these Number Groups are specifically related to Prime numbers.

The 'problem' with the outlying numbers 2 and 3 will be dealt with shortly.

1	5
7	11
13	17
19	23
25	29
31	35
37	41
43	47
49	53
55	59
61	65
67	71

By removing columns 2, 3, 4 & 6 and leaving only columns 1 & 5, we are left with only Primes (red) and some 'Other' numbers (black).

These 'Other' numbers are immediately identifiable as either:

Products of Prime Numbers   or   Exponents of Prime Numbers

$5 \times 7 = 35$ $\qquad$ $5^2 = 25$
$5 \times 11 = 55$ $\qquad$ $7^2 = 49$ etc.
$5 \times 13 = 65$ etc.

These 'Other' numbers are currently classified as Composite Numbers but should rightly be re-classified as Prime Composites and thought of exclusively as part of the Prime Number system, not the Composite Number system as they currently are.

COMPOSITE NUMBERS.

In order to further cement my case that Mod 9 is useful at inferring information not otherwise obtainable, we must look to the other side of the coin if you like and examine the 'not Prime' numbers or what are known as the Composite numbers.

Composite Numbers have multiple factors, not just itself and 1, as Primes do.

For example, 12 is a Composite Number and has factors of 1, 2, 3, 4, 6 and 12, while 13 is Prime as its only factors are 1 and 13.

As I examined the sequence of Composite Numbers as is currently posited by academia, Mod 9 analysis revealed a sequence that comes out like this:-

4 6 8 0 1 3 5 6 7 0 2 3

4 6 7 8 0 1 3 5 6 7 8 0 2 3

4 6 8 0 1 3 4 5 6

Can you see that these numbers are attempting to organise themselves into a repeating sequence but it's not quite happening?

I immediately realised that you simply have to include the numbers 2 and 3 traditionally thought of as Prime and then take out the Prime Composite Numbers, as described earlier.

So, for Primes 5 and greater, starting with the 5 x 5 = 25 and then 5 x 7 = 35 etc. we take these out and we should hopefully get some order appearing in the Composite Number Sequence.

This would prove that I was on the mark with the Prime Number analysis and my conjecture as to how to treat thos 'Other' numbers in Prime number columns.

Composite Numbers	Mod 9
4	4
6	6
8	8
9	0
10	1
12	3
14	5
15	6
16	7
18	0
20	2
21	3
22	4
24	6
25	7
26	8
27	0
28	1
30	3
32	5
33	6
34	7
35	8
36	0
38	2
39	3
40	4
42	6
44	8
45	0
46	1
48	3
49	4
50	5
51	6

Composite Numbers	Mod 9
2	2
3	3
4	4
6	6
8	8
9	0
10	1
12	3
14	5
15	6
16	7
18	0
20	2
21	3
22	4
24	6
26	8
27	0
28	1
30	3
32	5
33	6
34	7
36	0
38	2
39	3
40	4
42	6
44	8
45	0
46	1
48	3
50	5
51	6
52	7
54	0

CORRECT COMPOSITE NUMBER SEQUENCE.

In the table on the left you will see that I have added in numbers 2 and 3 and then excluded those numbers that are traditionally thought of as Composite numbers, but which should rightly be classed as a Prime Composite, for all Primes, greater than 5, starting with the 5^5 = 25 and then 5 x 7 = 35 etc.

Where there was no modulus before, there is now.

A 12 digit repeating sequence:-

2 3 4 6 8 0 1 3 5 6 7 0

This can be broken down into 3 columns of 4 to illustrate the Number Groups when reading vertically.

2 3 4 6

8 0 1 3

5 6 7 0

Further, when we break the sequence down into 2 columns of 6 we show the Number Groups again illuminated very well.

2 3

4 6

8 0

1 3

5 6

7 0

Prime Number Family	Modulo 9
1	1
5	5
7	7
11	2
13	4
17	8
19	1
23	5
25	7
29	2
31	4
35	8
37	1
41	5
43	7
47	2
49	4
53	8
55	1
59	5
61	7
65	2
67	4
71	8
73	1
77	5
79	7
83	2
85	4
89	8

CORRECT PRIME NUMBER FAMILY SEQUENCE.

Prime Numbers are in fact a very simple group and have a simple congruence exposed by analysis using Mod 9.

When properly classified we see a six number sequence that repeats Mod 9.

$$1\ 5\ 7\ 2\ 4\ 8$$

Again we see in this sequence the three dimensional numerical symmetry that shows the Number groups.

$$1\ 5$$

$$7\ 2$$

$$4\ 8$$

Prime Number Re-Classification.

Prime Numbers, by virtue of their indivisibility, are the building blocks of all material phenomena, or matter.

I had gleaned from the VBM research that as far as they are concerned the 9 0 represents Source, the central axis, and together with the 3 and 6, representing the polarity or boundaries, this number group represented magnetism. In their estimation the 1 2 4 8 7 5 that we see in the cellular mitosis sequence represents electricity. What they are saying, and I think they are right, is that the combination of the numbers 0 1 2 3 4 5 6 7 8 9 interact in such a way as to produce electro-magnetism.

In my theorised structure of the Universe, the Prime Numbers represent electricity as defined by the interaction of number groups 1 4 7 and 2 5 8, while the 3 6 9 are the containing magnetic infrastructure upon which electricity plays out her dance.

I believe we live in an electric universe, mathematically unfolding with great precision and accuracy through Time, that is totally governed by Prime Numbers and therefore it is Prime Numbers that we must understand thoroughly if we are to master an understanding of the universe and everything in it.

I want to argue vigorously that these previously referred to Prime Composite numbers, identified above, are really no different at all to these traditionally thought of Prime numbers, and actually hold the key to better understanding how Prime Numbers and the whole number system can work as a machine.

The Prime Composites are created exclusively by combinations only of each other. They are all, therefore, inherently Prime, in and of themselves, and nothing else. They are simply iterations of the Prime number sequence itself and all the possible combinations thereof. This allows and accounts for the infinite complexity of Physical Reality.

Reality is created by an infinite fractal expansion of Prime numbers that creates a grid, much like a spider that spins its web. I think that if we are to deny the above, it is surely like saying that what comes of a union between two human beings is not human somehow and that it could be defined as anything else.

Classification.

A Priori / Metaphysical Numbers.

If we think that all the other kinds of Primes greater than 5, so Seeds (those numbers traditionally thought of as Prime, so divisible only themselves and 1), Composites (Products of Prime Numbers) and Exponents (Primes multiplied by themselves), are Effects, then these, A Priori Numbers, 1 2 3 & 4 can be thought of as the Causes of those Effects.

Number 1.

The first number in this A Priori group is the Number 1 – which I am calling the Void.

Prime is omitted from the chosen name because immediately there is a question mark raised over even its strict belonging to a Prime number classification at all, A Priori or otherwise.

Even though 1 fits as a member of the 1 4 7 group, obviously; the key here is that the number 1 has an additional factor, exclusive to it as a Prime, other than 1 or itself.

That being -1.

$$-1^2 = (-1 \times -1) = 1$$

No other Prime number, A Priori or otherwise, can boast such a property and it is this, immediately, that puts it squarely into its own grouping within even the A Priori group of numbers.

Numbers 2 & 3.

The Composite Numbers 2 and 3, traditionally given the classification of Prime, are really the workhorses, the building blocks of the building blocks of all physical reality and the system of numbers.

All of physical creation comes down to the dance and weave of these two numbers, the 2 and the 3.

They should rightly be called the Causative or A Priori Composite Numbers within the A Priori Group of numbers 1 to 4.

Number 4.

4 is Composite also, and is important in that it is the first exponent being the first number with a square root - 2^2, and as such, introduces a new dimension or level of complexity by allowing for area. (One might argue that the first squared number could be said to be the number 1 also.)

Prime Seeds.

Prime Seeds are defined as being all those Prime numbers, 5 and above, traditionally thought of as Prime, i.e. divisible only by themselves and 1.

5 is the first of the Seed Primes that make up all matter in physical reality. It is the first effect created by the additive function of the 2 and 3.

5 is also directly connected to Phi or the Golden Mean, as expressed by the Fibonacci sequence, and which is found everywhere in nature and throughout the Human body.

It is connected thus:-

$$\Phi = \sqrt{\frac{5+\sqrt{5}}{5-\sqrt{5}}}$$

Original insight contributed by Erol Karazincir - pcerol@yahoo.com

Prime Composites & Exponents.

There are two types of these Prime Composite numbers, 'Pure Prime Composites' and 'Prime Exponents'.

'Prime Exponents' express the change in dimension of the 'Prime Seeds' whereas the Pure Prime Composites are simply the multiplication or product of Prime Seeds.

Visual Evidence.

As proof of the validity and worth of the above analysis and conjecture, please consider the following.

Below is a Square of 9 Number Spiral - drawn simply by placing the 1 in the middle, the 2 above, 3 to the left of the 2, 4 below the 3 etc. Spiralling around and around.

31	30	29	28	27	26	49
32	13	12	11	10	25	48
33	14	3	2	9	24	47
34	15	4	1	8	23	46
35	16	5	6	7	22	45
36	17	18	19	20	21	44

If we expand the spiral of numbers and highlight only the prime numbers traditionally thought of as prime, then we arrive at something called Ulam's Spiral which is show below:-

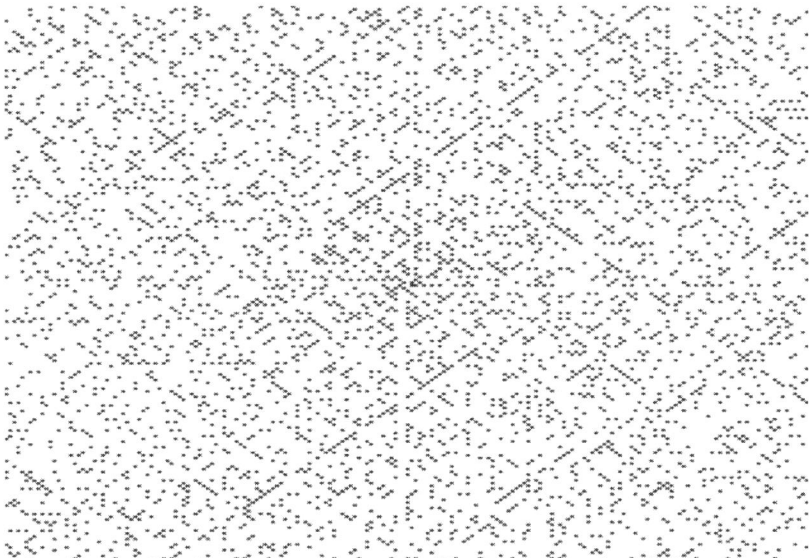

http://en.wikipedia.org/wiki/Ulam_spiral

The numbers, traditionally called Prime, produce, a grainy picture, for sure, but one which definitely shows preponderance by the Prime Numbers to appear on certain clear diagonals. Now, if we add in the Prime Composites and Exponents as I suggest, then we shall see something much more defined and quite incredible….

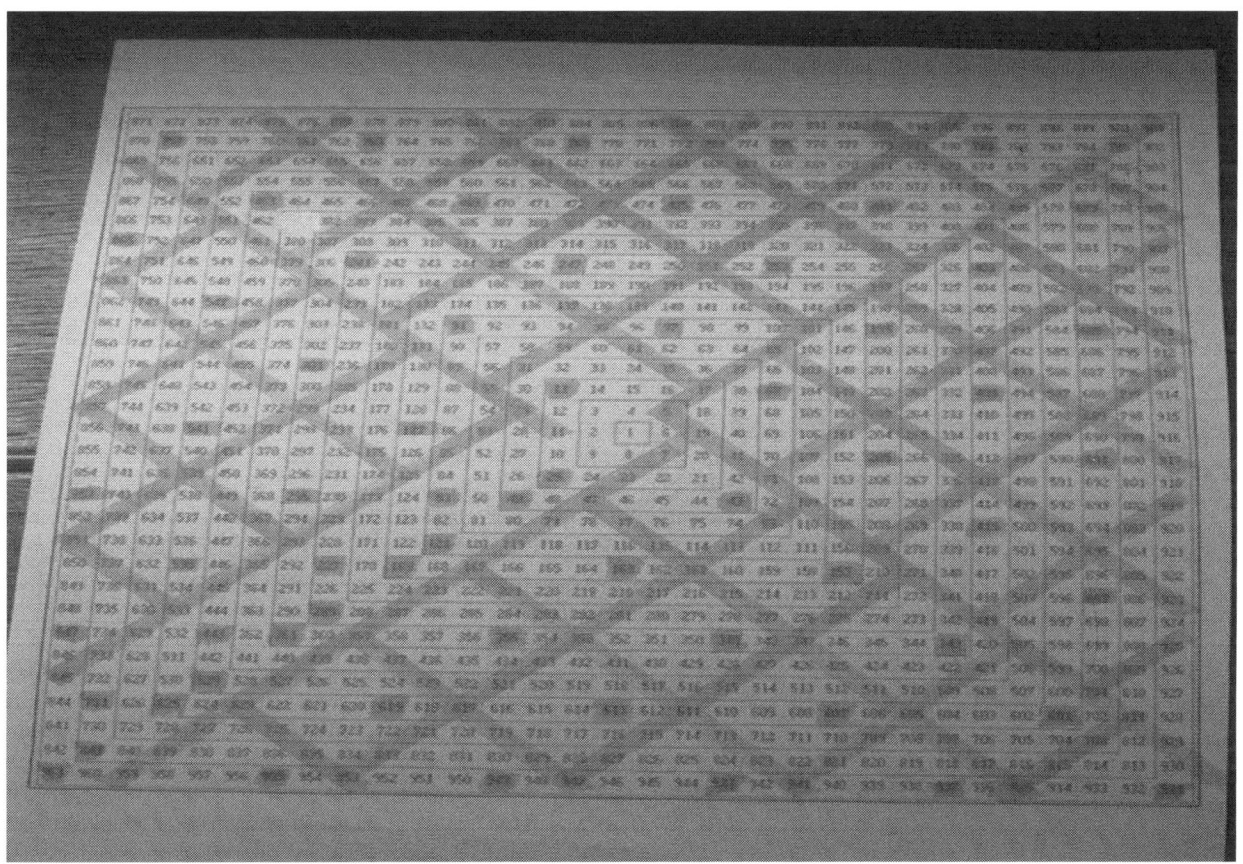

If Ulam's Spiral is important, then the amazing coherence and congruence displayed here must surely be of significant worth?

The grid or lattice is made up of diamonds that consist of 4 by 4 prime number nodes along the sides and 7 on the diagonals, containing a total of 25 numbers in each. 12 numbers are shared with other adjacent sectors and 13 numbers could be considered to be 'contained', one of which is a single Prime or Prime Composite and that I have coloured pink. Each diamond is constructed as either a pair of Golden Triangles or a Golden, or Square, Rhombus.

Each diamond contains a single Prime, that I have coloured pink, which strangely occurs in each diamond at the location of an eye, with the effect that the diamonds look like little fish swimming in a shoal, in an anti-clockwise direction. Clearly delineated are 4 definite quadrants. These can be visualised as the interacting 4 vortices or 2 pairs of poles that can be found in Walter Russell's cosmogony. North - South, and East - West. Next, please notice, on the South West diagonal, an intriguing and mysterious absence of this coherence, so striking elsewhere. Here we find a curious bunch of numbers highlighted in pink also, yet with a demonstrably even and predictable periodic occurrence. If you join these dots you would conceivably have the effect of a snake slithering up or down a pyramid edge. We know that some pyramids are built to exhibit this precise effect at solstices and equinoxes. The South West diagonal is dominated by the dimension expansion phase for numbers, and is announced by the presence of the Prime Exponent Numbers in the form of 5^2 for 25, 7^2 for 49, 11^2 for 121, 13^2 for 169, 17^2 for 289 etc. on up through the powers, for example 5^3 for 625 and on and on.

Square Pyramid Number Sequence.

If we imagine the square of nine number spiral in three dimensions we can see that the structure of Prime numbers is Square Pyramidal like the Great Pyramid at Giza.

Term	Sequence	Mod 9
1	1	1
2	5	5
3	14	5
4	30	3
5	55	1
6	91	1
7	140	5
8	204	6
9	285	6
10	385	7
11	506	2
12	650	2
13	819	0
14	1015	7
15	1240	7
16	1496	2
17	1785	3
18	2109	3
19	2470	4
20	2870	8
21	3311	8
22	3795	6
23	4324	4
24	4900	4
25	5525	8
26	6201	0
27	6930	0
28	7714	1
29	8555	5
30	9455	5
31	10416	3
32	11440	1
33	12529	1
34	13685	5
35	14910	6
36	16206	6
37	17575	7
38	19019	2
39	20540	2
40	22140	0

Square Pyramid Number Sequence.

1 5 5 3 1 1 5 6 6
7 2 2 0 7 7 2 3 3
4 8 8 6 4 4 8 0 0

Here we see a 27 digit repeating sequence for the Mod 9 results.

Here again we see the Number Groups emerge in the columns exhibiting the three dimensional numerical symmetry.

Interestingly the 3 Number Groups are balanced with each appearing exactly 3 times.

Where we find a 27 digit sequence in the Mod 9 sequence my feeling is that the sequence tells us something about the properties of Light.

Why this may be the case will become clearer in the last part of the book.

In conclusion, it seems absolutely clear to me that a re-classification of Prime and Composite numbers is in order. A Mod 9 analysis of numbers illuminates the perfect congruence that exists in the system of Prime and Composite Numbers when correctly classified.

Chapter 6: Numbers and the Structure of Reality.

During the early part of this research period I spent some time in Australia with old friends Tim & Liz Farrell, my studies were still in relatively nascent form, and we had been discussing the Harmonic Grid I use to time global asset markets and my Mod 9 analysis, and how I felt that the two were directly connected in that they both expose an order one otherwise cannot see.

In this section I advance what resembles the structure for a type of string theory to explain the structure of temporal quanta which contains and expresses physical reality.

As I remember, one night, early on in my stay, Tim had a dream that burst out of him in the morning and he drew what resembled the picture below to describe what he saw.

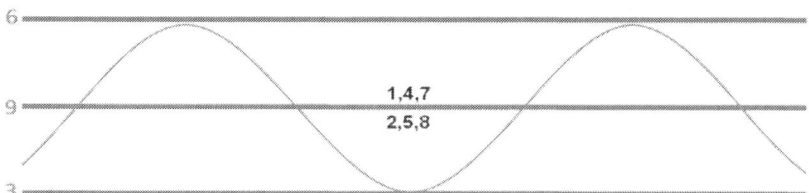

It had the 9 0 serving as the central axis that the 1 2 4 5 7 8 spiral around, bounded by the magnetic poles of the 3 and 6.

This cemented something in my mind and instant connections were made and new horizons opened up in my understanding. It was a critical moment that led to the following offering which can be applied to the DNA Helix.

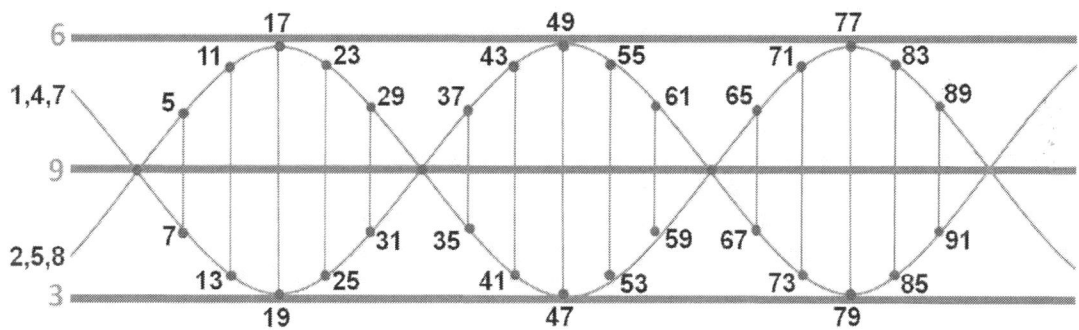

Central Axis = 9
Boundaries = 3 & 6
Primes = 1,4,7, & 2,5,8

The 9 is the Central Axis - The 9 Forms.

The 3 and 6 are the Boundaries - The 3 and 6 Contain and Confine.

The 1 4 7 and 2 5 8 Number Groups, are Controlled by Prime Numbers

The Prime Number Strings spiral, from opposite directions, around the Central Axis, the 9, within the confines of the 3 and 6.

In Terms of an Apple.

The 0 9 is represented by the core and the attachment to the tree.

The 3 and 6 are the boundaries that determine the size of the apple.

The 1 4 7 and 2 5 8 make up the flesh of the apple as it is formed / coalesced / integrated by spiraling from opposite directions around the core.

So it is with all 'things' in Physical Manifestation.

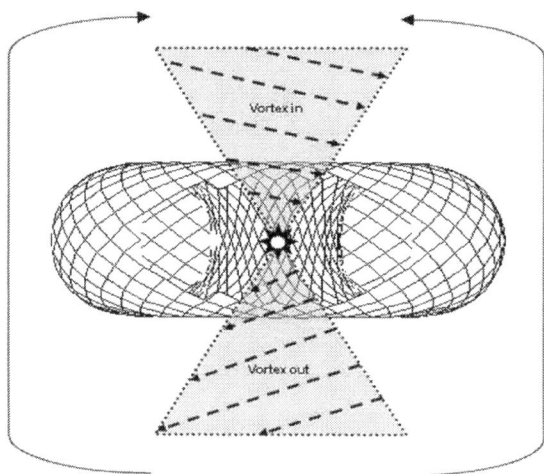

The 3 6 and 9 are the conduit for the 1 4 7 and 2 5 8 in the same way as the pipework in your house is the conduit that causes water to be available to you at different locations within it.

In this case though, the pipes are obviously not carrying water to all points, they are carrying, or providing the potential for, Electricity, to play out its dance.

I provide this metaphor mainly so that you may appreciate that the plumbing has nothing directly to do, or in common, with, the water, per se, other than its ability to provide a conduit for its delivery.

It is sort of the same with Electricity except, as we shall see in the next part, Magnetic Light and Electricity, it turns out, are the reciprocal of each other, which, of course, pipes and water are not.

Octaves of Perception.

During the course of my research I discovered the work of the late, great, Walter Russell who wrote that:-

'The known octaves which lie in the range of perception are five and one half. These begin with the third, or hydrogen octave, and end with the uranium group which are isotopes of Actinium and Tomium in the last octave.

The invisible octaves of finely divided matter of space are three and one half in number.

These octaves are beyond our range of perception, but they are not beyond our knowing.' - A New Concept of the Universe - Walter Russell.

When we examine these numbers we see:-

Visible Spectrum = 5.5 Octaves or 5.5 x 12 = 66 notes

Invisible Spectrum = 3.5 Octaves or 3.5 x 12 = 42 notes

5.5 + 3.5 = 9 Octaves x 12 = 108 notes

When we look at the relationship between these two numbers we see:-

5.5 / 3.5 = 11 / 7 = 1.571428 R - Electricity

1 4 7 and 2 5 8 Number Groups that control Prime Numbers.

3.5 / 5.5 = 7 / 11 = 0.636363 R - Magnetism

3 6 9 Number Group.

Electricity is numerically shown to be the Inverse or Reciprocal of Magnetism.

Russell continues - 'The Electric we can perceive through the Electric senses of our Electric bodies while the Magnetic we cannot directly perceive, yet can know.

Electricity is the only force which God makes use of to create this universe and the only two "tools" God makes use of for creating His universe of matter and motion are two pairs of opposed spiral vortices.

One of these opposite pairs meets at apices at wave amplitudes to create spheres of matter and the other opposed pair meets at cone bases upon wave axes to void both matter and motion'.

1 4 7 & 2 5 8

The Prime Number, Electric, Number Groups Are Opposed Spiral Vortices

'These two pairs of opposed electric spiral vortices are the basic units which construct all matter. Together they form the electric waves of motion which create the various pressure conditions which are needed to produce the many seemingly different elements of visible and invisible matter.

Electricity is divided into two equal-and-opposite forces which thrust away from each other to build this polarized universe. When inability to thrust away from each other takes its sequential turn in the pulse of the universal heartbeat, depolarization voids all opposition. Thus this universe consists of cycles of life followed by death - growth followed by decay, and generation followed by radiation - each expressed simultaneously and repeated sequentially forever without end.

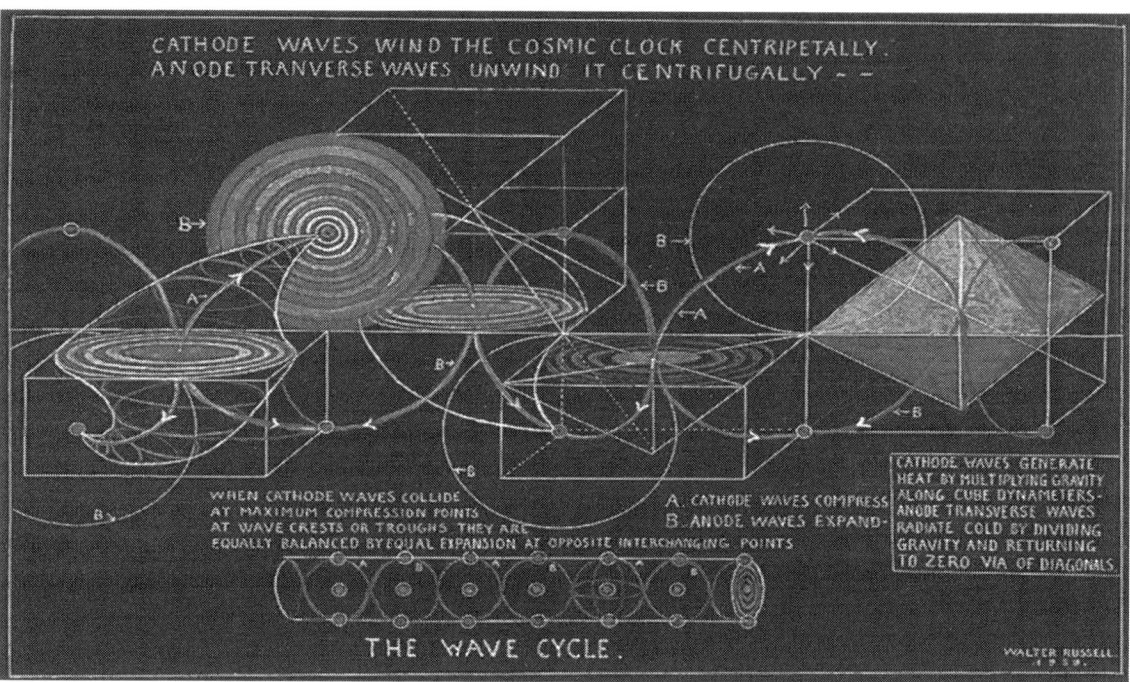

That which science calls magnetism, and believes to be a force which has the power of lifting tons of steel, is God's still Light which balances and controls the equality of electric division, but electricity alone performs all of the work of this universe.

The magnetic Light which controls the universal balance performs no work whatsoever.

A bar magnet picks up nails because of the electric current which divided that steel into its activated polarized conditions, and not because of its focal poles of stillness which center its two activities even though the electric current has been withdrawn. The steel retains its electric activity for long periods and acts as though the current still remained.

Magnetic Light control might be likened to the rudder of a ship which controls the direction of the ship's motion without in any way motivating that motion. It might again be likened to the fulcrum which extends its power of expression through motion to a lever, without in any way acting to motivate that expressed motion of the lever.

God's still magnetic Light is the fulcrum of this creating universe. Electricity is the two-way lever which extends from that fulcrum to give the universe its pulsing heartbeat of simulated life-death sequences.

Wherever God's Light appears in matter, there stillness centers motion, but there is no motion at that point. The center of gravity in a spherical sun or earth is one locatable point where God's Light is.

Likewise, the two still centers of north and south spiral vortices are other locatable balancing points of control.

Likewise, the shaft which connects all pairs of opposed poles is an extension of stillness from the zero of wave beginnings to the zero of wave amplitudes, and the return of motion to the zero of its beginnings in the stillness of its homing place on its wave axis.'

This is a universe of Light-at-rest from which two opposed lights-of-motion appear to manifest the IDEA which is eternally sealed in the Light-at-rest.

All matter is electric. Electricity conditions all matter under the measured control of the ONE MAGNETIC LIGHT which forever balances the TWO electrically-divided, conditioned lights of matter and space.

All matter is but pressure-conditioned motion.

Varying pressure conditions yield varying states of motion.

Varying states of motion are what science misinterprets as the elements of matter.

Varying pressures in a wave are tonal. In each octave wave there are four pairs of tones, each of which has the same relative position in its octave color spectrum as it has in its octaves of chemical elements.

Waves are, therefore, electric pressure-conditioned octaves of tones.'

The Divided Light

'In the light of the Creator's Mind is DESIRE to dramatize his ONE IDEA by dividing its one unconditioned, unchanging unity of balance and rest into pairs of oppositely-conditioned units, which must forever interchange with each other to seek balance and rest.' - A New Concept of the Universe - Walter Russell.

Walter Russell's Octave Wave Theory of the Chemical Elements.

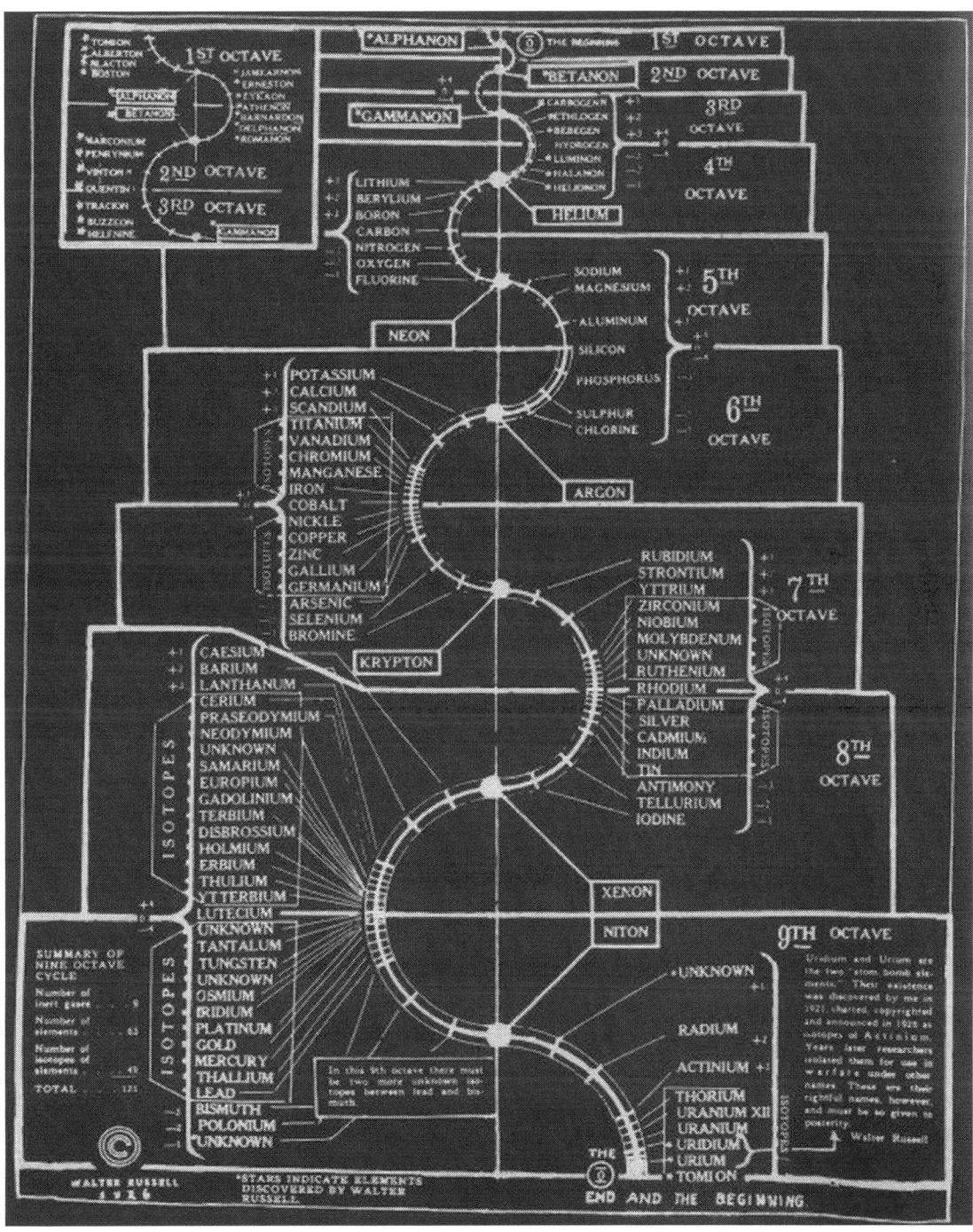

Chapter 7: The Vortex Theory of Atoms.

Collaborator Talal Ghannam PhD has written a truly brave and important book - The Mystery of Numbers. He has very kindly allowed me to extract from the book at will and reproduce here, which sets the scene for a greater understanding of the numerical aspect of particle physics.

His whole book is about the Number Groups 1 4 7, 2 5 8 and 3 6 9 as seen from the point of view of a PhD Physicist and as such our work touches on many of the same discoveries and themes albeit using differing nomenclature.

In his book, I was delighted to discover that for millennia a vortex theory of atoms had been generally accepted and that a substance called Ether or Aether permeated the universe through which Light was able to propagate freely, just as waves propagate through water. Many great scientists including Maxwell, Descartes, Einstein, Tesla and Faraday believed in the Ether.

That was, up until 1887, when the Michelson - Morley experiment seemed to disprove this theory, though even Michelson, one of the named architects of this experiment, refused to give up his belief in the ether. Others have also pointed to problems with the experiment but the Vortex theory of atoms was consigned to the dustbin.

I think it needs to be taken out again and dusted down because for me this theory maps perfectly to the number system exposed throughout my work and that of those upon whose shoulders I stand.

Talal writes that "Tesla believed that the ether is made of carriers (negative and positive) immersed in an insulating fluid and are always in a whirling motion forming kind of 'micro helices'.

The properties of this ether vary corresponding to movement, mass, and the existence of electromagnetic fields. In fact he believed that all the energy that exists in matter is received from the surrounding ether. Hence the ether is more like an energy repository capable of imparting its energy to all forms of matter, living and non-living. The reason behind the inertia of masses was believed to be due to the interaction between the ether and mass. However, once a mass starts moving within the ether, the charges of the atoms that form this mass will start creating rotary electromagnetic fields creating holes within the ether and forming what is called 'tube of force' that are free from any ether inside them. This will eliminate the existence of the ether inside the mass, and consequently make the movement of mass smoother and less resistive.

JJ Thomson's thoughts about these moving tubes of force can be summarized in the following points:

1 - that the tubes exist everywhere in space, either on closed circuits or terminating on atoms.
2 - that electric force becomes perceivable only when electric tubes have greater tendency to lie in one direction.
3 - that a beam of light is a group of electric tubes moving at the speed of light.
4 - if moving tubes entering a conductor are dissolved in it, mechanical momentum will be given to the conductor. This momentum is stored in a unit volume of the field being proportional to the vector product of the electric and magnetic vectors.

The ether was also used to explain the forces of nature such as:- electricity and magnetism, which are nothing but disturbances through the ether; and gravity, which is nothing but the pressure of the ether flow. The ether was also used to describe the elementary particles and atoms that the universe is made of. This theory was called - The Vortex Theory of Atoms.

First, let us explain what we mean by a vortex. A vortex is a spinning, funnel shaped, 3 dimensional spiral motion of any kind of matter and fluid, just like a tornado, and where energy is directed in a rotational configuration and concentrated at the center of the vortex as shown overleaf.

The basic idea of the vortex theory of atoms is that the subatomic particles, the electron, the proton, the neutron etc., are nothing but spirals, or more precisely, self-sustaining vortices of energy, having toroid-like shapes, similar to a doughnut, spinning in viscosity-free ether medium.

The direction of the spinning determines whether the etheric energy is flowing in or out of the particle, and whether this particle is positive or negative in terms of charge. In fact, mass, charge, magnetism etc. are all nothing but different states of the ether depending on its concentration, rotation, speed, etc."

This theory was championed by many of the most brilliant scientific minds of the 19th and 20th century, including Lord Kelvin (1824 - 1907), Hermann Von Helmholz (1921 - 1984) James Clark Maxwell (1831 - 1879), Nikola Tesla (1856 - 1943), and John W Keely (1827 - 1898)

Actually, Keely discovered that the proton is made of 3 sub-particles, long before modern physicists had proposed the quark model. He even went further by claiming that each of these sub-particles (the quarks) is made of another set of 3 particles, and so on, all being vortices in the ether. He claimed to have controlled 27 levels of these triple sub-particles.

Nikola Tesla, who is one of, if not the, most brilliant minds in the field of electromagnetism, also believed strongly in Vortex Theory. He envisioned the atom being differentiated from the ether by a spinning motion, just like a whirl of water in a calm lake.

Some scientists postulate that the flow of the ether through atomic particles and atoms is necessary for them to keep their spin and energy. They even argue that the flow of this ether through the stars and planets is necessary for them to keep their internal molten cores alive. Moreover, spirals and vortices permeate everything in nature, from the smallest to the largest. So why not also in the atomic world? "- The Mystery of Numbers - Talal Ghannam

Numbers & Particle Physics.

Just as the Numbers and Octaves pair, so too do the elementary particles!

A particle and its pair have exactly the same mass and fall downwards in the Earth's gravitational field. They are effectively mirrors of each other.

"The most important condition when recognising a particle and anti-particle is that when they combine, they should annihilate each other with a burst of complete pure energy, in other words a burst of self-mirrored photons' - Talal Ghannam - The Mystery of Numbers.

Particle	Anti-Particle
Neutron	Anti-Neutron
Proton	Anti-Proton
Electron	Positron
Neutrino	Anti-Neutrino

The Photon is considered its own Anti Particle which I liken to the 0 9 Number Pair relationship so we have a clear rationale for comparing the Numbers and Particles simply if we only knew which numbers to apply to which particle.

Talal proposes the following answer which I think is right.

Particle	Number	Number	Anti-Particle
Neutron	1	8	Anti-Neutron
Proton	2	7	Anti-Proton
Electron	3	6	Positron
Neutrino	4	5	Anti-Neutrino

So now we can attribute each number to an atomic particle or anti-particle. This will be important later on.

Talal then goes on to test these numbers against important reactions in the subatomic world, namely the B- Decay, the B+ Decay, and the electron capture 'K capture' reaction.

Due to the laws of conservation of energy, these reactions must tally up and we can see that they do if we use Mod 9.

B- Decay.

The Neutron spontaneously decays into the three subatomic particles, proton, electron and anti-neutrino.

Neutron decays to Proton + Electron + Anti - Neutrino

1 decays to   2 + 3 + 5 = 10 or 1 Mod 9 - Check

B+ Decay.

The energy induced decay of the Proton.

Energy + Proton decays to Neutron + Electron + Neutrino

9 + 2 decays to 1 + 6 + 4

11 = 11 - Check

Here we must also check energy with a value of zero:-

0 + 2 decays to 1 + 6 + 4

2 = 11 or 2 Mod 9 - Check

'K capture'.

Where, under energy induced conditions, a Proton will absorb an Electron and turn into a Neutron.

Energy + Proton + Electron turns into a Neutron + Neutrino

9 + 2 + 3 turns into 1 + 4

14   =   5 Mod 9 = 5 - Check

Here too we must check when Energy is zero:-

0 + 2 + 3 turns into 1 + 4

5   = 5 Check

'When a Proton meets an Electron they bond to each other, forming either a dipole or a hydrogen atom or even a neutron as in the K-capture, depending on the energy of the bonding' - Talal Ghannam

This in my mind is the spark of creation and is supportive of the idea that the Number 5 is central to creation and the first Prime in physical reality.

Proton - Value 2 + Electron Value 3 = 5

A meeting between a Particle and its Mirror results in a Photon - 1 and 8, 2 and 7, 3 and 6, 4 and 5 all add to 9 making this the number of the Photon, or Light.

Number 9 - The Number of Energy.

"It is interesting to see how similar the effect Number 9 has on the first eight numbers to the effect energy has on particles, which we explain as follows:

Firstly: when 9 is added to one of the 8 numbers it raises it numerically higher, e.g. 9 + 2 = 11, which is just what energy does to particles, it raises them into higher energetic states.

Secondly: adding 9 to a number will not change the digital root of the number (Talal's way of doing Mod 9), in the same way that energy will not change the nature of the particle; an electron stays an electron, it just gains more energy. (Nevertheless, in the extreme case when energy is too much, it can break some particles apart like in the B+ decay. But this is because these particles are composite to start with, made from so called quarks.)

Thirdly: adding 9 to these numbers will change their parity from even to odd and vice versa, e.g. 9 + 2 = 11 (odd). Amazingly, this is very similar to what happens in atoms, as when an electron absorbs a photon, the electron will want to go up into a higher state. However, the parity of this new state must be opposite to that of the original state; odd if it is even and vice versa, otherwise the transition is considered prohibited with zero probability. It is like saying that when an electron absorbs energy, its parity is reversed. This parity reversal is not restricted to electrons only, but applies to all other particles.

Hence, it seems that the correspondence between our nine numbers and the subatomic world is almost exact". - The Mystery of Numbers - Talal Ghannam.

Chapter 8: Canonical Numbers & Mod 99.

In light of Talal's work, it is proposed that where the Number Pairs are undivided, so 27 as opposed to 297 for example, they are at their most potent at the atomic level. When they come together, 'undiluted' so to speak, they produce a photon.

As examples, 3663, 4545, 5544, 3681 and 7227 are all canonical numbers as they are balanced in their number pairings.

As examples, 3143, 7203, 4288, 8093 and 7117 would not be canonical numbers as they are not balanced in their number pairings.

However, where 0 or 9 appear with an otherwise number pair balanced number that too would fall within the classification for a canonical number.

As examples, 639, 396, 270, 279 and 207 would all be considered canonical numbers.

After canonical two digit numbers like 27, numbers which include a 9 or 0 would be the next most pure interaction.

When looking at numbers in general, canonical or not, we can now translate these into particle interactions.

As an example 217 indicates a combination of Proton, Neutron and Anti Proton.

2 Digit Numbers & Mod 99.

Given that Talal's findings at the particle physics level show the number pairs and how when they interact they produce physical reactions, it played into something that revealed itself to me very slowly as I perused a large wall containing all the various number sequences looking for more patterns that might reveal themselves.

Earlier I showed the quick way to get the Mod 9 result for any given number. What I didn't realise initially, but later discovered through experimentation, was that if we wanted to find the Mod 99 result and therefore finish with a two digit number, we could do something similar by taking the digits from any given number, adding them in pairs, working backwards.

For example:-

1234 would be 34 + 12 = 46 Mod 99 or 12345 would be 45 + 23 + 1 = 69 Mod 99

On discovery of this, initially in the Hexagram, or 6 pointed star, sequence, it showed a repeating pattern as we shall see overleaf.

I then applied Mod 99 to all the important number sequences.

I am going to show that for example, the number 234567 is 'related' fractally to the number 36 in a way I think everyone might think crazy were it not for the irrefutable proof of its congruence which follows.

This is an aspect of the decimal system of numbers that is probably under-appreciated.

Hexagram Number Sequence.

The Hexagram number sequence is a key number sequence for my Integer Partition Theory, which occupies the third and final section of the book, and as such it is one which had been on my wall for a long time.

I applied the Mod 99 technique to this sequence first of all and was instantly rewarded with a geometric proof for what I was looking for.

The Numerical Universe

Term	Hexagram	Mod 9	Mod 99 1st Iteration	Mod 99 2nd Iteration
1	1	1	1	
2	13	4	13	
3	37	1	37	
4	73	1	73	
5	121	4	22	
6	181	1	82	
7	253	1	55	
8	337	4	40	
9	433	1	37	
10	541	1	46	
11	661	4	67	
12	793	1	100	1
13	937	1	46	
14	1093	4	103	4
15	1261	1	73	
16	1441	1	55	
17	1633	4	49	
18	1837	1	55	
19	2053	1	73	
20	2281	4	103	4
21	2521	1	46	
22	2773	1	100	1
23	3037	4	67	
24	3313	1	46	
25	3601	1	37	
26	3901	4	40	
27	4213	1	55	
28	4537	1	82	
29	4873	4	121	22
30	5221	1	73	
31	5581	1	136	37
32	5953	4	112	13
33	6337	1	100	1
34	6733	1	100	1
35	7141	4	112	13
36	7561	1	136	37
37	7993	1	172	73
38	8437	4	121	22
39	8893	1	181	82
40	9361	1	154	55
41	9841	4	139	40
42	10333	1	37	37

The Hexagram Sequence.

On the left we see the Hexagram sequence of numbers.

We see the repeating Mod 9 sequence in the third column showing 1 4 1 every three terms.

Then we have the first iteration of applying the Mod 99 technique and where the total is more than 99 the second iteration shows the resulting number giving the final Mod 99 result.

For example, on the 12th Term of the sequence 793, the first iteration gives a total of 93 + 7 = 100 and then applying the rule again gives us 00 + 1 = 1.

The analysis shows a clear 33 number palindromic sequence that repeats indefinitely.

The 33 number repeating sequence is centred on the 17th term by the number 49.

1 13 37 73 22 82 55 40 37 46 67 1 46 4 73 55

49

55 73 4 46 1 67 46 37 40 55 82 22 73 37 13 1

The number 49 is of particular interest appearing at this central space for this particular geometric sequence. Clearly 49 = 7 x 7 or 7^2 reinforcing this idea that so much about the natural world focuses on the interaction of the numbers 2 and 3 with the number 7. What is even more interesting and intriguing however is that the human eye can sense only one octave of the electro-magnetic spectrum, the 49th.

The Electro-magnetic (EM) Spectrum is the range of all possible electro-magnetic radiation. Electro-magnetic radiation can be divided into octaves, as sound waves are, adding up to a total of 81 octaves. It is interesting to note that there are only 81 stable chemical elements. Food for thought when considering Walter Russell's Octave Wave Theory. Light is Sound at a higher oscillation.

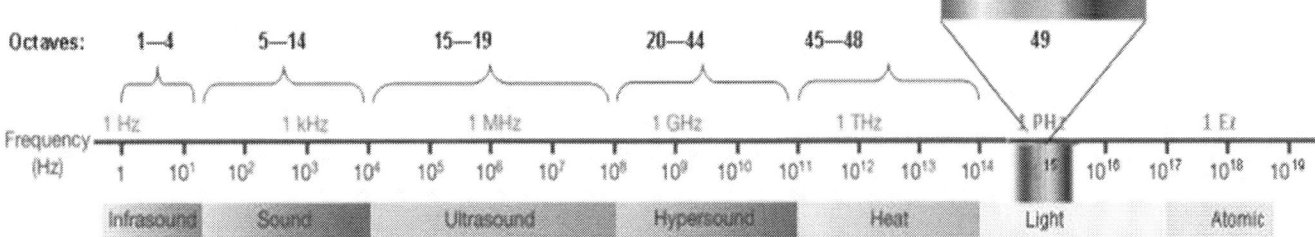

http://www.cocreatorsworld.com/science-of-harmony/the-forty-nine-octaves-of-sound-and-light/

Platonic Solids - Mod 99.

Tetrahedral	Mod 99	Octahedral	Mod 99	Cube	Mod 99	Dodecahedral	Mod 99	Icosahedral	Mod 99
1	1	1	1	1	1	1	1	1	1
4	4	6	6	8	8	12	12	20	20
10	10	19	19	27	27	48	48	84	84
20	20	44	44	64	64	124	25	220	22
35	35	85	85	125	26	255	57	455	59
56	56	146	47	216	18	456	60	816	24
84	84	231	33	343	46	742	49	1330	43
120	21	344	47	512	17	1128	39	2024	44
165	66	489	93	729	36	1629	45	2925	54
220	22	670	76	1000	10	2260	82	4060	1
286	88	891	99	1331	44	3036	66	5456	11
364	67	1156	67	1728	45	3972	12	7140	12
455	59	1469	83	2197	19	5083	34	9139	31
560	65	1834	52	2744	71	6384	48	11480	95
680	86	2255	77	3375	9	7890	69	14190	33
816	24	2736	63	4096	37	9616	13	17296	70
969	78	3281	14	4913	62	11577	93	20825	35
1140	51	3894	33	5832	90	13788	27	24804	54
1330	43	4579	25	6859	28	16264	28	29260	55
1540	55	5340	93	8000	80	19020	12	34220	65
1771	88	6181	43	9261	54	22071	93	39711	12
2024	44	7106	77	10648	55	25432	88	45760	22
2300	23	8119	1	12167	89	29118	12	52394	23
2600	26	9224	17	13824	63	33144	78	59640	42
2925	54	10425	30	15625	82	37525	4	67525	7
3276	9	11726	44	17576	53	42276	3	76076	44
3654	90	13131	63	19683	81	47412	90	85320	81
4060	1	14644	91	21952	73	52948	82	95284	46
4495	40	16269	33	24389	35	58899	93	105995	65
4960	10	18010	91	27000	72	65280	39	117480	66
5456	11	19871	71	29791	91	72106	34	129766	76
5984	44	21856	76	32768	98	79392	93	142880	23
6545	11	23969	11	35937	99	87153	33	156849	33

None of the Platonic Solid Number sequences under the Mod 99 microscope exhibit moduli, with the exception of the Cube where we see a 33 Modulus, where the 2 digit sequence repeats or wraps around and begins again.

This is of critical importance and something that sets the Cube apart from its Platonic brothers and sisters.

1 8 27 64 26 18 46 17 36 10 44 45 19 71 9 37 62 90 28 80 54 55 89 63 82 53 81 73 35 72 91 98 99

However, I noticed that if we exclude the final 99 that punctuates the sequences, we see the number pairs, each two digit pair adding to 99 to show a unique type of numerical symmetry as we move towards the centre pairing.

01 08 27 64 26 18 46 17 36 10 44 45 19 71 09 37

98 91 72 35 73 81 53 82 63 89 55 54 80 28 90 62

Number Groups Analysis - Mod 99.

I then applied the 2 Digit Technique to these specific combinations of numbers to see:-

3 6 9(0).

$$369 + 396 + 963 + 936 + 639 + 693 = 3996 = 135 = 36 \text{ Mod } 99$$

$$72 + 99 + 72 + 45 + 45 + 99 = 432 = 36 \text{ Mod } 99$$

$$360 + 306 + 63 + 36 + 630 + 603 = 1998 = 117 = 18 \text{ Mod } 99$$

$$63 + 9 + 63 + 36 + 36 + 9 = 216 = 18 \text{ Mod } 99$$

1 4 7.

$$147 + 174 + 741 + 714 + 471 + 417 = 2664 = 90 \text{ Mod } 99$$

$$48 + 75 + 48 + 21 + 75 + 21 = 288 = 90 \text{ Mod } 99$$

2 5 8.

$$258 + 285 + 528 + 582 + 825 + 852 = 3330 = 63 \text{ Mod } 99$$

$$60 + 87 + 33 + 87 + 33 + 60 = 360 = 63 \text{ Mod } 99$$

The sums of each of the Number Group combinations in their raw form reduce to 36, 63 and 90 which is really elegant being the magnetic infrastructure which contains the electric aspect.

When Mod 99 is applied to each of the combinations then we see:-

3 6 9 Number Group sums to 432, the frequency for Pythagorean Tuning A of Middle C.

3 6 0 Number Group sums to 216, 1/10th the mean diameter of the Moon and an octave below the 3 6 9 Number Group.

2 5 8 Number Group sums to 360, the degrees contained in any circle or quadrilateral.

1 4 7 Number Group sums to 288, the dimensions of New Jerusalem, a number whose importance will be fully revealed a little later on in the book.

Anthony Morris

Chapter 9: Recurring Decimals.

During an early investigation into the reciprocals of Prime Numbers I was inspired to produce an analysis of the numbers 0 - 99 and their divisive effect on all other numbers. I strongly suspect that recurring decimals are evidence and indicative of geometry and this analysis was essential in light of what is to come later on in the book. Therefore, when we see a 3 digit recurring decimal, I am thinking that represents a triangle. A 4 digit recurring decimal would represent a square, 5, a pentagon, 6 a Hexagon etc.

Below is the table of the analysis of the first 99 numbers and how many digits their recurring decimals cause. There are some surprising observations as you will see.

RECURRING DECIMAL ANALYSIS

None	1	2	4	5	8	10	16	20	25	32	40	50	64	80		
1 Dig	3	6	9	12	15	18	24	30	36	45	48	60	72	75	90	96
2 Dig	11	22	33	44	55	66	88	99								
3 Dig	27	37	54	74												
5 Dig	41	82														
6 Dig	7	13	14	21	28	35	39	42	52	56	63	65	70	77	78	84	91
8 Dig	73															
9 Dig	81															
12 Dig	26															
13 Dig	79															
14 Dig	53															
15 Dig	31	62	93													
16 Dig	17	34	51	68	85											
18 Dig	19	38	57	76	95											
21 Dig	43	86														
22 Dig	23	46	69	92												
28 Dig	29	58	87													
33 Dig	67															
35 Dig	71															
41 Dig	83															
42 Dig	49	98														
44 Dig	89															
46 Dig	47	94														
58 Dig	59															
60 Dig	61															
96 Dig	97															

27 & 37 are unique root numbers for triangles - 3 digit recurring decimals.

7 & 13 are unique root numbers for Hexagons - 6 digit recurring decimals with two dimensional numerical symmetry.

21 stands out as the only producer of a 6 digit recurring decimal which lacks two dimensional numerical symmetry.

41 uniquely produces a 5 digit recurring decimal - a pentagon.

73 uniquely produces a 8 digit recurring decimal with two dimensional numerical symmetry - an octagon.

81 uniquely produces a 9 digit recurring decimal with three dimensional numerical symmetry - a nonagon.

77 produces is a 6 digit recurring decimal with two dimensional numerical symmetry - a hexagon. Ordinarily we might expect a Line, being a multiple of 11.

49 produces a 42 digit recurring decimal with two dimensional numerical symmetry when all other multiples of 7, except 7 x 7 produce hexagons.

n/101 = 4 digit recurring decimals - Squares.

n/137 = 8 digit recurring decimals - Octagons.

All the numbers that produce 6 digit recurring decimals exhibit the two dimensional numerical symmetry showing the number pairs.

Number pairs always add up to 9 so the total of the two halves placed one above the other will always be 999 for a six digit recurring decimal. For example:-

$$1/7 = 0.142857 \text{ Recurring}$$

$$1\ 4\ 2$$

$$8\ 5\ 7$$

$$\mathbf{9\ 9\ 9}$$

All that is, except one.

Numbers divided by the number 21 produce 6 digit recurring decimals which uniquely lack this two dimensional numerical symmetry.

1/ 21 = 0.047619 Recurring

0 4 7

6 1 9

BUT THERE IS NO 2 OR 3 DIMENSIONAL SYMMETRY HERE!

The two dimensional numerical symmetry always totals 999 but here we see:-

0 4 7

6 1 9

6 6 6

666 - A well-known number with devilish connotations!

End of First Section.

Much of what has been set out in this section will, I hope, help to make sense and give a deeper appreciation of the remaining two parts of the book. I apologise if it has been a bit dry, the later sections of the book, I hope, will be much more interesting for those not obsessed with numbers and patterns.

The main purpose of this section has been to show the patient reader how all natural integer and geometric sequences, essential in the universe, operate in pairs and groups when we use modular arithmetic to show this order that cannot otherwise be appreciated from the seeming chaos of numbers.

I have demystified the Prime Number puzzle and shown how musical octaves behave and introduced a numerical structure of all things that exist in physical reality.

In addition, through the work of Talal, I have shown the reader how these Number Pairs can be applied to the world of particle physics and armed with this understanding have realised the importance of Mod 99.

Part 2: DNA & Amino Acids - The Game Unpacked.

AAA	AAC	AAT	AAT	AAG	ACA	ACC	ACT
ACG	ATA	ATC	ATT	ATG	AGA	AGC	AGG
CAA	CAC	CAT	CAG	CCA	CCC	CCT	CCG
CTA	CTC	CTT	CTG	CGA	CGC	CGT	CGG
TAA	TAC	TAT	TAG	TCA	TCC	TCT	TCG
TTA	TTC	TTT	TTG	TGA	TGC	TGT	TGG
GAA	GAC	GAT	GAG	GCA	GCC	GCT	GCG
GTA	GTC	GTT	GTG	GGA	GGC	GGT	GGG

Chapter 1: Introduction to DNA.

This part of the book came about courtesy of a challenge from good friend Neil Shotton who I had showed my early findings in the first part to. He could see that there may be something interesting in what I was doing but told me that I needed to employ / apply it somewhere tangible and useful, like DNA.

Again, this isn't something I knew anything about but I set about researching the subject to see what could possibly be unearthed.

Very briefly Deoxyribonucleic acid (DNA) is a molecule that carries the genetic instructions used in the growth, development, functioning and reproduction of all known living organisms.

DNA codes for the 20 Amino Acids that must be present wherever Life of any kind is found. Plants, Animals and Humans.

These 20 Amino Acids are coded for using a combination of 3 out of 4 possible Bases which produce what is called a Codon.

The Bases are:-

Adenine (A)   Cytosine (C)   Guanine (G)   Thymine (T)

Any 3 from 4 gives 4 x 4 x 4 = 64 Possibilities

For example, the Codon AAA codes for the amino acid called Lysine while AAC would code for Asparagine.

61 of these 64 Codons encode Amino Acids while the other 3 are 'Stop' codons which terminate the growing chain of Amino Acids at that point.

The 4 bases are paired and exclusive to each other, so that:-

Adenine links with Thymine   Cytosine links with Guanine

My mind turned immediately to the ancient game of Chess which is played out on an 8 x 8 square.

Having played a bit of chess as a kid, I also knew that there are only 20 opening moves in Chess which also seemed an interesting coincidence. 16 possible Pawn moves and 4 possible Knight moves.

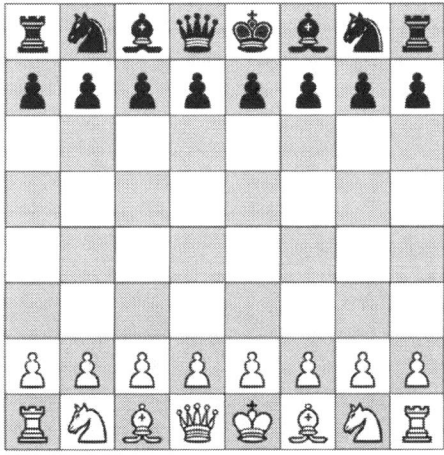

Immediately I thought to lay the Codons out on the chessboard as shown opposite.

AAA	AAC	AAT	AAT	AAG	ACA	ACC	ACT
ACG	ATA	ATC	ATT	ATG	AGA	AGC	AGG
CAA	CAC	CAT	CAG	CCA	CCC	CCT	CCG
CTA	CTC	CTT	CTG	CGA	CGC	CGT	CGG
TAA	TAC	TAT	TAG	TCA	TCC	TCT	TCG
TTA	TTC	TTT	TTG	TGA	TGC	TGT	TGG
GAA	GAC	GAT	GAG	GCA	GCC	GCT	GCG
GTA	GTC	GTT	GTG	GGA	GGC	GGT	GGG

Below we can see which Codon codes for which amino acid.

Lysine	Asparagine	Asparagine	Lysine	Threonine	Threonine	Threonine	Threonine
Isoleucine	Isoleucine	Isoleucine	Methionine	Arginine	Serine	Serine	Arginine
Glutamine	Histidine	Histidine	Glutamine	Proline	Proline	Proline	Proline
Leucine	Leucine	Leucine	Leucine	Arginine	Arginine	Arginine	Arginine
Serine	Serine	Serine	Serine	STOP	Tryptophan	Cysteine	Cysteine
Phenylalanine	Phenylalanine	Leucine	Leucine	STOP	STOP	Tyrosine	Tyrosine
Glutamic Acid	Aspartic Acid	Aspartic Acid	Glutamic Acid	Alanine	Alanine	Alanine	Alanine
Valine	Valine	Valine	Valine	Glycine	Glycine	Glycine	Glycine

With the exception of Tryptophan which only appears once, all other amino acids are coded for, between 2 and 6 times.

I felt that seeing the Amino Acids in this way would give the most accurate rendering of the whole 'game' of DNA and I could not have been more fortunate in the results of doing so.

Each Amino Acid has a known and undisputed chemical formula. I was the type of kid who when I got the game of Monopoly I wanted to find out exactly how much money was possible to accrue and so in this analysis I wanted to count the total number of nucleons (Protons and Neutrons) that this group of amino acids required.

Armed with the chemical formulas it was then straightforward to make a count of these nucleons as when one examines each of the chemical elements under a microscope we can see a specific number of nucleons in each.

In the analysis that follows, all I have done is to simply count up the Protons and Neutrons for each amino acid, on a per square / codon basis, for each of the 64 Codons / Squares on the board.

Chemical Structure.

Amino Acids are fortunately not very complex chemical equations and use just 5 chemical elements for their construction.

Chemical Element	Protons	Neutrons
Hydrogen	1	0
Carbon	6	6
Nitrogen	7	7
Oxygen	8	8
Sulphur	16	16

Amino Acid Components.

There are 2 Component Parts to an Amino Acid. The Standard Block and the Side Chain.

Standard Block. - Common to 19 of the 20 Amino Acids (the odd man out is Proline, though the molecular / atomic count remains unaltered).

Side Chain. - This is where all the 'magic' happens that gives each of the Amino Acids their 'individuality'.

The Standard Block can be thought of as a household drill, on to which may be placed different shaped 'drill bits' to achieve a desired effect. These 'drill bits' are the Side Chains of the Amino Acids.

As an example, the most simply constructed Amino Acid, Glycine, consists of simply the Standard Block with a single Hydrogen atom attached.

Chapter 2: DNA Analysis.

Understanding the Tables.

In order to understand the following tables what I have done is to look at each group of sixteen, Base sorted codons, so the 16 codons beginning with A for Adenine and then the 16 codons beginning with C for Cytosine, T for Thymine and G for Guanine.

As an example, in the Side Chain Analysis section we look at the formulas for just the Side Chain aspect of the amino acid, coded for example by, AAA - an Adenine led Codon - which codes for an amino acid called Lysine. I then add up all the Protons and Neutrons (Nucleons) used by this group of 16 Adenine A led codons and then their Thymine, Guanine and Cytosine counterparts, to produce the table below.

SIDE CHAIN	H	P	N	Sum	C	P	N	Sum	N	P	N	Sum	O	P	N	Sum	S	P	N	Sum	Total P	Total N	Nucleon Total
ADENINE	108	108	0	108	45	270	270	540	10	70	70	140	8	64	64	128	1	16	16	32	528	420	948
THYMINE	72	72	0	72	51	306	306	612	1	7	7	14	6	48	48	96	2	32	32	64	465	393	858
A + T	180	180	0	180	96	576	576	1152	11	77	77	154	14	112	112	224	3	48	48	96	993	813	1806
	H	P	N	Sum	C	P	N	Sum	N	P	N	Sum	O	P	N	Sum	S	P	N	Sum	Total P	Total N	Nucleon Total
CYTOSINE	120	120	0	120	58	348	348	696	18	126	126	252	2	16	16	32	0	0	0	0	610	490	1100
GUANINE	60	60	0	60	26	156	156	312	0	0	0	0	8	64	64	128	0	0	0	0	280	220	500
C + G	180	180	0	180	84	504	504	1008	18	126	126	252	10	80	80	160	0	0	0	0	890	710	1600
TOTALS	360	360	0	360	180	1080	1080	2160	29	203	203	406	24	192	192	384	3	48	48	96	1883	1523	3406

KEY - H-Hydrogen P-Proton N-Neutron C-Carbon N-Nitrogen O-Oxygen S-Sulphur.

So, for example, reading left to right, the group of 16 Adenine (A) led codons use exactly 108 Hydrogen atoms (top left corner under H).

I have then noted the Proton (P) and Neutron (N) Count in the next cells and a Sum of Nucleons column before we come to Carbon (C) where we see that the Group of 16 Adenine led codons uses exactly 45 Carbon atoms, 270 Protons and 270 Neutrons as Carbon has 6 of each and 45 x 6 = 270 of each for a total of 540 Nucleons.

Next we see the group uses 10 Nitrogen (N), 8 Oxygen (O), 1 Sulphur (S) and so on. I look at the total number of Protons, which in the case of Adenine, is 528 and then Neutrons, totalling 420, for a grand total of 948 Nucleons.

We can then see that Thymine (T) led codons use exactly 72 Hydrogen atoms, 51 Carbon atoms, 1 Nitrogen, 6 Oxygen and 2 Sulphur.

The next row A + T is the totals for the 32 Adenine and Thymine led codons, so for Hydrogen we see a total of 180 Hydrogen and 90 Carbon etc.

For the Cytosine and Guanine base pairing, we see C + G showing a total of 180 Hydrogen, 84 Carbon 18 Nitrogen 10 Oxygen and no Sulphur for a total of 890 Protons and 710 Neutrons for a total of 1600 Nucleons.

I have analysed the numerical attributes of the Side Chain and the Standard Block individually and then looked at the combined totals. First we will begin with the Side Chain Analysis.

Chapter 3: Side Chain Analysis.

In my view, this is where all the magic happens, where all the metaphorical drill bits for the drill that is the Standard Block are formed that create the diversity of Amino Acids that underpin all biology.

SIDE CHAIN	H	P	N	Sum	C	P	N	Sum	N	P	N	Sum	O	P	N	Sum	S	P	N	Sum	Total P	Total N	Nucleon Total
ADENINE	108	108	0	108	45	270	270	540	10	70	70	140	8	64	64	128	1	16	16	32	528	420	948
THYMINE	72	72	0	72	51	306	306	612	1	7	7	14	6	48	48	96	2	32	32	64	465	393	858
A + T	180	180	0	180	96	576	576	1152	11	77	77	154	14	112	112	224	3	48	48	96	993	813	1806
	H	P	N	Sum	C	P	N	Sum	N	P	N	Sum	O	P	N	Sum	S	P	N	Sum	Total P	Total N	Nucleon Total
CYTOSINE	120	120	0	120	58	348	348	696	18	126	126	252	2	16	16	32	0	0	0	0	610	490	1100
GUANINE	60	60	0	60	26	156	156	312	0	0	0	0	8	64	64	128	0	0	0	0	280	220	500
C + G	180	180	0	180	84	504	504	1008	18	126	126	252	10	80	80	160	0	0	0	0	890	710	1600
TOTALS	360	360	0	360	180	1080	1080	2160	29	203	203	406	24	192	192	384	3	48	48	96	1883	1523	3406

KEY - H-Hydrogen P-Proton N-Neutron C-Carbon N-Nitrogen O-Oxygen S-Sulphur.

360 Hydrogen Atoms.

The total number of Hydrogen atoms utilised in the Side Chain across the 4 bases A C G and T totals 360 exactly.

Obviously this is inextricably linked to the most important of geometries, as far as life is concerned.

The 360 degrees of the circle! Fascinating!

I propose that Hydrogen now needs to be thought of in terms of circles and angles where perhaps Hydrogen content within compounds controls the angles.

This aspect I hope will become clearer a little further on.

180 Carbon Atoms.

Exactly half the count of Hydrogen and now lending itself to the idea of a semi-circle of 180 degrees.

Again, I was amazed, and now delighted to see the Octave Relationship between Carbon and Hydrogen proposed in the phenomenal work of Walter Russell, whose Octave Wave Theory we looked at in the first section of the book.

His work also illuminates the Octave relationship between Carbon and Hydrogen as depicted in the chart opposite.

Walter Russell's Octave Wave Theory.

To my knowledge there is only one philosopher ever to have assigned an octave relationship between Hydrogen and Carbon, that being Walter Russell. In the diagram below, we can see Hydrogen at the beginning of the 4th Octave with Carbon at the beginning of the 5th, separated exactly by an Octave.

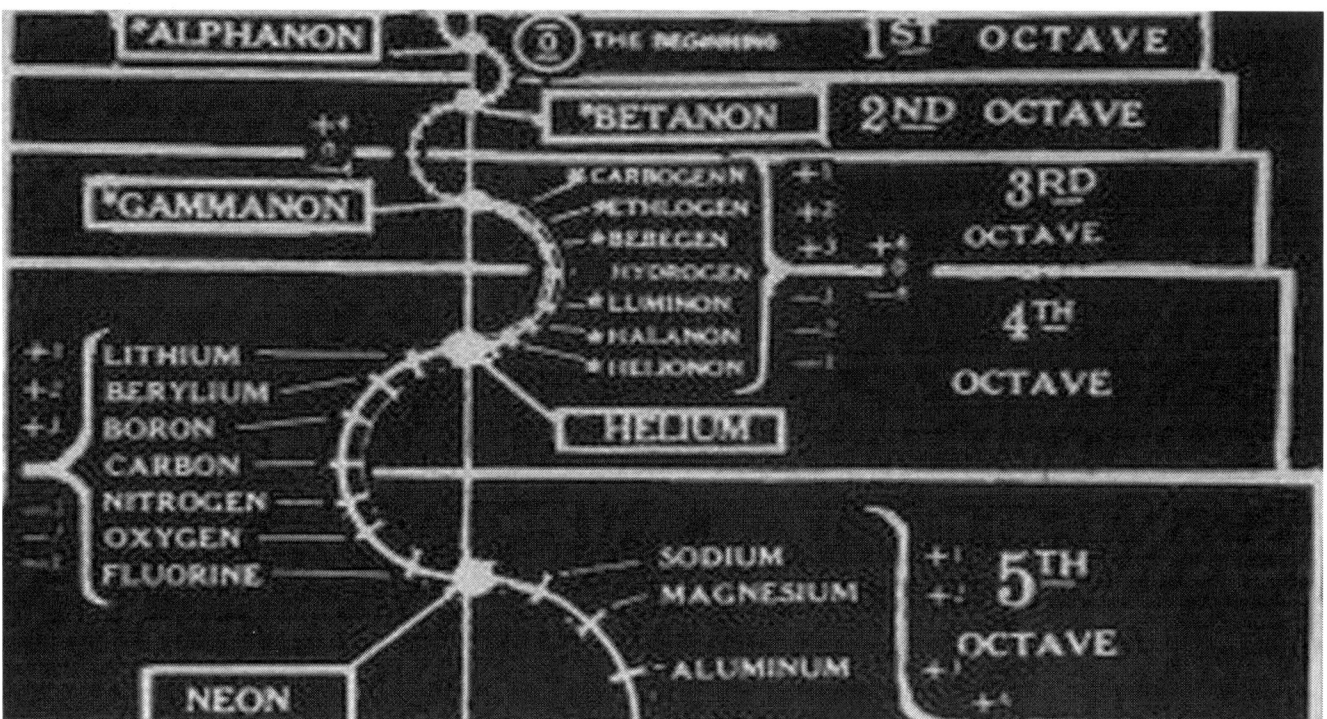

Adenine and Thymine.

The Hydrogen count for Thymine led codons is 72 while its pair Adenine has a Hydrogen count of 108 illuminating an important musical ratio.

$$108 : 72 \;=\; 3 : 2 \;=\; \text{The Perfect Fifth}$$

The Pivot or centre point of the Hydrogen count between Adenine 108 & Thymine 72 is at 90.

$$90 \times 6/5 = 108 \qquad 90 \times 4/5 = 72$$

6:5 is the Minor Third **5:4 is the Major Third**

Cytosine & Guanine.

In the Cytosine & Guanine pairing we see that the total Hydrogen count is in the relationship.

$$120 : 60 \;=\; 2 : 1 \;=\; \text{The Octave}$$

The Pivot for the Hydrogen Count between Cytosine 120 and Guanine 60 is at 90.

$$90 \times 4/3 = 120 \text{ and } 90 \times 2/3 = 60$$

3:2 is the Perfect Fifth **4:3 is the Perfect Fourth**

Side Chain Connection to the Number Groups.

Earlier I showed some analysis of the permutations of the Number Groups, where coincidentally I had noted that:

3 6 9

$$369 + 396 + 963 + 936 + 639 + 693 = 3996 = 108 \times 37$$

The Total Hydrogen Count for Adenine led codons is 108.

2 5 8

$$258 + 285 + 528 + 582 + 825 + 852 = 3330 = 90 \times 37$$

The Hydrogen Count Pivot between Adenine and Thymine led codons is 90.

1 4 7

$$147 + 174 + 741 + 714 + 471 + 417 = 2664 = 72 \times 37$$

The Total Hydrogen Count for Thymine led codons is 72.

Number 37.

The number 37, you will remember, is one of only two numbers, with 27, to produce 3 digit recurring decimals as their effect on all other numbers by division, triangles.

Both these numbers will be shown to be pivotal later on.

Chapter 4: Standard Block Analysis.

STANDARD BLOCK	H	P	N	Sum	C	P	N	Sum	N	P	N	Sum	O	P	N	Sum	S	P	N	Sum	Total P	Total N	Nucleon Total
ADENINE	64	64	0	64	32	192	192	384	16	112	112	224	32	256	256	512	0	0	0	0	624	560	1184
THYMINE	52	52	0	52	26	156	156	312	13	91	91	182	26	208	208	416	0	0	0	0	507	455	962
A + T	116	116	0	116	58	348	348	696	29	203	203	406	58	464	464	928	0	0	0	0	1131	1015	2146
	H	P	N	Sum	C	P	N	Sum	N	P	N	Sum	O	P	N	Sum	S	P	N	Sum	Total P	Total N	Nucleon Total
CYTOSINE	64	64	0	64	32	192	192	384	16	112	112	224	32	256	256	512	0	0	0	0	624	560	1184
GUANINE	64	64	0	64	32	192	192	384	16	112	112	224	32	256	256	512	0	0	0	0	624	560	1184
C + G	128	128	0	128	64	384	384	768	32	224	224	448	64	512	512	1024	0	0	0	0	1248	1120	2368
TOTALS	244	244	0	244	122	732	732	1464	61	427	427	854	122	976	976	1952	0	0	0	0	2379	2135	4514

KEY - H-Hydrogen P-Proton N-Neutron C-Carbon N-Nitrogen O-Oxygen S-Sulphur.

Again we see the Octave Relationship between Hydrogen and Carbon in the Totals for the 4 Bases.

244 Hydrogen v 122 Carbon

Prime Numbers 37 and 61.

STANDARD BLOCK	H	P	N	Sum	C	P	N	Sum	N	P	N	Sum	O	P	N	Sum	S	P	N	Sum	Total P	Total N	Nucleon Total
TOTALS	244	244	0	244	122	732	732	1464	61	427	427	854	122	976	976	1952	0	0	0	0	2379	2135	4514

All of the numbers found in the Standard Block are related to only two numbers, 37 again, and 61.

61.

Hydrogen Total is 244 = 4 x 61   Hydrogen Protons total 244 = 4 x 61   Hydrogen Neutrons is 0.

Carbon Total is 122 = 2 x 61   Carbon Protons and Neutrons each total 732 = 12 x 61

Nitrogen Total is 61 = 1 x 61   Nitrogen Protons and Neutrons each total 427 = 7 x 61

Oxygen Total is 122 = 2 x 61   Oxygen Protons and Neutrons each total 976 = 16 x 61

Total Protons is 2379 = 39 x 61   Total Neutrons is 2135 = 35 x 61   Total Nucleons is 4514 = 74 x 61

37.

STANDARD BLOCK	Total P	PIVOT	Total N	Nucleon Total
ADENINE	624	592	560	1184
THYMINE	507	481	455	962
Total Nucleons				2146
CYTOSINE	624	592	560	1184
GUANINE	624	592	560	1184
Total Nucleons				2368
TOTALS	2379	2257	2135	4514

The Pivot between Protons and Neutrons in Adenine led Codons is 592 = 16 x 37

The Pivot between Protons and Neutrons in Thymine led Codons is 481 = 13 x 37

The Pivot between Protons and Neutrons in Cytosine led Codons is 592 = 16 x 37

The Pivot between Protons and Neutrons in Guanine led Codons is 592 = 16 x 37

The Pivot between the Total Number of Protons and Neutrons is 2257 = 61 x 37

The Total Number of Nucleons for Adenine, Cytosine and Guanine led Codons is 1184 = 32 x 37

The Total Number of Nucleons for Thymine led Codons is 962 = 26 x 37

As I have shown, the Standard Block Totals are dominated by multiples of the prime numbers 37 and 61.

Interestingly, these numbers that are connected via their figuracy, producing the 4th and 5th Centred Hexagons respectively.

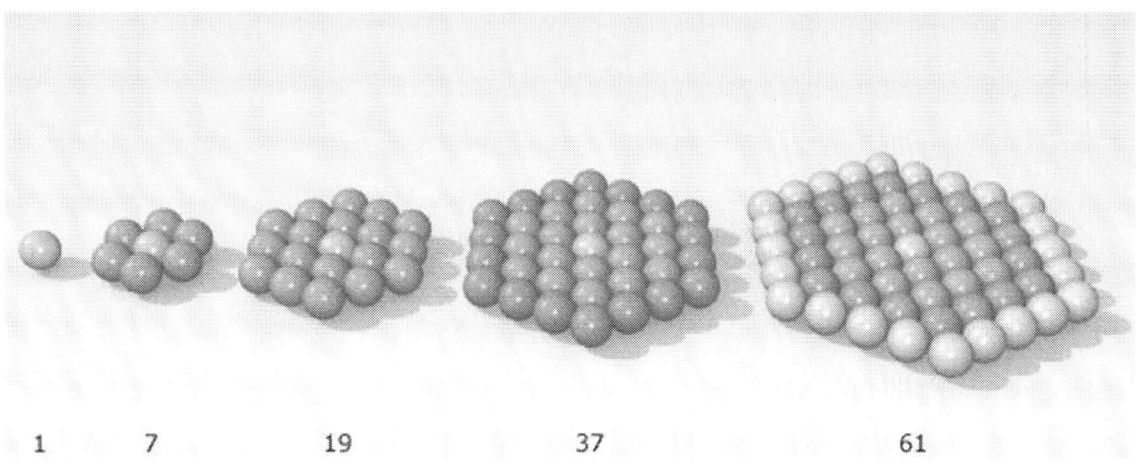

http://www.nextnumber.com/show?1D

Chapter 5: Combined Analysis.

COMBINED	H	P	N	Sum	C	P	N	Sum	N	P	N	Sum	O	P	N	Sum	S	P	N	Sum	Total P	Total N	Nucleon Total
ADENINE	172	172	0	172	77	462	462	924	26	182	182	364	40	320	320	640	1	16	16	32	1152	980	2132
THYMINE	124	124	0	124	77	462	462	924	14	98	98	196	32	256	256	512	2	32	32	64	972	848	1820
A + T	296	296	0	296	154	924	924	1848	40	280	280	560	72	576	576	1152	3	48	48	96	2124	1828	3952
	H	P	N	Sum	C	P	N	Sum	N	P	N	Sum	O	P	N	Sum	S	P	N	Sum	Total P	Total N	Nucleon Total
CYTOSINE	184	184	0	184	90	540	540	1080	34	238	238	476	34	272	272	544	0	0	0	0	1234	1050	2284
GUANINE	124	124	0	124	58	348	348	696	16	112	112	224	40	320	320	640	0	0	0	0	904	780	1684
C + G	308	308	0	308	148	888	888	1776	50	350	350	700	74	592	592	1184	0	0	0	0	2138	1830	3968
TOTALS	604	604	0	604	302	1812	1812	3624	90	630	630	1260	146	1168	1168	2336	3	48	48	96	4262	3658	7920

KEY - H-Hydrogen P-Proton N-Neutron C-Carbon N-Nitrogen O-Oxygen S-Sulphur.

7920.

In the bottom right hand corner you will see that the total number of Protons and Neutrons utilised by the entire 'Game of Amino Acids', that must be present for life to be perpetuated, when all are added together, is 7920 exactly.

Why is that interesting?

Well, because it really looks like we plants animals and humans are living in just the right place, numerically speaking.

Why?

7920 British Miles is the Mean Diameter of the Earth!

I know many people will immediately say – wait – the mile is an arbitrary Man-made thing so this must just be some sort of coincidence – I disagree completely, the origin of the foot and the mile goes back deep into the mists of time and I believe that the imperial measurement system was either discovered, or somehow delivered into awareness at some point in great antiquity, and not invented, in the same way that Man discovered Music.

From Ancient Metrology.

7920 feet = 1 League   7920 inches = 1 Furlong   12 Furlongs = 1 League

1 Link = 7.92 inches   100 Links = 1 Chain   10 Chains = 1 Furlong

Note that 7920 inches = 1 Furlong and 7920 miles = Earth's Mean Diameter which means that 1 inch is to a Furlong as one Mile is to the Earth's Diameter.

The Great Pyramid at Giza.

The Great Pyramid of Giza, in Egypt, fascinates many of us. It is an astonishing construction built with an incredible accuracy, far beyond any achievable by our current technology.

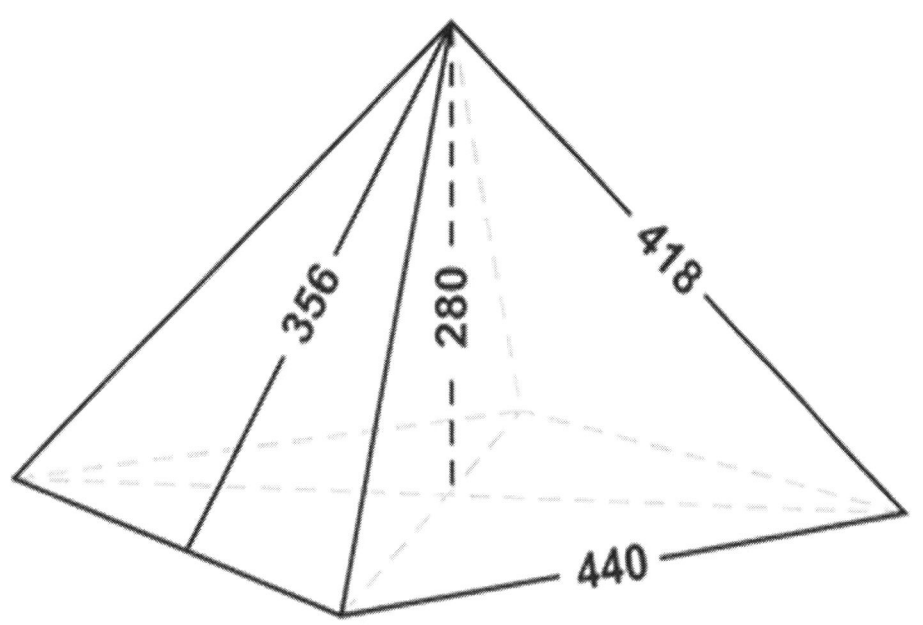

Measured in Egyptian Cubits, the ratio of base to height is 440 / 280 = 11/7 which is a good approximation for Pi / 2. This is the same relationship we saw in Walter Russell's Octave Wave Theory because 5.5 / 3.5 is the same as 11 / 7.

$$11! / 7! = 11 \times 10 \times 9 \times 8 = 7920$$

11! (11 Factorial) = 11x10x9x8x7x6x5x4x3x2x1 Divided by 7! (7 Factorial) = 7x6x5x4x3x2x1

New Jerusalem.

Here we return to the work of the late, great, John Michell, from his book The Dimensions of Paradise. The vision of the Holy City that occurred to St. John on the island of Patmos is described in Revelation 21.

'And I saw a new heaven and a new earth: for the first heaven and the first earth were passed away; and there was no more sea. And I John saw the holy city, New Jerusalem, coming down from God out of heaven, prepared as a bride adorned for her husband.

As proof that the New Jerusalem was no mere illusion or imaginative fancy, St. John gives its actual measured dimensions.

And he that talked with me had a golden reed to measure the city, and the gates thereof. And the city lieth foursquare, and the length is as large as the breadth; and he measured the city with the reed, twelve thousand furlongs. The length and the breadth of it are equal. And he measured the wall thereof, an hundred and forty four cubits, according to the measure of a man, that is, of an angel.

Evidently the New Jerusalem took the form of a cube with each of its twelve sides measuring 12,000 furlongs, each of its six faces 144 million square furlongs, and with a volume of 1.728×10^9 cubic furlongs.

Somehow associated with it is a wall of 144 cubits. However the foot is defined, a furlong is 660 feet and a cubit is 1 ½ feet, so the two measures are on different scales.

St. John in Revelation 11 is told, "Rise and measure the temple of God, and the altar, and them that worship therein". For that purpose the angel gave him a "reed like a rod", in Greek καλαμος δμοιος ραβδω.

The total numerical value of the letters in that phrase is 1,729, which by the conventions of Gematria is equivalent to 1,728 or 12^3, and 1.728 feet is the length of the Egyptian Royal Cubit (equal to 1 ½ Egyptian feet) by which the New Jerusalem is measured.

The difference in scale between the 12,000 furlongs and the 144 cubits of the New Jerusalem indicates that it represents both the macrocosm, the order of the heavens and the constitution of human nature. Both are measured by the sacred units that apply to the astronomical as well as to the human scale and thus unite the two.

When the dimensions of the New Jerusalem are made commensurable as 12 furlongs and 14,400 cubits, the geometric ground plan of the City becomes visible, for:-

14,400 cubits = 14,400 x 1.728 = 24,883.2 feet and 12 furlongs = 12 x 660 = 7920 feet

The significance of these measures is that a circle of diameter 7.920 feet has a circumference of 24,883.2 feet.

In ancient sacred or cosmological arithmetic, the π ratio between the diameter and the circumference of a circle was made rational as a simple fraction, the most convenient being 22/7.

For numerical reasons, others were also used including 864/275. This is slightly more accurate than 22/7, and it makes the above calculation exact, for 7920 x 864/275 = 24,883.2.

Thus the basic plan of the New Jerusalem is a square of 12 furlongs containing a circle of circumference 14,400 cubits.

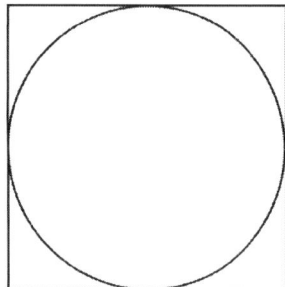

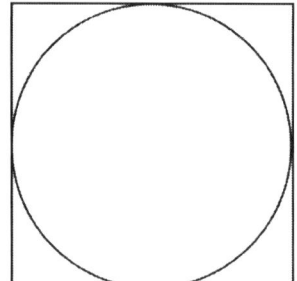

The New Jerusalem

Diameter = 7920 feet
Circumference = 24883.2 feet
Perimeter of Square = 31680 feet

The Earth

Diameter = 7920 miles
Circumference = 24883.2 miles
Perimeter of Square = 31680 miles

Measured in miles instead of feet, the wall of the New Jerusalem forms a plan of the earth, for the earth's mean diameter is some 7920 miles and its circumference through the poles is 24883.2 miles. The significance of this number is that $248832 = 12^5$.

The New Jerusalem in Stonehenge.

Michell continues - 'Another occurrence of the New Jerusalem diagram in England is in the ground plan of Stonehenge. There is evidence of an ancient connection between that old monument and Glastonbury Abbey, in that the main axis of Glastonbury, from St. Benedict's Church down the length of the abbey and along Dod Lane toward a former beacon site at Gare Hill in Wiltshire, points at Stonehenge some forty miles to the east. In their respective lore there is only one tradition of a link between the two places. Both are named in the old Welsh Triads as sites of the Perpetual Choirs of Britain, a relic of the Orphic rule, which preceded the Druid regime, when the order of society was maintained by music'.

"The present demonstration is of the pattern behind the Stonehenge ground plan, in every essential respect it is identical with that of Glastonbury and with the New Jerusalem diagram.

Briefly described, the temple of Stonehenge consists of two concentric stone circles enclosing two U-shaped stone arrangements, all contained within a circular bank and ditch.

Opposite we see how perfectly the Stonehenge ground plan correlates with the New Jerusalem diagram superimposed.

The square is that of St. John's city, or the twelve hides of Glastonbury, 7,920 feet wide, and the circle within it is the wall 14,400 cubits around, all the dimensions being reproduced on a scale of 1:100.

The outer circle has the same perimeter as the square, 316.8 feet. Stonehenge is thus founded on the classic image of sacred geometry, the squared circle representing the reconciliation of opposites, which is the common feature of temples and foundation myths the world over.

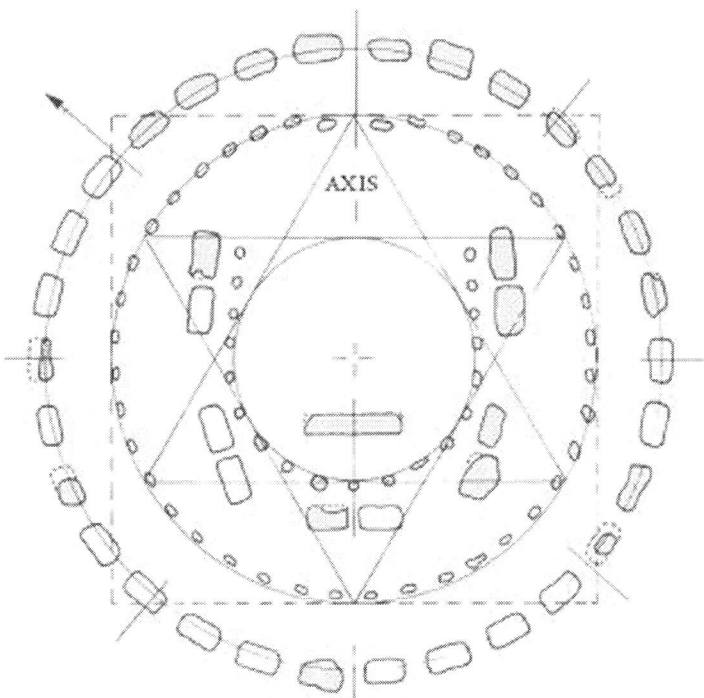

Above we see the ground plan of Stonehenge overlaid by the New Jerusalem diagram, which evidently determined the form of the temple. The mean circumference of the outer sarsen circle with lintels is 316.8 feet, a hundredth part of 6 miles, and a square with perimeter of 316.8 feet would contain a circle enclosing the bluestone ring, diameter 79.20 feet. The nineteen stones of the inner U-shaped structure enclose a circle of diameter 39.6 feet'.

The Dimensions of Paradise by John Michell published by Inner Traditions International and Bear & Company, ©2008. All rights reserved. http://www.Innertraditions.com Reprinted with permission of publisher.

We shall see 3168 again in no uncertain terms, later in the book.

Platonic Solids & 7920.

A little research into the numbers involved here turned up our friend 7920 again; this time as the sum of the angles in the faces of the first 4, classically elemental, Platonic Solids.

Platonic Solid	Element	Physical Faces	Face Angles	Sum Face Angles	Sum of Angles
Tetrahedron	Fire	4	60	180	720
Octahedron	Air	8	60	180	1440
Hexahedron	Earth	6	90	360	2160
Icosahedron	Water	20	60	180	3600
	Total	38		900	**7920**
Dodecahedron	Time	12	108	540	6480
	Total	50		1440	14400

Could it be that there is a link between the Platonic Solids and DNA or is this all just some mad coincidence?

Food for thought certainly, so I continued to look deeper.

Platonic Solids & Number Pairs.

When I looked at the proportions each element would occupy as a percentage of the four Earthly elements, I discovered that they rather elegantly show the Number Pairs which would seem to shine further light onto their inherent nature.

Platonic Solid	Proportion	Number Pairs	Element
Tetrahedron	720 / 7920	0.09090909	Fire
Octahedron	1440 / 7920	0.18181818	Air
Hexahedron	2160 / 7920	0.272727272	Earth
Icosahedron	3600 / 7920	0.454545454	Water
	2880 / 7920	0.36363636	

Interestingly, I found that the face angle sums, in quotient to the Dodecahedron which showed the numbers 1,2,3 and 5.

Platonic Solid	Proportion	Numbers	Element
Tetrahedron	720 / 6480	0.11111111	Fire
Octahedron	1440 / 6480	0.22222222	Air
Hexahedron	2160 / 6480	0.33333333	Earth
Icosahedron	3600 / 6480	0.55555555	Water
	2880 / 6480	0.444444444	

Then I looked at the face angles of each Platonic Solid, in quotient to the total.

Platonic Solid	Proportion	Numbers	Element
Tetrahedron	720 / 14400	0.05	Fire
Octahedron	1440 / 14400	0.1	Air
Hexahedron	2160 / 14400	0.15	Earth
Icosahedron	3600 / 14400	0.25	Water
	Total	0.55	
Dodecahedron	6480 / 14400	0.45	Time

The sum of the first four elemental solids as a percentage of all five including the Dodecahedron we see that they make up 55% of the total.

55 is a very interesting number and one that is literally central to my Integer Partition Theory detailed in the final part of the book.

55 is the sum of the numbers 1 to 10 and also the 10th number in the Fibonacci sequence, the spiralling sequence of growth.

$$1 + 2 + 3 + 4 + 5 + 6 + 7 + 8 + 9 + 10 = \mathbf{55}$$

$$1\ 1\ 2\ 3\ 5\ 8\ 13\ 21\ 34\ \mathbf{55}$$

When we look at the sum of the Dodecahedron face angles relative to the four elemental solids we see that:-

$$6480 / 7920 = \mathbf{0.81\ Recurring}$$

81 is an interesting number because all the elements in the periodic table above 83 are radioactive and subject to decay while numbers 43 and 61, Technetium and Promethium are not stable and do not exist in Nature, meaning that there are only 81 stable chemical elements and you will remember that there are a total of 81 octaves in the electro-magnetic spectrum.

In addition I showed in the chapter on Music that:-

$$1/81 = 0.123456789$$

Kepler & the Mysterium Cosmographicum.

It has become apparent during the course of my research that there are theories from long ago that appear to have been debunked and which have been summarily discarded when in fact they are very likely correct or perhaps misapplied. Nowhere is this truer than with Johannes Kepler and his Mysterium Cosmographicum.

Kepler proposed that the distance relationships between the six planets known at that time could be understood in terms of the five Platonic solids, enclosed within a sphere that represented the orbit of Saturn.

He said that the Platonic Solids dictated the structure of the universe and reflect God's plan through geometry. His work was the first published defence of the Copernican heliocentric system. Kepler claimed to have had an epiphany on July 19, 1595, while teaching in Graz, demonstrating the periodic conjunction of Saturn and Jupiter in the zodiac: he realized that regular polygons bound one inscribed and one circumscribed circle at definite ratios, which, he reasoned, might be the geometrical basis of the universe. He found that each of the five Platonic solids could be uniquely inscribed and circumscribed by spherical orbs; nesting these solids, each encased in a sphere, within one another would produce six layers, corresponding to the six known planets - Mercury, Venus, Earth, Mars, Jupiter, and Saturn. By ordering the solids correctly - octahedron, icosahedron, dodecahedron, tetrahedron, cube - Kepler found that the spheres could be placed at intervals corresponding (within the accuracy limits of available astronomical observations) to the relative sizes of each planet's path, assuming the planets circle the Sun.

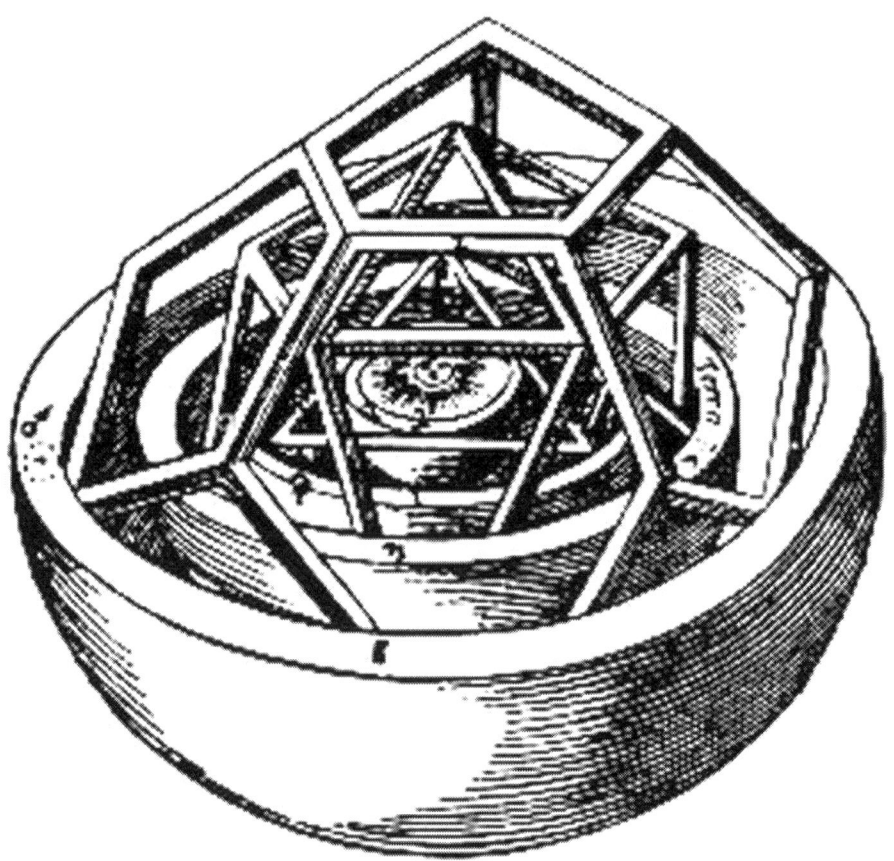

Public Domain, https://commons.wikimedia.org/w/index.php?curid=37301

Sporadic Groups: Mod 9 and the Monster.

To be fair, only specialists in this area of high conceptual mathematics will really appreciate this section. I do not profess to have any expertise in this area either but numerically it seems there are too many coincidences and I feel I must shine a light on them.

What follows is pure intuitive conjecture and I am inviting collaboration from experts to root out the truth of the matter.

7920 - Matthieu Group M11.

In Group Theory, a discipline within Mathematics, a Sporadic Group is one of the 26 exceptional groups found in the classification of Finite Simple Groups.

A simple group is a group G that does not have any normal subgroups except for the subgroup consisting only of the identity element, and G itself. The classification theorem states that the list of finite simple groups consists of 18 countably infinite families, plus 26 exceptions that do not follow such a systematic pattern. These are the sporadic groups. They are also known as the sporadic simple groups, or the sporadic finite groups.

Table of Sporadic Groups.

Name	Order	Factorization
Mathieu group M_(11)	7920	$2^4 \cdot 3^2 \cdot 5 \cdot 11$
Mathieu group M_(12)	95040	$2^6 \cdot 3^3 \cdot 5 \cdot 11$
Janko group J_1	175560	$2^3 \cdot 3 \cdot 5 \cdot 7 \cdot 11 \cdot 19$
Mathieu group M_(22)	443520	$2^7 \cdot 3^2 \cdot 5 \cdot 7 \cdot 11$
Janko group J_2=HJ	604800	$2^7 \cdot 3^3 \cdot 5^2 \cdot 7$
Mathieu group M_(23)	10200960	$2^7 \cdot 3^2 \cdot 5 \cdot 7 \cdot 11 \cdot 23$
Higman-Sims group HS	44352000	$2^9 \cdot 3^2 \cdot 5^3 \cdot 7 \cdot 11$
Janko group J_3	50232960	$2^7 \cdot 3^5 \cdot 5 \cdot 17 \cdot 19$
Mathieu group M_(24)	244823040	$2^{10} \cdot 3^3 \cdot 5 \cdot 7 \cdot 11 \cdot 23$
McLaughlin group McL	898128000	$2^7 \cdot 3^6 \cdot 5^3 \cdot 7 \cdot 11$
Held group He	4030387200	$2^{10} \cdot 3^3 \cdot 5 \cdot 2 \cdot 7^3 \cdot 17$
Rudvalis Group Ru	1.45926E+11	$2^{14} \cdot 3^3 \cdot 5^3 \cdot 7 \cdot 13 \cdot 29$
Suzuki group Suz	4.48345E+11	$2^{13} \cdot 3^7 \cdot 5^2 \cdot 7 \cdot 11 \cdot 13$
O'Nan group O'N	4.60816E+11	$2^9 \cdot 3^4 \cdot 5 \cdot 7^3 \cdot 11 \cdot 19 \cdot 31$
Conway group Co_3	4.95767E+11	$2^{10} \cdot 3^7 \cdot 5^3 \cdot 7 \cdot 11 \cdot 23$
Conway group Co_2	4.23054E+13	$2^{18} \cdot 3^6 \cdot 5^3 \cdot 7 \cdot 11 \cdot 23$
Fischer group Fi_(22)	6.45618E+13	$2^{17} \cdot 3^9 \cdot 5^2 \cdot 7 \cdot 11 \cdot 13$
Harada-Norton group HN	2.73031E+14	$2^{14} \cdot 3^6 \cdot 5^6 \cdot 7 \cdot 11 \cdot 19$
Lyons Group Ly	5.17652E+16	$2^8 \cdot 3^7 \cdot 5^6 \cdot 7 \cdot 11 \cdot 31 \cdot 37 \cdot 67$
Thompson Group Th	9.07459E+16	$2^{15} \cdot 3^{10} \cdot 5^3 \cdot 7^2 \cdot 13 \cdot 19 \cdot 31$
Fischer group Fi_(23)	4.08947E+18	$2^{18} \cdot 3^{13} \cdot 5^2 \cdot 7 \cdot 11 \cdot 13 \cdot 17 \cdot 23$
Conway group Co_1	4.15778E+18	$2^{21} \cdot 3^9 \cdot 5^4 \cdot 7^2 \cdot 11 \cdot 13 \cdot 23$
Janko group J_4	8.67756E+19	$2^{21} \cdot 3^3 \cdot 5 \cdot 7 \cdot 11^3 \cdot 23 \cdot 29 \cdot 31 \cdot 37 \cdot 43$
Fischer group Fi_(24)^'	1.25521E+24	$2^{21} \cdot 3^{16} \cdot 5^2 \cdot 7^3 \cdot 11 \cdot 13 \cdot 17 \cdot 23 \cdot 29$
Baby Monster group B	4.15478E+33	$2^{41} \cdot 3^{13} \cdot 5^6 \cdot 7^2 \cdot 11 \cdot 13 \cdot 17 \cdot 19 \cdot 23 \cdot 31 \cdot 47$
Monster group M	8.08017E+53	$2^{46} \cdot 3^{20} \cdot 5^9 \cdot 7^6 \cdot 11^2 \cdot 13^3 \cdot 17 \cdot 19 \cdot 23 \cdot 29 \cdot 31 \cdot 41 \cdot 47 \cdot 59 \cdot 71$

Perhaps, most significantly, in terms of advancing a radical way of understanding the Game of DNA and the Group of Amino Acids, is the fact that 7920 is also the number of M11, in the Matthieu Group, the smallest of the 20 Sporadic Groups, the largest being the Monster.

I had first come across the Sporadic Groups very early on in my research, courtesy of my father having introduced me to the book 'SYMMETRY OF THE MONSTER', written by a fellow Garrick Club member, Mark Ronan.

Before I had even begun the book I had noticed Appendix 4 in the back which lists 'The 26 Exceptions' which is a list of the 26 exceptional symmetry atoms – these so-called Sporadic Groups.

I had just discovered the virtues of Mod 9 analysis revealing information that could not otherwise be gleaned from Fibonacci and Prime Number sequences so I immediately applied Mod 9 to all of the numbers associated with these exceptional symmetries only to find they all returned a result of 9, or 0, (Mod 9), but with one exception.

J1 in the Janko Group = 175560 = 33 Mod 99 = 6 Mod 9.

I e-mailed Professor Ronan immediately to ask him if there was anything unusual about this exception, J1 in the Janko Group, and he responded right away that 'funny I should ask, but in mathematical circles, J1 is known as the 'mechant nain' or 'naughty child' of mathematics.

Now, I knew nothing about Sporadic Groups or exceptional symmetry atoms but Mod 9 made J1 stick out like a sore thumb.

The secret of J1 may lie in correctly interpreting and understanding what exactly the effect is, of having 3 as a factor, just once. Every other exceptional symmetry atom has a minimum of 3^2 (3 squared), or in other words, 3 shows up as a factor on at least 2 occasions. This of course ensures that the Mod 9 result will always be a 9 (or 0) as 3^2 x anything at all Mod 9 = 9 or 0.

The Monster.

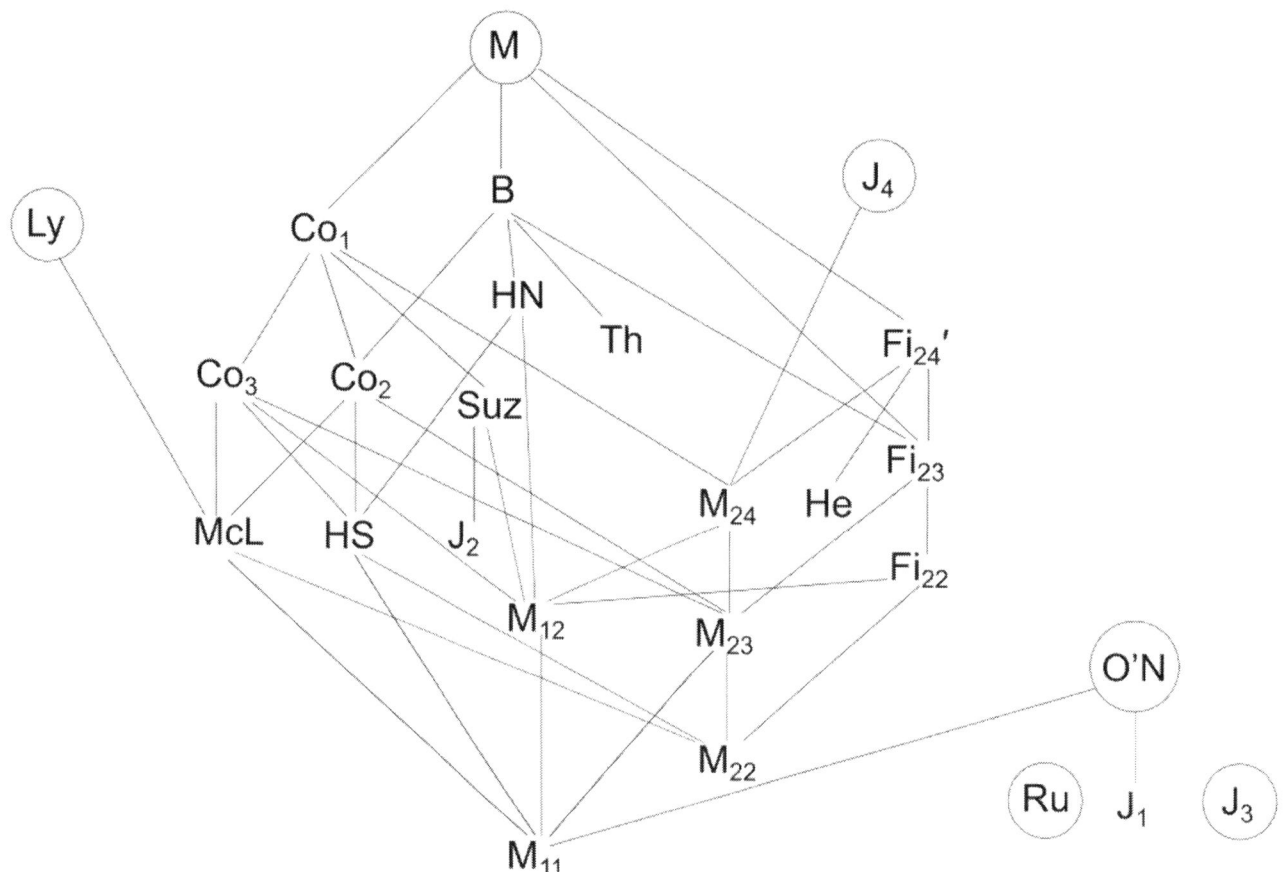

https://en.wikipedia.org/wiki/File:SporadicGroups.svg

Of the 26 sporadic groups, interestingly, exactly 20 can be seen inside the Monster group as subgroups or quotients of subgroups.

The 6 exceptions are J1, J3, J4, O'N, Ru and Ly. These 6 groups are sometimes known as 'the pariahs'.

It is also very interesting in light of many of other research, that the Monster, the largest of these exceptional symmetry groups, only uses 15 of the first 20 prime numbers, as its factors, ranging from 2 to 71.

There are 20 Amino Acids and I think that each one is formulated around a specific Prime Number, the understanding of which could be profoundly useful.

Because it is not strictly a group of Lie type, the Tits group is sometimes regarded as a sporadic group, in which case the sporadic groups number 27, another number central to my Integer Partition Theory, as we shall see, and that represents Light.

Matthieu Groups Comparison.

Here I have just done a quick analysis of the relationships between the Matthieu Groups.

Name	Order	Factorization
Mathieu group M_(11)	7920	2^4·3^2·5·11
Mathieu group M_(12)	95040	2^6·3^3·5·11
Mathieu group M_(22)	443520	2^7·3^2·5·7·11
Mathieu group M_(23)	10200960	2^7·3^2·5·7·11·23
Mathieu group M_(24)	244823040	2^(10)·3^3·5·7·11·23

95040 M12 / 7920 M11 = 12   **10200960 M23 / 443520 M22 = 23**

244823040 / 10200960 = 24   **443520 M22 / 7920 M11 = 56**

Supersingular Prime Numbers.

The Supersingular Primes are exactly the set of primes that divide the group order of the Monster Group.

Monster group M	8.08017E+53	2^(46)·3^(20)·5^9·7^6·11^2·13^3·17·19·23·29·31·41·47·59·71

There are exactly 15 Supersingular Primes:

5 of the first 20 Prime Numbers are not used and are not Supersingular Primes

37   43   53   61   67

Note - Total of First 20 Prime Numbers 2 – 71 is 261 + 378 = 639.

3960.

3960 Miles is the Mean Radius of the Earth.

The Magnetic Number Group 3960 appears three times in the DNA analysis.

First, as the Pivot between the total Protons and Neutrons utilised in the whole 'Game of Life'.

Combined Analysis	Total Protons	Pivot	Total Neutrons	Nucleon Total
Adenine + Thymine	2124		1828	3952
Cytosine + Guanine	2138		1830	3968
Totals	4262	3960	3658	7920

Protons 4262   Pivot 3960   Neutrons 3658

Second, as the Pivot between the total number of Protons and Neutrons utilised by the Base Pairs.

Combined Analysis	Total Protons	Total Neutrons	Nucleon Total
Adenine + Thymine	2124	1828	3952
Pivot			3960
Cytosine + Guanine	2138	1830	3968

Adenine & Thymine 3952   Pivot 3960   Cytosine & Guanine 3968

Third, as the Pivot between the Side Chain and the Standard Block Nucleon total.

	Total Protons	Total Neutrons	Nucleon Total
Standard Block	2379	2135	4514
Pivot			3960
Side Chain	1883	1523	3406

Side Chain 3406   Pivot 3960   Standard Block 4514

Here, we have seen that the Earth emerges as the main player when one looks at the whole game of DNA and unpacks all the pieces to count up all the Protons and Neutrons.

Chapter 6: Caveat and First 'Proof'.

When I had first added up the numbers it was quickly pretty clear to me that if I could find just 2 more Protons, the numbers would come out perfectly in line with many of the numbers that stand out in the rest of my work. All my analysis in this section of the book has assumed that there were in fact 2 Hydrogen atoms 'missing' from the aggregated 64 Codons.

I had no idea how or where I was going to find them but I knew they had to be there somewhere for everything to drop into place. Because I was looking for 2 protons, this had to mean looking for 2 Hydrogen atoms as all the other atoms contribute neutrons and too many protons anyway.

This meant that my answer could only possibly be those amino acids that are coded for exactly twice, leaving the possibilities for finding these 2 Hydrogen atoms as:-

Asparagine   Aspartic Acid   Glutamine   Lysine   Glutamic Acid

Histidine   Phenylalanine   Tyrosine   Cysteine

I was certain that these 2 Hydrogen atoms were on the C and G side, mainly because the total atoms on the C & G side including these 2 errant Hydrogen atoms are such stand out rounded and clearly related numbers in 1100 and 550 respectively.

This assumption then cuts the possibilities down to:-

Histidine   Glutamine   Glutamic Acid   Aspartic Acid

A little research and I knew it must be Histidine because:-

Histidine can bind a proton to the non-bonded electron pair of its ring nitrogen to become a weak acid at low pH. The pK of the acid is 6.0 so that at neutral pH, Histidine is about 90% in the basic form with about 10% still in the acid form.

Histidine is the only amino acid that has a functional group that titrates in the physiological pH range. It is a polar Atom, Hf<O that tends to favour the formation of helical structure. Depending upon its form, which depends on the localized pH of its environment, Histidine can serve as both a proton donor and accepter. The non-bonded electron pair of the basic form are always available for metal chelation. This versatility has been utilized and Histidine is quite often found at the active site of enzymes and as a point of attachment for metal containing group. http://class.fst.ohio-state.edu/fst605/lectures/Lect9.html

As a result, I have adjusted the Hydrogen atomic count for Histidine to 6 from 5 and as there are 2 occurrences, which gives me the 2 Hydrogen atoms I was looking for.

Conclusions & Feedback.

This paper would seem to show the existence of a primordial, numerical and mathematical, geometric and ultimately musical process, inherent to the 'Game of Life' and its origins, and further the ordering of the Universe in which we exist. Not a single Hydrogen atom, nor even a subatomic particle can be missing or displaced between bases without the entire symmetrical, musical and numerical structure being lost completely.

To say these numbers are random is akin to saying that Hamlet is the possible result of a monkey with a typewriter or of tossing however many letters up in the air and having them fall into place to produce the play.

With such an obvious numerical, musical and geometric structure is it really likely that Life evolved stochastically from some primordial soup? Isn't it more likely that DNA is delivered as a whole 'machine' created through resonance?

Does this now mean that it is much more likely that the origin of Life is the result of the inherent canonical, numerical, geometric and musical template that pervades the created Universe and has done since the moment of its inception, if there ever was one?

Feedback.

I was very fortunate to have Jean Claude Perez, who is a leading light in the area of numerical analysis of DNA, review the nascent version of this research in early December 2012 and he gave it an unequivocal thumbs up with a 'that is very impressive..!!!' opening statement in his response to my email to him. - http://creationwiki.org/Jean-claude_Perez

Jean Claude introduced me to Jordi Sola-Soler, a PhD Biomedical Engineer, Technical University of Catalonia, who wrote to me saying - 'I really think it contains a different and fruitful view, from which many interesting interpretations of the Genetic Code may arise' Jordi has a great website http://www.sacred-geometry.es

Part 3: Integer Partition Theory.

'Behind it all is surely an idea so simple, so beautiful, that when we grasp it - in a decade, a century, or a millennium - we will all say to each other, how could it have been otherwise? How could we have been so stupid for so long?' - John Archibald Wheeler.

'Measure what is measurable, and make measurable what is not so. The Universe cannot be read until we have learnt the language and become familiar with the characters in which it is written. It is written in mathematical language, and the letters are triangles, circles and other geometrical figures, without which means it is humanly impossible to comprehend a single word.' - Galileo Galilei.

'Space-time, in small enough regions should not only be merely "bumpy", not only erratic in its curvature; it should fractionate into ever-changing multiply-connected geometries' - John Archibald Wheeler.

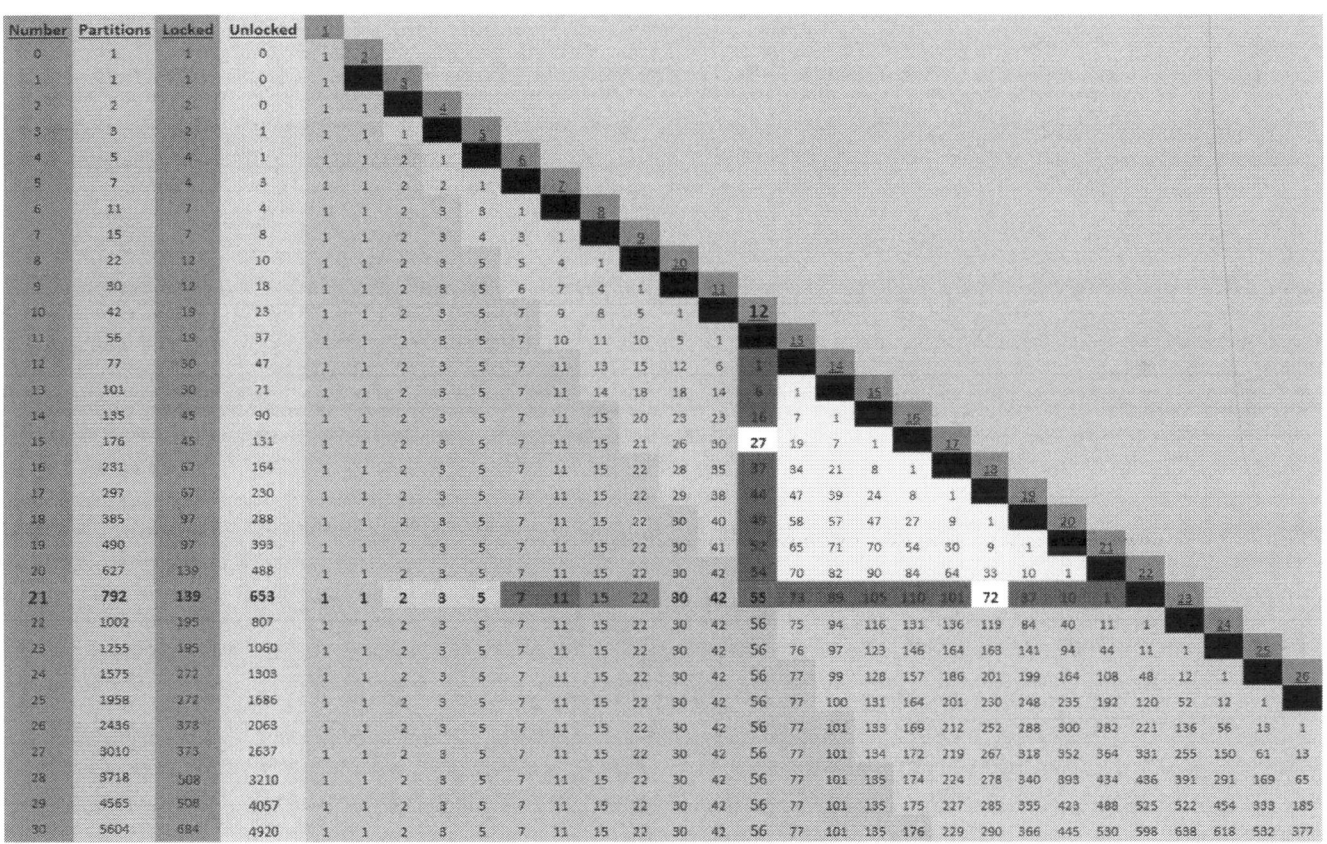

Chapter 1: Introduction.

The Integer Partition Table.

Totals	Partition of the number K / K approaches infinity / Stage X=(K - #spaces + 1)	K is a finite number	1	2	3	4	5	6	7	8	9	10	11	12	13	14	15	16	17	18	19	20	21	22	23	24	25	26	27	28	29	30	31	32	33	34
1	1	0																																			
1	2	1	1																																		
2	3	2	1	1																																
3	4	3	1	1	1																															
5	5	4	1	1	2	1																															
7	6	5	1	1	2	2	1																													
11	7	6	1	1	2	3	3	1																												
15	8	7	1	1	2	3	4	3	1																											
22	9	8	1	1	2	3	5	5	4	1																										
30	10	9	1	1	2	3	5	6	7	4	1																										
42	11	10	1	1	2	3	5	7	9	8	5	1																									
56	12	11	1	1	2	3	5	7	10	11	10	5	1																								
77	13	12	1	1	2	3	5	7	11	13	15	12	6	1																							
101	14	13	1	1	2	3	5	7	11	14	18	18	14	6	1																						
135	15	14	1	1	2	3	5	7	11	15	20	23	23	16	7	1																					
176	16	15	1	1	2	3	5	7	11	15	21	26	30	27	19	7	1																				
231	17	16	1	1	2	3	5	7	11	15	22	28	35	37	34	21	8	1																			
297	18	17	1	1	2	3	5	7	11	15	22	29	38	44	47	39	24	8	1																		
385	19	18	1	1	2	3	5	7	11	15	22	30	40	49	58	57	47	27	9	1																	
490	20	19	1	1	2	3	5	7	11	15	22	30	41	52	65	71	70	54	30	9	1																
627	21	20	1	1	2	3	5	7	11	15	22	30	42	54	70	82	90	84	64	33	10	1															
792	22	21	1	1	2	3	5	7	11	15	22	30	42	55	73	89	105	110	101	72	37	10	1														
1002	23	22	1	1	2	3	5	7	11	15	22	30	42	56	75	94	116	131	136	119	84	40	11	1													
1255	24	23	1	1	2	3	5	7	11	15	22	30	42	56	76	97	123	146	164	163	141	94	44	11	1												
1575	25	24	1	1	2	3	5	7	11	15	22	30	42	56	77	99	128	157	186	201	199	164	108	48	12	1											
1958	26	25	1	1	2	3	5	7	11	15	22	30	42	56	77	100	131	164	201	230	248	235	192	120	52	12	1										
2436	27	26	1	1	2	3	5	7	11	15	22	30	42	56	77	101	133	169	212	252	288	300	282	221	136	56	13	1									
3010	28	27	1	1	2	3	5	7	11	15	22	30	42	56	77	101	134	172	219	267	318	352	364	331	255	150	61	13	1								
3718	29	28	1	1	2	3	5	7	11	15	22	30	42	56	77	101	135	174	224	278	340	393	434	436	391	291	169	65	14	1							
4565	30	29	1	1	2	3	5	7	11	15	22	30	42	56	77	101	135	175	227	285	355	423	488	525	522	454	333	185	70	14	1						
5604	31	30	1	1	2	3	5	7	11	15	22	30	42	56	77	101	135	176	229	290	366	445	530	598	638	618	532	377	206	75	15	1					
6842	32	31	1	1	2	3	5	7	11	15	22	30	42	56	77	101	135	176	230	293	373	460	560	653	732	764	733	612	427	225	80	15	1				
8349	33	32	1	1	2	3	5	7	11	15	22	30	42	56	77	101	135	176	231	295	378	471	582	695	807	887	919	860	709	480	249	85	16	1			
10143	34	33	1	1	2	3	5	7	11	15	22	30	42	56	77	101	135	176	231	296	381	478	597	725	863	984	1076	1090	1009	811	540	270	91	16	1		
12310	35	34	1	1	2	3	5	7	11	15	22	30	42	56	77	101	135	176	231	297	383	483	608	747	905	1060	1204	1291	1297	1175	931	603	297	96	17	1

https://en.wikipedia.org/wiki/File:Integer_Partition_Table.png

Integer Partition Theory proposes a radically 'new' way of looking at and understanding the Universe but one that maps to some of the theory discussed by leading scientific academics today.

I say 'new' because it seems obvious to me now that we must have had a far greater understanding of the workings of the Universe in great antiquity and it is entirely possible that the Integer Partition Table and my discoveries within it may have been understood perfectly at one time. Everything is cyclical and this information was always destined to re-appear.

The critical relationship between number and geometry cuts to the heart of what this part of the book is about, a link that ought to be embedded in every young child.

Perhaps the greatest mathematician of all, the self - taught Indian, Ramanujan, (about whom the recent film ' The Man Who Knew Infinity is based on), won acclaim at the beginning of the 20th century for, among other things, his discovery of certain mathematical patterns in the Partition Sequence and it is in the Integer Partition Table, that represents how each of the Partition sequence numbers are arrived at, that I show the way in which Number can be seen to be at the root of all Art, an idea that is in alignment with the perennial philosophy advanced by Plato and Pythagoras, with creation the result of some ultimate master artist's work.

The Integer Partition Table is proffered as the algorithm for a particle engineered 'machine' that inherently organises and unfolds number in a geometric way, integrating Light and Vibration, the building blocks of the universe.

In short, the theory I am advancing can be boiled down to - Spherical Waves of Vibration, represented by the Number 37, Condition, Cubed Light Particles, represented by the Number 27, to create the design template for all of what is described by physical reality and are the conditions for Life, all biology, to exist at all.

Everything that exists anywhere is made of Light that has been Conditioned by Vibration.

Introducing Partitions.

A Partition is very simply a way of writing a positive number so 1 2 3 4 etc. as a sum of other positive numbers, where the order of the parts is not significant. In layman's terms, how many ways can I make the number 4 using the number 4 or smaller numbers, i.e 3, 2 or 1?

For example, the Number 4 can be partitioned in 5 ways.

One way using the 4: 4.

One way using the 3: 3 + 1.

Two ways using the 2: 2 + 2 and 2 + 1 + 1.

One way using the 1: 1 + 1 + 1 + 1.

(4)   (3+1)   (2+2   2+1+1)   (1+1+1+1)

Apart from giving the number of partitions for each number, this number can be broken down into what are called 'Stage Values'. A Stage Value shows how many times each of the building block numbers can be used, so the Stage Values for the Number 4 are broken down as:

Stage 1: 1 for the 4   Stage 2: 1 for the 3   Stage 3: 2 for the 2   Stage 4: 1 for the 1 = 5 Total

So it is written in the table as Number 4 Partitions 5 Stage Values 1 1 2 1

This produces the table of numbers shown opposite about which I will explain more shortly.

Let's take another example, the Number 5, can be partitioned in 7 ways.

5   41   (32   311)   (221   2111)   11111

1   1   2   2   1   Stage Values

Surprisingly, such a seeming elementary and almost recreational indulgence requires some profound mathematics for its study and there is much more to the Partition Table than at first meets the eye. In fact, it is my belief that All the Secrets of the Universe may be deciphered when a full understanding of this algorithm, and the mechanics of its unfolding, are properly understood.

Rather as I have done with my paper on Prime Numbers, and DNA and Amino Acids, what I have done is to look at the 'game' of Number Partitions with fresh, unconditioned eyes, to discover some radically new thoughts about how the 'Game of Partitions' might work and how integral they actually might be to the structure and creation of Physical Reality.

Partitioning numbers, for me, is akin to peeling the layers of an onion to inspect the guts of the number we are scrutinising.

It shows every geometric form that a number has the potential to morph into and in turn suggests the particle interaction necessary. Some numbers are particularly revealing when you open them up and see how they work, as I hope to convey.

Now, because of my previous years of research I was incredibly luckily armed with exactly what to unconsciously look for in terms of the numbers that might make my ears prick up and it really did not take me but a moment to spot some glaringly obvious numbers that burst out at me from the Integer Partition Table.

I could hardly believe my eyes or my luck!

The Integer Partition Table - Key.

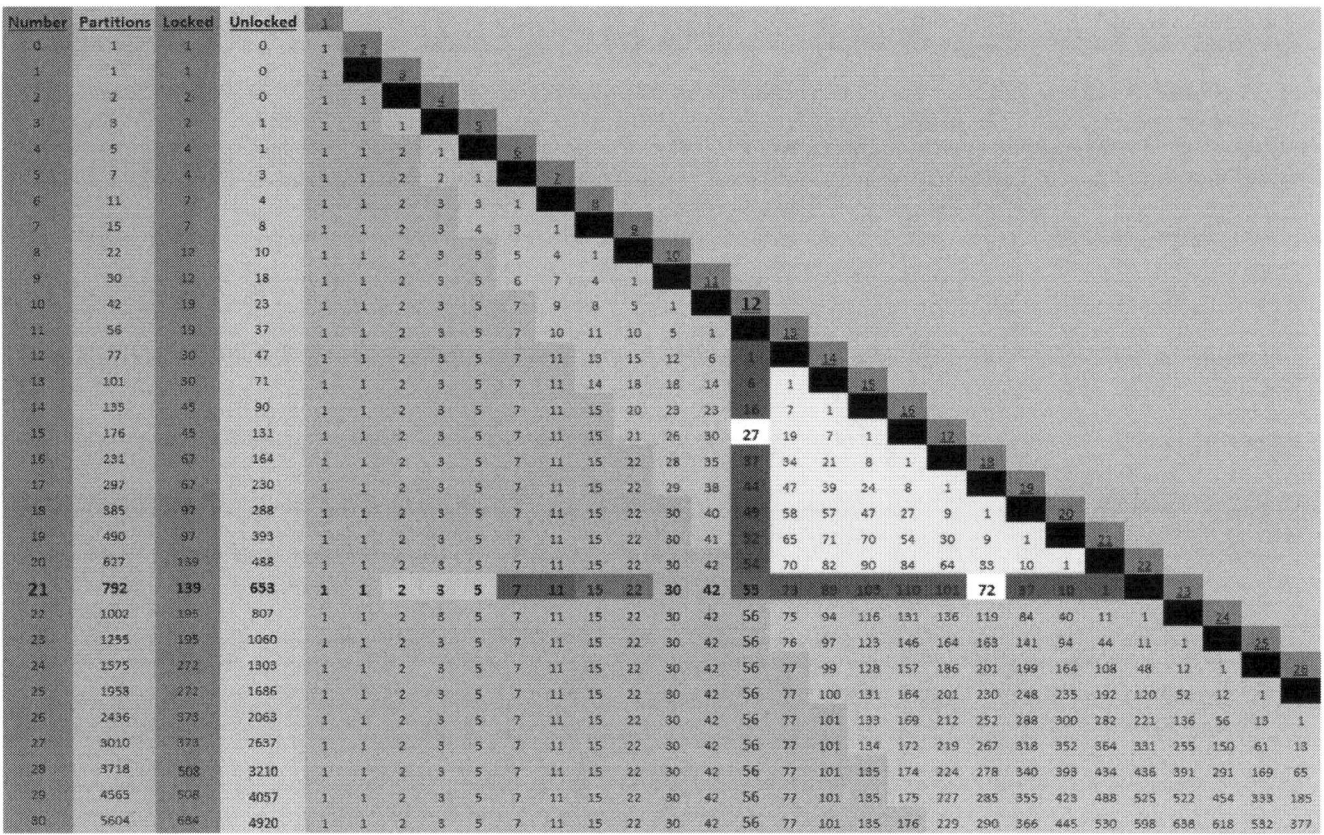

The above representation of the Partition Table is designed to illustrate the machine that creates and perpetuates the universe as per my embryonic Integer Partition Theory.

Above we can see the Numbers 1 to 30 in the first column – the number of partitions for that number in column 2 followed by their respective locked and unlocked potential totals and finally their respective Stage Values, as produced using the method just described.

The light blue shaded cell numbers represent the Stage Values for every number that has taken its place in the Partition Sequence. I call these 'Locked' potentials. Column 3 is the total of these Locked potentials for each number. As an example, we can know that the 12th Stage Value for the number 1 million will still only be 56, it never changes after a given point.

The orange shaded cell numbers represent the Stage Values for every number that can be said to be on its way to taking its place in the Partition Sequence. I call these 'Unlocked' potentials and the total of these, for each number, may be found in Column 4.

The Red shaded cell numbers are what I call the 'Hologram Projector Infrastructure Numbers' the fulcrum of which is the Olive coloured Number, 55. The area coloured yellow within the boundaries of the Red Hologram Projector structure I am calling the 'Holographic Plane'.

The Multi-coloured numbers preceding the Number 55 make up the Dimensionless Dimension – giving a total of 137, the mysterious number of Kabbalah and famously associated with what is called the Fine Structure Constant, the greatest mystery in physics, obsessed over by some of the leading minds of the 20th Century, and the force which essentially holds everything together.

137 is the source vibration or frequency that when joined together with cubed particles of light, using the algorithm of Number as it unfolds according to the Partition Table, produces the whole physics of the universe and biology, emanating from what I propose is the Carbon Central Axis, what the ancients described as the 'Tree of Life', represented by the Number 55.

The Numerical Universe

Chapter 2: 792 Partitions of Number 21.

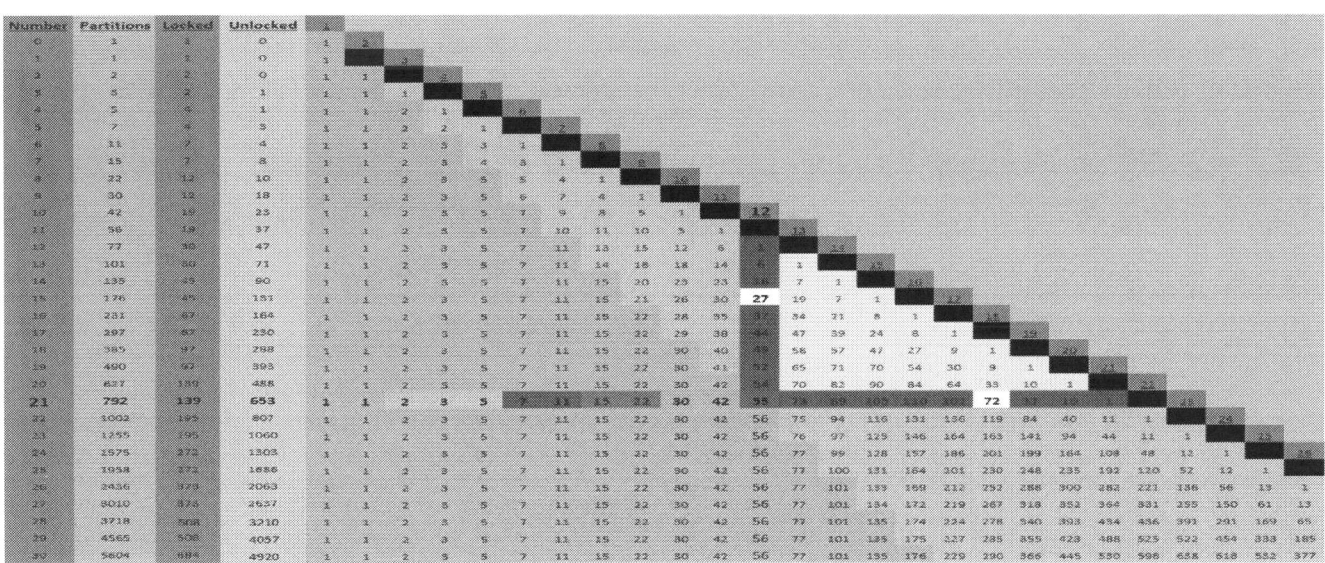

As a result of my previous research, practically the moment I saw this table, I was instantly attracted by the number 792 and then more so as I saw it was the total number of partitions for the number 21. To recap briefly:-

792.

A study of ancient metrology allowed me to know that 792 was certainly represented amply:-

1 League = 7920 Feet   1 Furlong = 7920 Inches   1 Link = 7.92 Inches   1 Chain = 792 Inches

7920 Miles = Mean Diameter of Earth.

7920 is the total number of Protons and Neutrons needed to create the 'Game of Life'.

7920 is the sum of the face angles of the first four Platonic Solids.

7920 is the order of the smallest of the Sporadic Groups, Matthieu 11.

Being a very keen cricketer I also knew that 792 was the number of inches that was the measure of the stumps from each other:

22 yds = 66 feet x 12 = 792 inches.

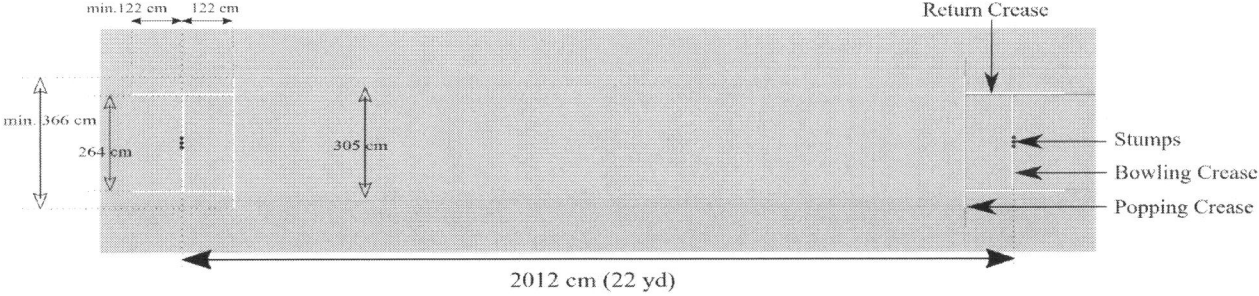

https://en.wikipedia.org/wiki/Cricket#/media/File:Cricket_pitch.svg

21.

If 792 leapt out at me, 21 was almost easier to recognise and the two found together were instantly intriguing and compelling.

Earlier, I showed an investigation into recurring decimals where I discovered that 21 is unique among numbers in terms of its effects, by division, on all other numbers, but I had no idea or frame of reference for how it could be of importance when I first discovered it.

1/ 21 = 0.047619 Recurring

0 4 7
6 1 9

BUT THERE IS NO 2 OR 3 DIMENSIONAL SYMMETRY HERE!

The two dimensional numerical symmetry always totals 999 but here we see:-

0 4 7
6 1 9
6 6 6

21, uniquely, produces 6 digit recurring decimals, Hexagons, lacking this, otherwise, ubiquitous, 'Numerical Symmetry'.

So what?

Well, as it turns out, asymmetry is critical to any good theory because two Japanese scientists and a Tokyo-born American shared the 2008 Nobel Prize for physics for helping explain why the universe is asymmetrical and that if it was not it would not be fit for life.

"The Nobel committee lauded Yoichiro Nambu, now of the University of Chicago, and Makoto Kobayashi and Toshihide Maskawa of Japan for work that helped show why the universe is made up mostly of matter and not anti-matter via processes known as broken symmetries. The fact that our world does not behave perfectly symmetrically is due to deviations from symmetry at the microscopic level," the committee said. This broken symmetry allowed particles of matter to outnumber particles of anti-matter. This is lucky for all living things - because if the universe were symmetrical, anti-matter would be constantly meeting matter, and exploding". - http://www.canada.com/calgaryherald/news/story.html?id=31b1ced8-0b8c-4874-b654-cefe493e2321

Combined Analysis	Total Protons	Total Neutrons	Nucleon Total
Adenine + Thymine	2124	1828	3952
Pivot			3960
Cytosine + Guanine	2138	1830	3968

This inherent asymmetry is fascinating especially as it appears in my DNA paper as the asymmetry in the total number of Nucleons used by Adenine & Thymine led codons, being 3952, and by Cytosine & Guanine led Codons, being 3968.

Each are 8 away from the Pivot, the axis or mid-point of which is 3960 - the mean radius of the Earth in miles.

From the Number theoretical point of view, 8 obviously fits well because the numbers available to us using Mod 9 where 9 and 0 are self-similar are 8.

Interestingly, this is rather close to the bulge in the equator of the Earth.

Chapter 3: Origins - Integer Partition Theory.

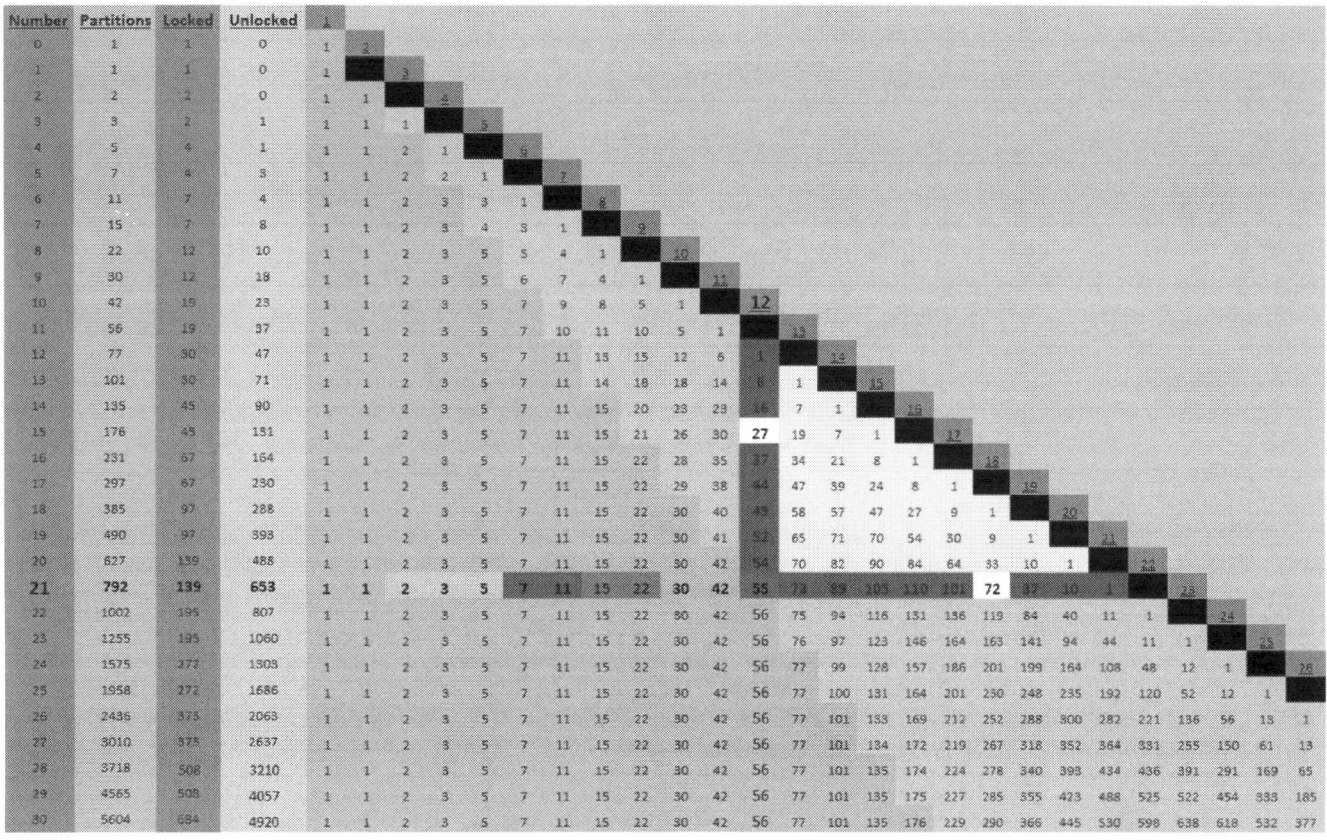

Reading horizontally along the Number 21 to take in all the different stage values, there before me were so many of the important numbers pervading various areas of my research, in particular 37, 55, 72 and 73. The others all had interesting properties that I felt sure would reveal their importance in good time with further investigation.

Then it all came together in my mind in a perfect moment. I looked at the numbers vertically, for the 12th Stage, 12 being the mirror of 21, and saw the Number 55 as the 12th stage for the Number 21. I knew 55 as the sum of the first ten numbers and also the tenth in the sequence of the biologically important Fibonacci numbers.

I quickly saw that the Number 27 at the 12th Stage Value for the Number 15 would map perfectly to the 72 on the 18th Stage Value for the Number 21, through a rotation of 90 degrees.

I saw all the geometric integrations taking place for each of the numbers that were vertically and horizontally connected on what I call the 'Hologram Projector Infrastructure'.

21 is the mirror of 12 obviously, but I knew from my Prime Number Sequence that there were more mirrors here:

12th Prime is 37

21st Prime is 73

12 and 21   37 and 73

I realised immediately that I was looking at something really very special, thanks to some extraordinary work done by Vernon Jenkins that I had literally only just stumbled across and skimmed, containing his work on opening line of Genesis in the Bible and the importance of the figuracy of number and the integration of geometries.

Anthony Morris

37 / 55 / 73

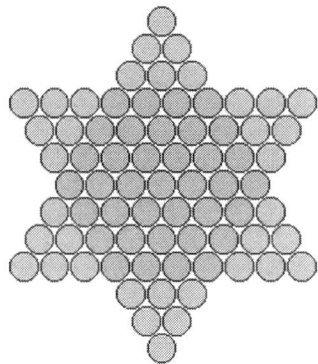

Depicting the related hexagon/hexagram pair 37/73
as the intersection/union of 55-as-triangle
with an inverted copy of itself

http://homepage.virgin.net/vernon.jenkins/Symb.htm

What I could see in my mind was the integration of Spherical Sound Waves, horizontally on the Number 21, that condition Cubes of Light Particles, vertically on Stage 12, and that inevitably produces what I am calling the 'Holographic Plane' of physical reality.

I believe the Integer Partition Table shows the exact algorithm that Numbers utilise to build the projection of Physical Reality.

There is only Vibration (nothingness, energy, frequency) and Light - the unfolding and self-ordering and replicating numbers are the algorithm - the machine.

This describes Man perfectly too because we have the Light of our Mind and the energy, frequency or vibration which comes from our intention.

The intention of your attention, with repetition, is what causes effects to occur and the mastery or manifestation of anything, just like the universe does it.

The Numerical Machine.

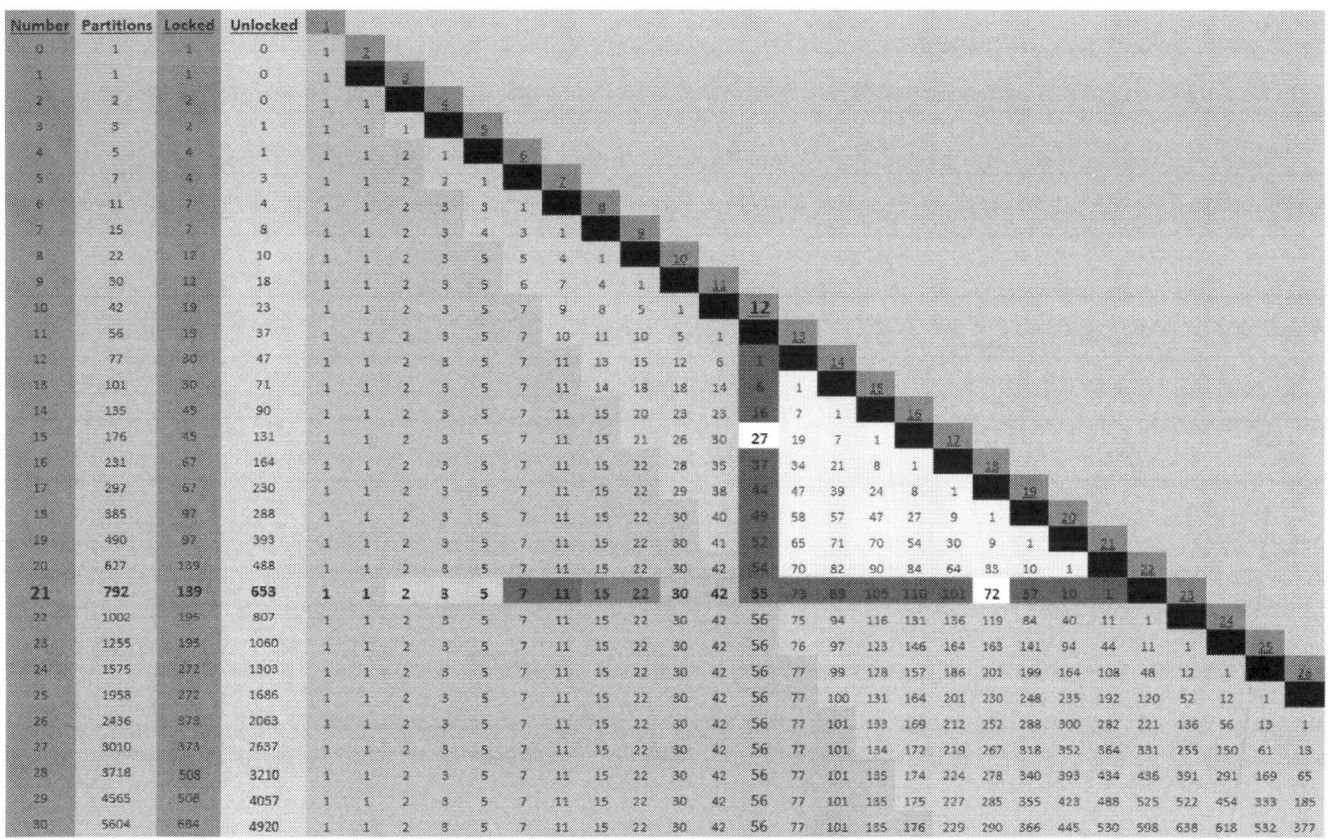

Integer Partition Theory proposes that there is a numerical, geometric and musical machine that creates, organises and perpetuates the universe.

The Hologram Projector principally consists of just 2 numbers and their mirrors that uniquely create triangles as a result of their effect on all other numbers.

27 & 37

72 & 73

Light Particles are represented by the Number 27.

27 - The Cause

Integer Partition Theory proposes that Cubes of Light Particles, represented by the Number 27, are conditioned by Spherical Waves of Vibration, represented by the Number 37, to form the whole of the universe.

37   (55)   73

Boundary   Pivot   Boundary

Conditioned Light Particles are represented by the Number 72.

72 - The Effect

Chapter 4: Locked Potentials for the Number 21.

Locked v Unlocked Potentials.

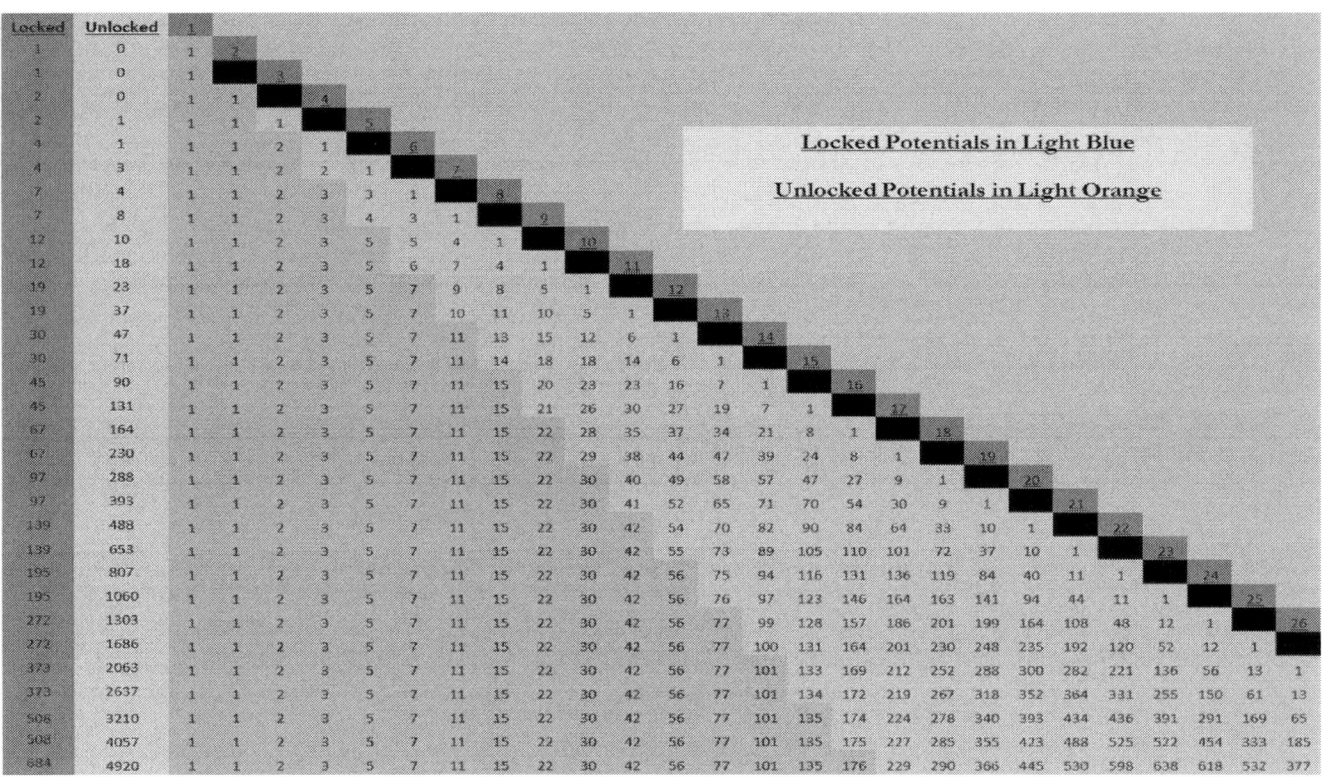

Above you will see that some of the Stage Value numbers are shaded in stair step fashion. These 'Stage Values' seem to 'lock' into place so that, for example, all numbers 8 and above, will have a 5th Stage Value of 5.

Light Blue Shaded Numbers I call 'Locked Potentials'
Light Orange Shaded Numbers I call 'Unlocked Potentials'

I call a Stage 'Unstable or 'Unlocked" if it has not yet reached its maximum 'Locked Potential' as a number that takes its place as part of the Partition Sequence.

When one considers the table from a vertical perspective, looking down the page, these 'Unlocked' Stage Value Numbers could be said to be on their way to taking their place in the Partition Sequence. They have not yet completed their journey if you like. They could be said to be transitioning, or transforming into the finished product.

All numbers greater than 2 could be said to have a Locked and an Unlocked Component.

137.

Looking horizontally at the 'locked potentials', so the shaded numbers, for the Number 21, and if we leave 0, 1 and 2 aside, on the basis that they have no unlocked element, (they are whole in and of themselves), I was stunned to find that the first 9 terms for numbers in the sequence are then:-

| 1 | 1 | 2 | 3 | 5 | 7 | 11 | 15 | 22 | 30 | 42 |

$$2 + 3 + 5 + 7 + 11 + 15 + 22 + 30 + 42 = 137$$

This number, 137, is absolutely central to the mystery of natural philosophy and to find this number embedded in such a way in the Integer Partition Table is extraordinary, especially in light of what you will be reading next. The positioning and unfolding could hardly be more perfect.

Many of the greatest minds of the 20th Century obsessed over this number.

Arnold Sommerfeld championed whole numbers with all the passion of the Kabbalists whose central number is also 137 and about which more will be said later. It was he who discovered 137 while trying to solve a puzzling feature of atoms.

'The fine structure of spectral lines, the characteristic combination of wavelengths of light emitted and absorbed by each chemical element - the fingerprint or DNA, as it were, of each wavelength of light. It was dubbed the 'fine structure constant' (which in fact equals 1/137, though for convenience physicists refer to it as 137). From the moment 137 first popped up in his equations, he and other physicists saw that its importance went far beyond the fact that it solved this puzzle. They quickly realised that this unique "fingerprint" was the sum of certain fundamental constants of Nature, specific quantities believed to be invariable throughout the universe, quantities central to relativity and the quantum theory.

But if this one number were so important, should it not be possible to deduce it from the mathematics of these theories? Disturbingly, no one could.

The fine structure constant turns out to be exquisitely tuned to allow life as we know it to exist on our planet. Perhaps it was not surprising then, that physicists began referring to 137 as a mystical number.

According to Bohr's theory, when an electron drops from a higher to a lower order it emits light which is recorded in the laboratory as a spectral line. Bohr was able to work out equations for these spectral lines that could be compared with the data obtained in the laboratory. By 1915 scientists had more accurate spectroscopes that enabled them to make closer inspection of the spectral lines. This revealed that many of the individual lines were in turn made up of several more closely spaced lines: they were said to have a fine structure. Certain spectral lines also split into several lines or multiplets when the atom was placed near a magnet, but the fine structure was always there. 'It was given by Nature herself without our agency,' Sommerfeld wrote.

Sommerfeld's primary contribution to atomic physics was his work on the fine structure problems. His brainwave was to apply relativity theory to Bohr's theory, changing the mass of the electron following Einstein's famous equation $E = MC^2$. The result was astounding: an extra term appeared in Bohr's equation for a single spectral line. This extra term made it possible to predict that certain lines would actually split and reveal their fine structure.

Sommerfeld called the quantity that set the distance between the split spectral lines in this extra term the 'fine structure constant' and designated it with the Greek letter Alpha. His equation was:

Alpha - Fine Structure Constant = 2 x Pi x e^2 / h x c

The fine structure constant is made of three fundamental constants: the charge of the electron e (1.60×10^{-19} coulombs - a coulomb is the unit of electric charge): the speed of light c (3×10^8 metres / second), which defines relativity theory: Planck's constant h (6.63×10^{-34} Joule-seconds), which defines quantum theory and determines the size of the grains into which the microscopic world is partitioned, be it grains of energy, mass or even space itself. Pi is the ratio of the circumference of a circle to its diameter (3.141529). The constants e, c and h had already been measured. Thus the discovery of the fine structure constant was a step toward the great goal of finding a theory that would unite the domains of relativity and quantum theory, the large and the small, the macrocosm and the microcosm.

There was one extraordinary feature of the fine structure constant. The three fundamental constants that make it up have dimensions - such as space and time - and therefore depend on the units in which they are measured, whether metric, imperial, or some other. So although they would certainly play an essential part in a relativity or quantum theory formulated by physicists on a planet in another galaxy, they might not have precisely the same values as they have on Earth.

But when they come together to form the fine structure constant, something extraordinary happens. All of their units cancel out and as a result the fine structure constant is a pure number without any dimensions. No matter what number system this will always be true. Sommerfeld calculated it as 1/137 or 0.00729 - a rather unexciting way of expressing such a momentous result.' Arthur I Miller - 137 - Jung, Pauli and the Pursuit of a Scientific Obsession.

1/137 = 0.00 72 99 27 Recurring.

137 creates 8 digit recurring decimals at its effect on all other numbers by division, a property shared with the number 73 and which links these two numbers deeply. It also displays the numerical symmetry associated with all even length recurring decimals.

$$0072$$

$$9927$$

$$9999$$

37 is the 12th Prime Number   73 is the 21st Prime Number

12 + 21 = 33 - The 33rd Prime Number is 137

Note also that the Sum of the 3 causal Hologram Projector Numbers:-

$$27 + 37 + 73 = 137$$

The product of these numbers produces a rather beautiful palindromic result deeply connected to those numbers that are proposed in Integer Partition Theory to represent Cause and Effect.

$$27 \times 37 \times 73 = 72\ 9\ 27$$

Locked Potentials for Number 21 - Mod 9 & 99.

| 1 | 1 | 2 | 3 | 5 | 7 | 11 | 15 | 22 | 30 | 42 |

Interestingly the Mod 9 total for this sequence is:-

$$2 + 3 + 5 + 7 + 11 + 15 + 22 + 30 + 42 = 137 = 38\ Mod\ 99$$

Mod 9 $2 + 3 + 5 + 7 + 2 + 6 + 4 + 3 + 6 = 38$

This is interesting as 38 is the Magic Constant for the Magic Hexagon, the only one there is.

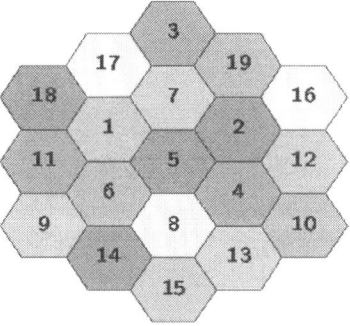

It is the only Hexagon where the numbers can be arranged so as to create what is called a magic constant, that is, whichever way you add it up, it comes to the same sum. In the case of the Magic Hexagon, the Magic Constant is 38.

19 & 20 - 137.

I also noticed that the difference between the total number of partitions for the number 19 and 20, 490 and 627 respectively, giving 627 - 490 = 137. Perhaps a coincidence, but I really doubt it, mainly because a great deal in Nature seems to be organised in to a 19 +1 structure.

For example, in the areas of biochemistry and nuclear chemistry, Peter Plichta PhD writes in his excellent book, God's Secret Formula, that of the 20 Amino Acids common to all biology, there are 19 left-oriented amino acids, and one amino acid which can have both a left and right orientation. He also writes that the total number of pure isotopes in all the stable elements is exactly 20. Of these 20 pure isotopes, the lowest element, Beryllium is the only one with an even atomic number, being 4. This is followed by the atomic numbers 9, 11, 13, 15, 21, 25, 27, 33, 39, 41, 45, 53, 55, 59, 65, 67, 69, 79, and 83, all of which are odd numbers.

1 + 19 Amino Acids and 1 + 19 Pure Isotopes.

137 - The Age of the Universe & Time.

Integer Partition Theory also proposes and predicts that the Universe cannot be any other age than it is. What I mean by that is that, in what we would count a billion years of our experience down here on Earth, if we measured the Universe at that point in Time, we would find that the universe is still exactly $1.370370370 \times 10^{11}$ years old. The universe has never been any other age. Time is an illusion; there is only now, a flickering of Now. How can I make such a seemingly absurd suggestion?

Einstein's well known equation $E = MC^2$ can be re-written as $M = E / C^2$ where E = the spherical vibration represented by the number 37 and its triangle - the perfect scalene triangle which is created as half of an equilateral triangle, and C^2 = the cubes of light particles represented by the number 27 and its triangle - the isosceles right angled triangle with 45, 45, 90 angles. More on these shortly.

M = 37 / 27 = 1.370 370 R

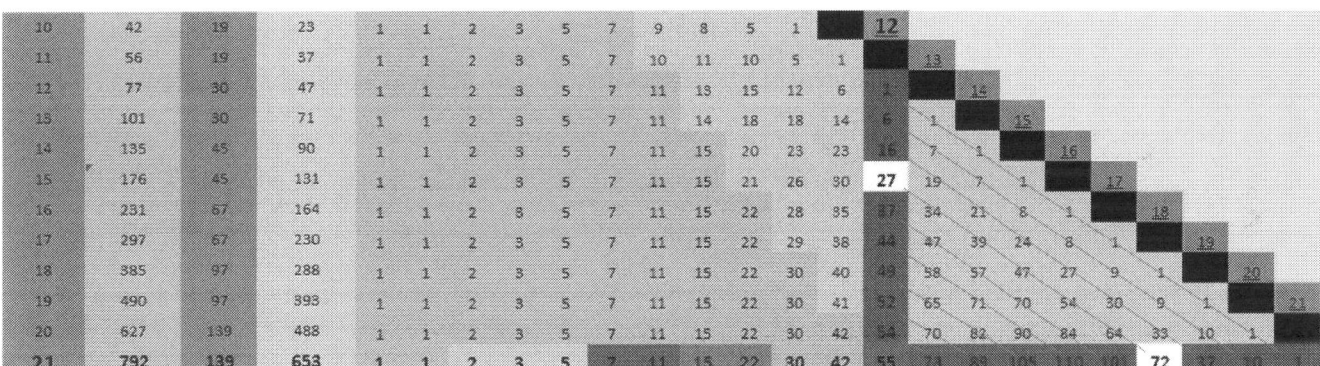

The 11 Dimensions.

The Central Axis - The Z Axis represented by the Number 55.

The Time Axis - The Y Axis - Vertical Red Hologram Projector Infrastructure Numbers on the 12th Stage Value.

The Space Axis - The X Axis - Horizontal Red Hologram Projector Numbers on the Number 21.

8 Dimensions - The Holographic Plane defined as the area contained by the Hologram Projector Infrastructure.

When we increase 1.370370R by 10^{11} - which essentially means 10 raised to the 11th dimension, we arrive at the age of the universe and the only age the universe could ever be, 13.7037037 billion years old.

What $M = E / C^2$ actually means is that Mass = SPHERICAL Energy / Frequency / Vibration Waves divided by CUBES of Light Particles. In this way Light is conditioned by Vibration and the system of, and inevitable unfolding of numbers are proposed as the algorithm that perpetuates and constitutes the hologram projection that is the experience of physical reality.

Caesium 137.

Talking of time, it is perhaps no coincidence that Caesium 137, with its 55 Protons no less, is the element we use to measure time. Scientists have to have an independent way of defining the second in terms of microwaves and the frequency of the microwaves emitted by Caesium are a near perfect match for the second.

'Caesium-based atomic clocks observe electromagnetic transitions in the hyperfine structure of caesium-133 atoms and use it as a reference point. The first accurate caesium clock was built by Louis Essen in 1955 at the National Physical Laboratory in the UK. They have been improved repeatedly over the past half-century, and form the basis for standards-compliant time and frequency measurements. These clocks measure frequency with an error of 2 to 3 parts in 1014, which would correspond to a time measurement accuracy of 2 nanoseconds per day, or one second in 1.4 million years. The latest versions are accurate to better than 1 part in 1015, which means they would be off by about 2 seconds since the extinction of the dinosaurs 66 million years ago, and has been regarded as 'the most accurate realization of a unit that mankind has yet achieved.'
http://en.wikipedia.org/wiki/Caesium

Caesium clocks are also used in networks that oversee the timing of cell phone transmissions and the information flow on the Internet.

137 - Chlorophyll & Hemoglobin.

There are, apparently coincidentally, 137 atoms in both Chlorophyll, found in plant life, and Hemoglobin, found in the blood.

These 137 atoms are organized into 6 Concentric Rings around a single Magnesium Atom with 12 equal radial branches out from the centre.

The chemical formulas are:-

Hemoglobin - C55 H72 O5 N4 Fe   **Chlorophyll - C55 H72 O5 N4 Mg**

Please note that there are 55 Carbon Atoms and 72 Hydrogen Atoms in each formula.

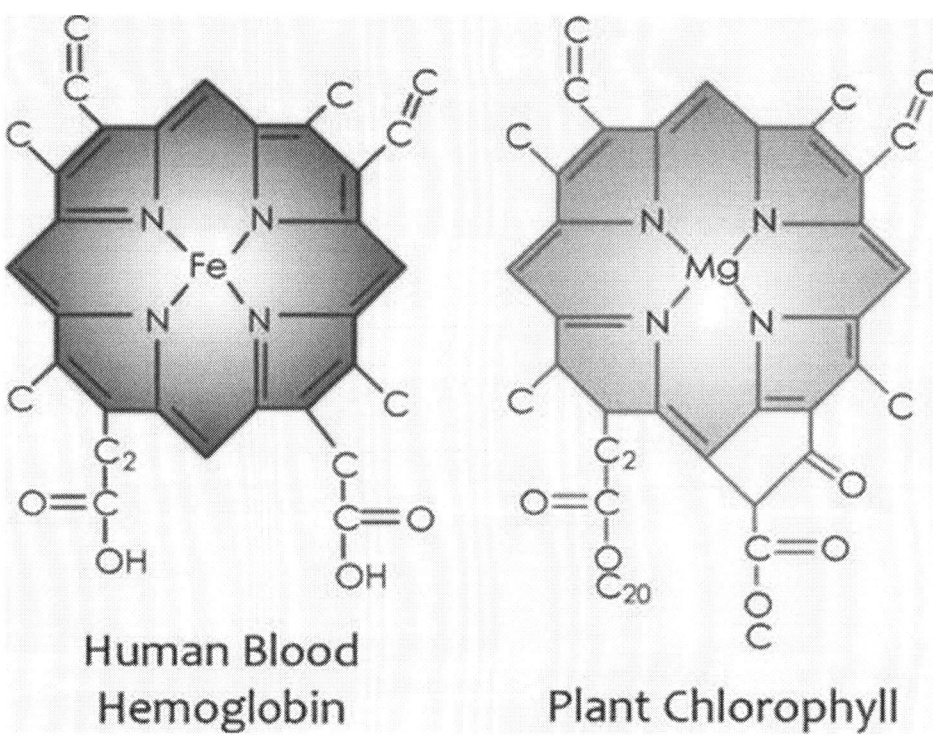

http://www.nature-education.org/cart2/chlorophyll-haemoglobin.jpg

137 & the SATOR Square.

Further evidence for 137 has turned up in a very old and mysterious puzzle called the SATOR Square.

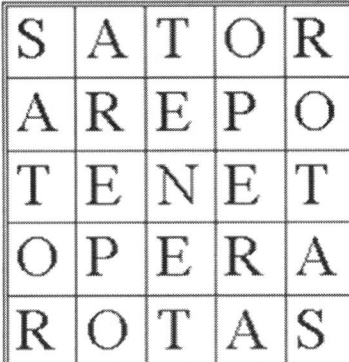

Image: Wikimedia Commons

The SATOR Square or ROTAS Square is a word square thought to contain a Latin palindrome featuring the words SATOR AREPO TENET OPERA ROTAS, in this or in the reverse order, written in a square so that they may be read top-to-bottom, bottom-to-top, left-to-right, and right-to-left. This 2000 year old enigma was found in the ruins of Pompeii and then all over the world. Its meaning has been the subject of a great deal of scholarly speculation with no clear consensus for a Latin translation or its meaning.

I don't think this has anything to do with Latin or the Romans and this is the reason that the meaning of this curio has remained beyond the reach of scholars.

History of the Region.

The largest of the Greek cities in the region was Neapolis (modern Naples), founded between the 7th and 6th centuries B.C. Other important Greek cities include Cumae and Pithicusae. As a result of the alliance between the city of Pompeii and the surrounding Greek colonies (referred to collectively as Magna Graecia), much of the early influence in Pompeian development is Greek. Most of the earliest regions of the city are clearly Greek in design.

Isopsephy.

Isopsephy is the Greek word for the practice of adding up the number values of the letters in a word to form a single number. Isopsephy is related to Gematria which is the same practice using the Hebrew Alphabet.

Original Greek Cipher used in Isopsephy

1 = α	10 = ι	100 = ρ
2 = β	20 = κ	200 = σ
3 = γ	30 = λ	300 = τ
4 = δ	40 = μ	400 = υ
5 = ε	50 = ν	500 = φ
6 = ς (ϝ)	60 = ξ	600 = χ
7 = ζ	70 = ο	700 = ψ
8 = η	80 = π	800 = ω
9 = θ	90 = ϟ	900 = ϡ

First, I used the original version of the Greek cipher, seen opposite, which translates numbers to letters and vice versa. Over the years some of these were taken out of use so it was important to use this original cipher and no more modern version. When we apply this cipher to the SATOR Square, we see the numbers in the table below.

__SATOR Square in Numbers.__

	S	A	T	O	R
	A	R	E	P	O
	T	E	N	E	T
	O	P	E	R	A
	R	O	T	A	S
	200	1	300	70	100
	1	100	5	80	70
	300	5	50	5	300
	70	80	5	100	1
	100	70	300	1	200
Total					
2514	671	256	660	256	671
		927	660	927	

I then looked at this same table of numbers in Mod 9. Extraordinarily, all the numbers I could hope to see, to be relevant to my Integer Partition Theory were just staring me in the face!

__SATOR Square in Numbers - Mod 9.__

	200	1	300	70	100
	1	100	5	80	70
	300	5	50	5	300
	70	80	5	100	1
	100	70	300	1	200
	2	1	3	7	1
	1	1	5	8	7
	3	5	5	5	3
	7	8	5	1	1
	1	7	3	1	2
Total					
93	14	22	21	22	14
		36	21	36	

Notice the Mod 9 results for each half total 36, being the number of constituents in the Holographic Plane, and that they straddle the 21, adding to 72, the number representing the effect of physical reality. 21 is the sum of the first six numbers and 36 the sum of the first eight which makes them both triangular numbers. 36 is also one of the few numbers to be both triangular and square.

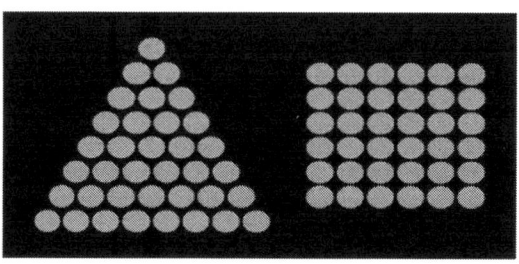

SATOR Square in Numbers - Mod 9 Emphasised.

2	1	3	7	1
1	1	5	8	7
3	5	5	5	3
7	8	5	1	1
1	7	3	1	2
14	22	21	22	14
	36	21	36	

The first numbers to leap out at me were those of The Fine Structure Constant Number, 137, associated with the greatest mystery in physics and what many consider to be the central question in natural philosophy. It was extraordinary to see how the 137 is encompassed by the Numbers 12 and 21 which are the key numbers in the Partition Table, where everything 'happens'.

Using the central 5 as the focus we see that 55 is represented and perhaps Integer Partition Theory offers the answer to the ultimate question: - Why 10? It is important to remember that the sum of numbers 1 to 10 is the triangular 55 and it is the 10th Fibonacci Number.

1 1 2 3 5 8 13 21 34 55

The Creator used the decimal system and Man has discovered it in the same way he discovered the principles of harmonics in Music. These are inherent systems which display the nature of Nature.

Now all this looks very good news for Partition Theory as it seemed clear to me that the SATOR Square is the creation cipher and of extraordinary and fundamental importance.

This idea was further cemented when I went a little 'off piste' in my thinking and decided to overlay a swastika symbol to the 5 x 5 SATOR Square. A 5 x 5 square is the smallest square that you can draw a swastika on to, and be able to see it fully defined. The result was frankly unbelievable in terms of its connection to my theory.

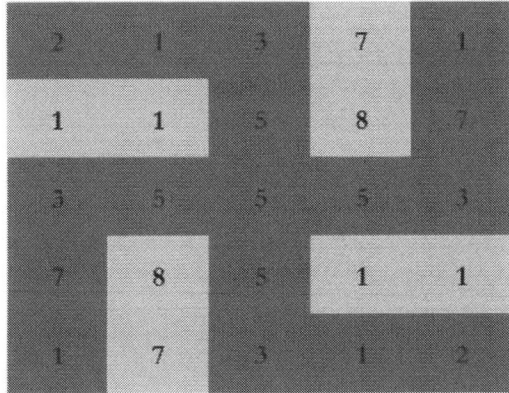

The total used by the whole Swastika symbol appears to be 59 if we count the central 5 only once but 64 if we count it twice, which in my view we must as it belongs to both arms of the swastika taken in isolation.

2 1 3 5 5 5 3 1 2 1 7 3 5 5 5 3 7 1 = 64

This number 64 immediately got my attention, the chess board again and being the number of Codons produced by the 4 bases in biological DNA.

Amazingly, when I took the Swastika arms apart I was rewarded with each Swastika arm totalling each of the 2 key numbers for Light and Vibration, 27 and 37 central to my Integer Partition Theory.

2	1	3	7	1	2	1	3	7	1
1	1	5	8	7	1	1	5	8	7
3	5	5	5	3	3	5	5	5	3
7	8	5	1	1	7	8	5	1	1
1	7	3	1	2	1	7	3	1	2

2 1 3 5 5 5 3 1 2 = 27   **1 7 3 5 5 5 3 7 1 = 37**

The logic of making use of the swastika as an overlay for the SATOR Square and the Greek origins of the SATOR Square was cemented by a picture sent to me by my friend Alex Syed who had been to Paestum, not far from Pompeii.

Paestum was a major ancient Greek city on the coast of the Tyrrhenian Sea in Magna Graecia. After its foundation by Greek colonists under the name of Poseidonia it was eventually conquered by the local Lucanians and later the Romans. The Lucanians renamed it to Paistos and the Romans gave the city its current name. Clearly the ancient Greeks made use of the swastika.

With the correct letter to number cipher applied, this famous square is revealed itself as a further numerical cipher, one that I believe is central to the creation of this Numerical Universe. This would seem to indicate that at some point in antiquity that the ancients may have known a great deal about the numerical, geometric and harmonic structure of the universe.

137 & Kabbalah

Last but not least is the connection between 137 and Kabbalah. Kabbalah is part of a wider body of Hermetic literature. Hermetic wisdom is credited to have come from the Egyptian god, Thoth, also known as Hermes Trismegistus by the Greeks.

'In Kepler's time Hermetic Literature was enthusiastically embraced as an antidote to the rational approach of Greek philosophy and science. It was full of mystery and magic and spoke in terms of a vital or living force at the heart of the cosmos.' - Arthur I Miller - 137 - Jung, Pauli and the Pursuit of a Scientific Obsession.

Central to Kabbalah are the Sephirot, meaning emanations, are the 10 attributes /emanations in Kabbalah, through which Ein Sof (The Infinite) reveals himself and continuously creates both the physical realm and the chain of higher metaphysical realms.

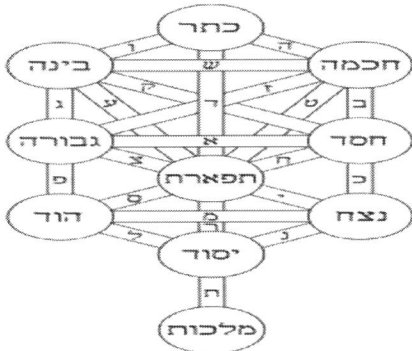

http://en.wikipedia.org/wiki/Sephirot

The Sephirot is normally shown as a Tree of Life with 10 branches - 5 pairs of opposites, just as the 10 numbers that we have are paired.

0 9   1 8   2 7   3 6   4 5

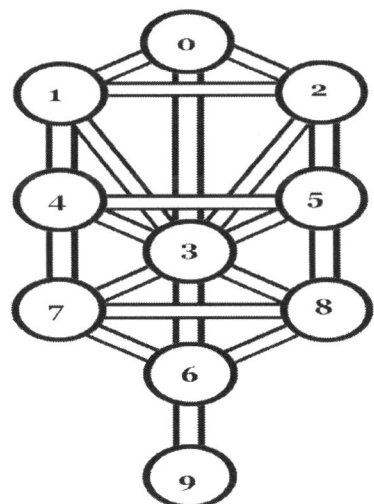

In my interpretation of it, the central aspect of the Sephirot represents the 3 6 9 (0) Magnetic Number Group with the Prime Number related Electric Number Groups 1 4 7 and 2 5 8 to the left and right respectively.

By 1500 Kabbalah had become part of Christian theology with many interested in Gematria which assigned numbers to letters of the Hebrew alphabet like the Greeks who called in Isopsephy.

The Hebrew word for Kabbalah is:-

קַבָּלָה
The Gematria Value is 137.

By about 1600, Christian Kabbalah was accepted, along with alchemy and astrology, as the paths to true wisdom. As with all things, beliefs have their day in the Sun and it wasn't long before a colder, materialistic science, which depended on the measuring of the inanimate, displaced all these once held paths to true wisdom.

Kabbalah had a great influence on Johannes Kepler who was inspired to ask himself why there were 6 planets and why they were ordered the way they were.

Fittingly, the Hebrew name for the biblical God of Israel is:-

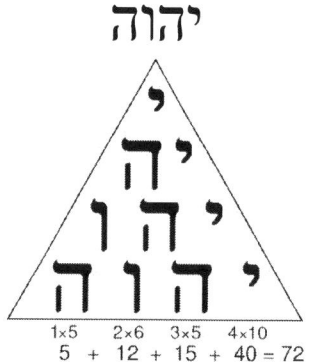

https://en.wikipedia.org/wiki/File:Tetragrammaton-Tetractys.png

Above we see יהוה represented in what is traditionally known as a Tetractys, which is essentially a 4 base triangle representing 10. There are some who believe that the Tetractys and its mysteries influenced the early Kabbalists.

Shown in this way we can see that by following the rules of Gematria the total value comes to 72.

Chapter 5: Key Numbers 12 and 21.

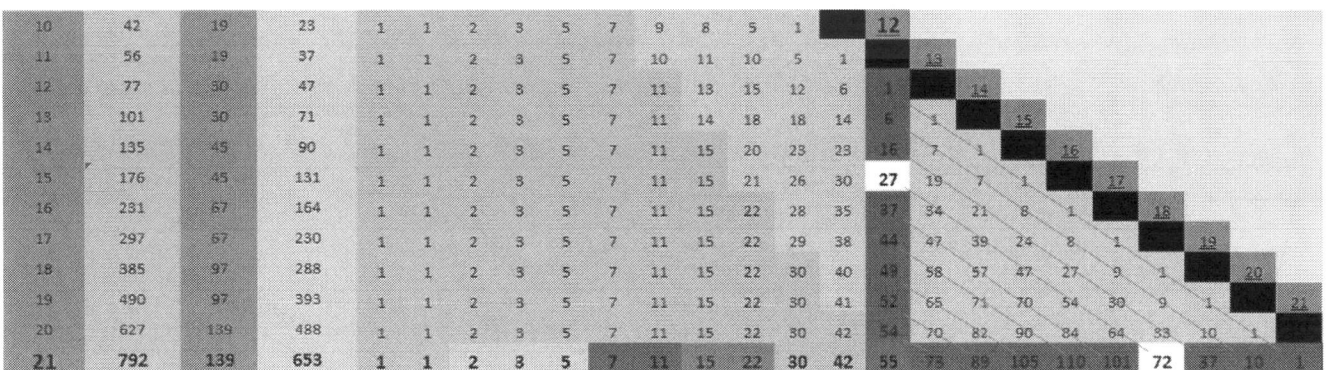

π - 12 and 21.

While toying around with the important numbers in my theory I came across something quite elegant:-

The total number of Partitions for the number 21 is 792.

The integration between Sound and Light happens on the 12th Stage Value of the Number 21.

$$792 / 12 \times 21$$

$$= 792 / 252$$

$$= 22 \times 36 / 7 \times 36$$

$$= 22 / 7$$

22/7 is the ancient approximation for π - (3.142857 R)

252 - 12 x 21.

$$12 \times 21 = 252$$

Interestingly:-

$$252 \times 252 = 441 \times 144$$

This is interesting when we examine the range of numbers:-

144 (+108) 252 (+189) 441

The Range is 441 - 144 = 297

297 = The number of Partitions for the Number 17

792 is the mirror of 297

This speaks to the way collaborator Michael Joyce sees things:-

'There is a characteristic unique to only Earth in our solar system, involving its diameter 7920 miles, number 72 and their reflected ones, 0297 and 27. When 7920 is divided by 72 and 0297 by 27, each makes 'ELEVEN', the Earth Number.'

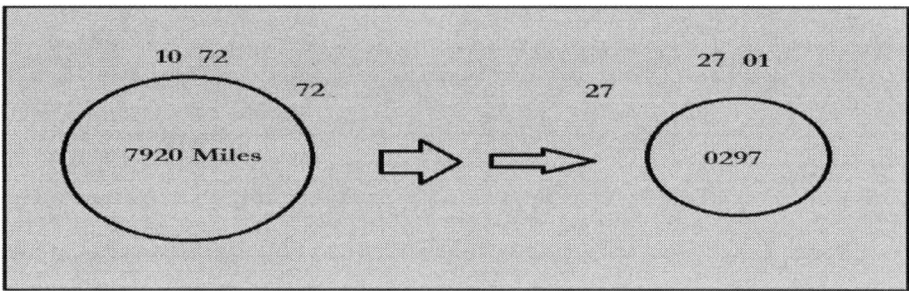

This would suggest TWO Earths, our physical one and an 'unseen' one.

252 is quite an interesting number in its own right because if you flip a coin 10 times in a row, there are exactly 252 ways in which it can turn out that you get exactly 5 heads and 5 tails.

For the maths boffins, 252 is the central binomial coefficient and is where the Ramanujan tau function is found.

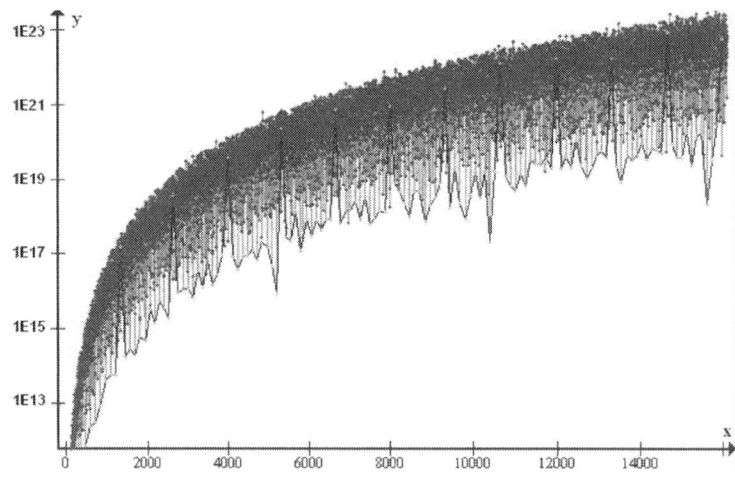

http://en.wikipedia.org/wiki/Ramanujan_tau_function

Values for n < 16,000 with logarithmic scale. The blue line picks only the values of *n* that are multiples of 121 which is 11 x 11.

12 + 21.

1221 = 33 x 37 which interestingly is the 21st Dodecagon number, a Dodecagon being a 12 sided polygon constructed of 12 Pentagons. The Dodecagon in this theory represents the containing geometry of TIME or the Ether. This is connected to the 3 6 9 (0) Number Group.

1221 & Planck Mass.

I wondered if I might find this number 1221 in Physics?

Well, remarkably, we see it in the Planck Mass which to 4 significant figures is 1.221×10^{19} GeV/c^2. This number is important for the relative masses.

'The Planck mass is nature's maximum allowed mass for point-masses (quanta). If two quanta of the Planck mass or greater met, they could spontaneously form a black hole whose Schwarzschild radius equals their Compton wavelength. Once such a hole formed, other particles would fall in, and the black hole would experience runaway, explosive growth (assuming it did not evaporate via Hawking radiation).

Nature's stable point-mass particles, such as electrons and quarks, are many, many orders of magnitude smaller than the Planck mass and cannot form black holes in this manner. On the other hand, extended objects (as opposed to point-masses) can have any mass.

Unlike all other Planck base units and most Planck derived units; the Planck mass has a scale more or less conceivable to humans. It is traditionally said to be about the mass of a flea, but more accurately it is about the mass of a flea egg.

The Planck mass can be derived approximately by setting it as the mass whose Compton wavelength and Schwarzschild radius are equal.

The Compton wavelength is, loosely speaking, the length-scale where quantum effects start to become important for a particle; the heavier the particle, the smaller the Compton wavelength.

The Schwarzschild radius is the radius in which a mass, if confined, would become a black hole; the heavier the particle, the larger the Schwarzschild radius.

If a particle were massive enough that its Compton wavelength and Schwarzschild radius were approximately equal, its dynamics would be strongly affected by quantum gravity. This mass is (approximately) the Planck mass.'
http://en.wikipedia.org/wiki/Planck_mass

Fibonacci & the Number 21.

The Fibonacci Sequence we saw earlier is traditionally taught as (0) 1 1 2 3 5 8 13 21 34 55 89 144 233 etc.

As we have seen, the difference between these numbers approximates to 1.618, which is known as the Golden Mean or Phi.

21 is the 8th Fibonacci number and thus could be said to complete the octave.

A common mistake is thinking that the ratios of consecutive Fibonacci numbers are the only numbers that can create the Golden Mean.

The fact is, ANY sequence of numbers, starting with ANY random two numbers, you will still eventually get to the Phi ratio if you divide the latter into the former.

The same is true even if you begin with a negative number.

$$-8 \ 4 \ -4 \ 0 \ -4 \ -4 \ -8 \ -12 \ -20 \ -32 \ -52 \ -84 \ -136$$

At this stage we already see that the terms are getting close to the Golden Mean or Phi value of 1.618:-

$$-136 \ / \ -84 = 1.619047$$

The golden mean ratio always reasserts itself to control the numbers - it is inherent to any sequence of numbers formed as an addition of previous two numbers.

However the 1 1 2 3 5 8 13 is the natural unfolding in the universe and is what makes the numbers in that sequence so special.

In pondering over this sequence, I wondered what terms could come before the zero and leave the positive side of the sequence unaffected and I saw a rather nice sequence:-

$$-21\ 13\ -8\ 5\ -3\ 2\ -1\ 1\ 0\ 1\ 1\ 2\ 3\ 5\ 8\ 13\ 21$$

There is a type of symmetry being displayed here that doesn't seem very normal because of the alternating positive and negative value numbers as we go beyond zero. The above sequence contains 17 Fibonacci terms. 17 having 297 partitions, the mirror of 792. 2 full octaves of eight terms either side of their fulcrum point of balance 0, beginning with -21 and 13 ending with 21. I feel this is pretty deep and essential.

To the left of zero, the total of the terms:-

$$-21\ 13\ -8\ 5\ -3\ 2\ -1\ 1 = -12$$

To the right of zero, the total or the terms:-

$$1\ 1\ 2\ 3\ 5\ 8\ 13\ 21 = 54$$

The Range is -12 to + 54 = 66.

66 is the Sum of the First 11 Numbers and Triangular.

Addition of all the terms of this sequence gives:-

$$-12 + 54 = 42\ (2 \times 21)$$

Multiplying

$$1 \times -1 \times 2 \times -3 \times 5 \times -8 \times 13 \times -21 = 65520 = 81\ \text{Mod}\ 99$$

There are 81 stable chemical elements.

Mod 9 Sequence Analysis.

$$-21\ 13\ -8\ 5\ -3\ 2\ -1\ 1\ 0\ 1\ 1\ 2\ 3\ 5\ 8\ 13\ 21$$

$$-3\ 4\ -8\ 5\ -3\ 2\ -1\ 1\ 0\ 1\ 1\ 2\ 3\ 5\ 8\ 4\ 3$$

Mod 9 Digit Total = 24

When we multiply the Mod 9 results we see:-

$$1 \times 1 \times 2 \times 3 \times 5 \times 8 \times 4 \times 3 = 2880$$

$$1 \times -1 \times 2 \times -3 \times 5 \times -8 \times 4 \times -3 = 2880$$

2880 is an important number we will visit later on.

12 and 21 – The Difference of Squares.

$$21^2 - 12^2 = 297\ \text{Number of Partitions for the number 17}$$

$$31^2 - 13^2 = 792\ \text{Partitions for 21 and the mirror of 297}$$

Chapter 6: Key Numbers: 27 & 37.

Triangles.

In the course of the research I had discovered that 27 & 37 are the only two digit Root Numbers that produce three digit recurring decimals as their effect on all other numbers by division with 3 digit Recurring Decimals representing triangles.

Partition Theory suggests there are in fact only two numbers required for creation - 27 and 37 - everything else is done with mirrors.

These are two seriously special numbers because 27 and 37 possess an incredibly intimate relationship via their reciprocals that no others boast.

$$1/37 = 0.027 \text{ Recurring} \qquad 1/27 = 0.037 \text{ Recurring}$$

Remember the reciprocal nature of Electricity and Magnetism as illuminated by Walter Russell's Octaves of Perception.

Everything seems to be a dance of the a priori composite numbers, 2 and 3, in combination with the 'virgin' number 7. These two numbers 27 & 37 produce 2 different types of triangles.

Plato said that Surface was made of triangles and squares and these form the central aspect of Integer Partition Theory.

Plato - The Triangles That Fill Up Space.

I must now return to another great work of John Michell, How the World is Made, to explain, far better than I ever could, why the triangle is so important in the grand scheme of things and to introduce Plato's work and ideas to the reader.

'The importance of the triangle in the structuring of space is demonstrated in the five regular figures of three dimensional or solid geometry by which Plato in Timaeus (54-57) symbolizes the five elements, earth, air, fire, water and the ether.

Three of these, the tetrahedron (4 sides), the octahedron (8 sides), and the icosahedron (20 sides), are made up of equilateral triangles and represent in order fire, air, and water. The two other solids are the cube or hexahedron with six square sides, the symbol of the Earth and the dodecahedron with twelve pentagonal sides which represents the ether and orders the whole universe.

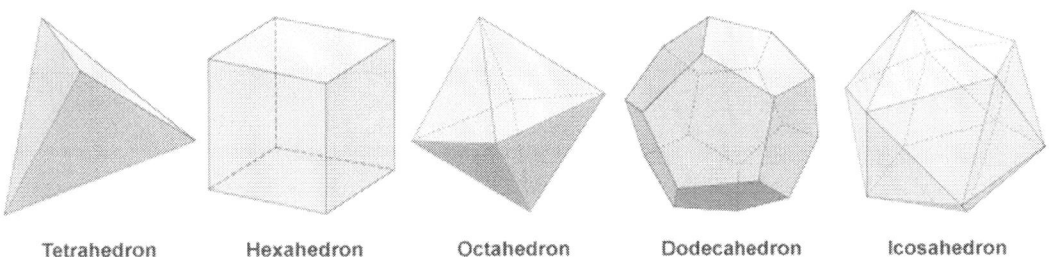

Tetrahedron   Hexahedron   Octahedron   Dodecahedron   Icosahedron

There are two kinds of right angle triangle.

One is the isosceles with two equal sides and the other, the scalene, with three unequal sides. The first is half a square bisected along its diagonal, with angles of 90, 45 and 45 degrees while the scalene has infinite variety of shapes.

The most perfect of these, said Plato, is half an equilateral triangle, with one side half the length of its hypotenuse and with angles of 90, 60, and 30 degrees' - 'How the World is Made' by John Michell.

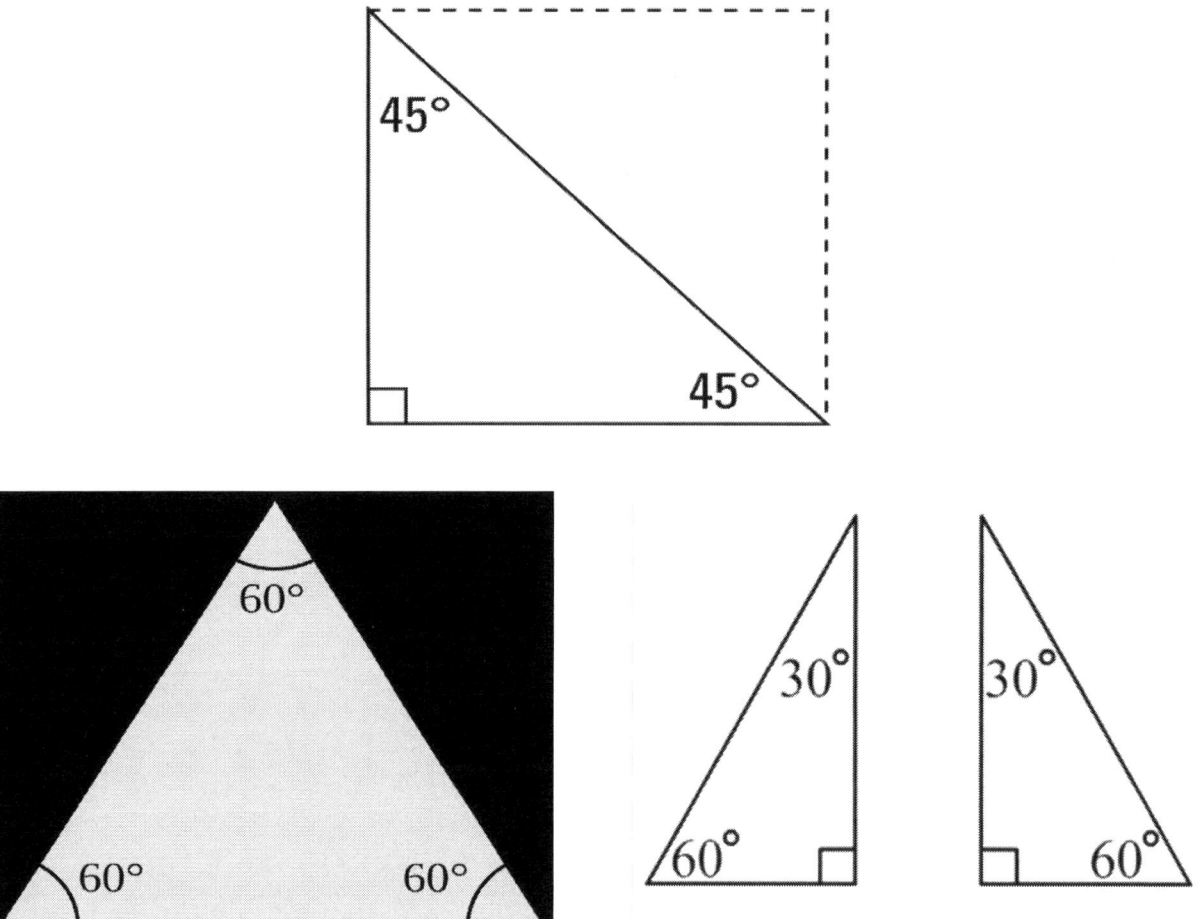

Back to Michell – 'These two triangles, each united with its pair, are the building blocks of creation, forming the first four regular solids which, in the geometer's cosmology, correspond to atoms.

As the basic components of fire, earth, air and water, they are in constant motion, and as they knock and jostle each other they are broken up into separate triangles, which are then sub-divided into similar shapes on different scales.

At the same time, the particles are reassembling, small triangles joining with others to make big ones and these combining to form new atoms or geometric solids.

The scalene triangles can become parts of tetrahedrons, octahedrons or icosahedrons, because they have sides made of equilateral triangles, whereas the isosceles can only fit into a cube, the element of the earth.

As an essay in physics, Plato's account is strikingly odd but he persists with it in detail, describing the interactions of the atoms and elements and illustrating the correspondences between earth, air, fire and water and the 4 geometric shapes.

For example, spicy dishes taste hot and burn like fire because the spice atoms are tetrahedrons, the smallest and most prickly of shapes. Sweet and creamy things, on the other hand, consist mainly of icosahedrons, whose rounded shape soothes and pleases the tongue.

That sounds plausible enough, but no one today would take Plato's account literally, and no one was ever meant to (I beg to differ on this point - AM) Plato is here elaborating on the traditional, geometric myth of creation.

With high wit and beautiful imagery he is leading his students into the intricacies of solid geometry, not boring them with shapes, numbers and figures but bringing life to the subject through the allegories attached to it.

A noticeable omission from Plato's story is the fifth regular solid, the dodecahedron.

It plays no part in the clashing and breaking up to which the other four are subjected. All that Plato says in Timaeus is that 'God used it for arranging the constellations of the whole heaven'.

Elsewhere (Phaedrus 110) he indicates that the dodecahedron is the Earth's essential form – its etheric envelope perhaps.

As a symbol the dodecahedron is uniquely significant, for with its 12 pentagonal faces, it provides an image of the ideal earth, where all twelve types of humanity dwell in harmony under the guidance of the twelve zodiacal gods.

The reality of this image, if only on the level of ideals, is apparent in the dodecahedron, the most perfect and noble polyhedron in geometry.' How the World is Made by John Michell.

The Two Triangles of Creation.

My theory is that the isosceles triangle being formed of the square, and being a cube number, is connected to Light and the number 27 which is 3^3 (three cubed) and that the equilateral derived scalene triangle is connected to Vibration and the number 37.

27 Triangle - Connections

Proton - Anti Proton Interaction to Create Light

Cubes   Hexagons   Light   Phi   Time   Isosceles   Pentagons

Dodecahedrons   Earth   41   Squares   Magnetism   101   7   13

37 Triangle - Connections

Electron – Anti Proton Interaction to Create Vibration

Equilateral   Scalene   Tetrahedrons   Octahedrons   Icosahedrons

Spheres   Vibration   Thought   Intention   Consciousness

Energy   Frequency   Fire   Air   Water   Electricity   73   137

Triangles in Mainstream Science.

Triangles have recently been found by mainstream science to be of profound importance - this paraphrased from
http://resonance.is/news/study-of-graphene-reveals-structure-of-space-time-2/

  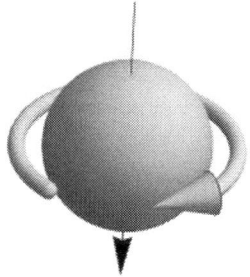

"Two Physicists at UCLA working with a highly conductive nanotech material called graphene recently released a new postulate on the origin of the spin of electrons. Professor Chris Regan and graduate student Matthew Mecklenburg discovered that they could accurately predict the spin of electrons if they divided space itself into a lattice of positions interconnected by triangles. "An electron's spin might arise because space at very small distances is not smooth, but rather segmented, like a chessboard," Regan said in an interview for the UCLA Newsroom article.

Space is usually considered infinitely divisible - given any two positions, there is always a position halfway between. But in a recent study aimed at developing ultra-fast transistors using graphene, researchers from the UCLA Department of Physics and Astronomy and the California Nano Systems Institute show that dividing space into discrete locations, like a chessboard, may explain how point-like electrons, which have no finite radius, manage to carry their intrinsic angular momentum, or 'spin.'

While studying graphene's electronic properties, they found that a particle can acquire spin by living in a space with two types of positions - dark tiles and light tiles. The particle seems to spin if the tiles are so close together that their separation cannot be detected. 'An electron's spin might arise because space at very small distances is not smooth, but rather segmented, like a chessboard,' Regan said.

Unveiling a concept that is at once novel and deceptively simple, Regan and Mecklenburg found that electrons' two-valued spin can arise from having two types of tiles - light and dark - in a chessboard-like space. And they developed this quantum mechanical model while working on the surprisingly practical problem of how to make better transistors out of a new material called graphene. This calculation involved understanding how light interacts with the electrons in graphene.

The electrons in graphene move by hopping from carbon atom to carbon atom, as if hopping on a chessboard. The graphene chessboard tiles are triangular, with the dark tiles pointing 'up' and light ones pointing 'down.' When an electron in graphene absorbs a photon, it hops from light tiles to dark ones. Mecklenburg and Regan showed that this transition is equivalent to flipping a spin from 'up' to 'down.'

In other words, confining the electrons in graphene to specific, discrete positions in space gives them spin. This spin, which derives from the special geometry of graphene's honeycomb lattice, is in addition to and distinct from the usual spin carried by the electron. In graphene the additional spin reflects the unresolved chessboard-like structure to the space that the electron occupies.

'My adviser was very excited to see that spin can emerge from a lattice. It makes you wonder if the usual electron spin could be generated in the same way. It's not yet clear if this work will be more useful in particle or condensed matter physics,' Regan said, 'but it would be odd if graphene's honeycomb structure was the only lattice capable of generating spin.'

I hope I will have convinced the reader that I am not alone in seeing the importance of the triangle.

27 & 37 - Time.

As I was looking up the precise length of a year in days I found the number:-

$$365.242199 \text{ days} = 1 \text{ Year}$$

For some reason I found myself, calculator in hand, and looked at the reciprocal to see:-

$$1 / 365.242199 \text{ days} = 0.00 \ \ 27 \ \ 37 \ \ 909$$

When I saw the numbers, I dropped the calculator!

27 and 37 are the only numbers needed to create Physical Reality because I am convinced that Plato was right, Surface is triangles.

27.37 - Uncanny Event - 6th November 2012.

The first person I told about my theory once it was first formed was good mate Neil Shotton who provided the inspiration to look at DNA and who had kindly come down to listen to me going on about a theory which rankles with his world view to put it mildly.

We were out fetching supper from Sainsbury's for that evening having just spent the afternoon with me while I was explaining much about my thinking and having isolated the numbers 27 and 37 as being central to the whole theory.

As the checkout total flashed out a final balance at £27.37 there was an involuntary burst of laughter from me which I quickly had to bring under control!

Neil was unaware, until we had returned to the car, what had happened, where I showed him the receipt.

Interesting synchronicity!

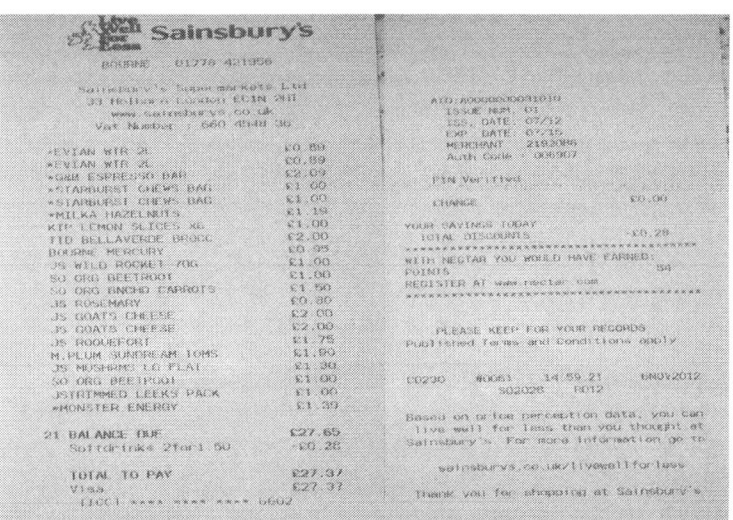

21 Items, no less.......

DNA & Amino Acids - 2737.

In addition to looking at the total number of Protons and Neutrons for the 64 Codons, I wondered what the total would be for just one of each of these Amino Acids common to all biology.

Amino Acid	Ticker	Mass	Hydrogen	Carbon	Nitrogen	Oxygen	Sulphur	Protons	Neutrons	Total	Freq	Total Protons & Neutrons
Glysine	GLY	75.0669	1	0	0	0	0	1	0	1	1	75
Alanine	ALA	89.0935	3	1	0	0	0	9	6	15	1	89
Serine	SER	105.0930	3	1	0	1	0	17	14	31	1	105
Proline	PRO	115.1310	6	3	0	0	0	24	18	42	1	116
Valine	VAL	117.1469	7	3	0	0	0	25	18	43	1	117
Threonine	THR	119.1197	5	2	0	1	0	25	20	45	1	119
Cysteine	CYS	121.1590	3	1	0	0	1	25	22	47	1	121
Isoleucine	ILE	131.1736	9	4	0	0	0	33	24	57	1	131
Leucine	LEU	131.1736	9	4	0	0	0	33	24	57	1	131
Asparagine	ASN	132.1184	4	2	1	1	0	31	27	58	1	132
Aspartic Acid	ASP	133.1032	3	2	0	2	0	31	28	59	1	133
Glutamine	GLN	146.1451	6	3	1	1	0	39	33	72	1	146
Lysine	LYS	146.1882	10	4	1	0	0	41	31	72	1	146
Glutamic Acid	GLU	147.1299	5	3	0	2	0	39	34	73	1	147
Methionine	MET	149.2124	7	3	0	0	1	41	34	75	1	149
Histidine	HIS	156.1552	6	4	2	0	0	44	38	82	1	156
Phenylalanine	PHE	165.1900	7	7	0	0	0	49	42	91	1	165
Arginine	ARG	174.2017	10	4	3	0	0	55	45	100	1	174
Tyrosine	TYR	181.1894	7	7	0	1	0	57	50	107	1	181
Tryptophan	TRP	204.2262	8	9	1	0	0	69	61	130	1	204
STOP		0.0000	0	0	0	0	0	0	0	0	0	0
TOTALS		2739.0169	119	67	9	9	2	688	569	1257	20	2737

The answer was none other than 2737!

Messenger RNA – 27:37.

Ribonucleic acid, or RNA is one of the three major biological macromolecules that are essential for all known forms of life (along with DNA and proteins). A central tenet of molecular biology states that the flow of genetic information in a cell is from DNA through RNA to proteins: 'DNA makes RNA makes protein'

Phenylalanine	UUU	AUU	Isoleucine
Phenylalanine	UUC	AUC	Isoleucine
Leucine	UUA	AUA	Isoleucine
Leucine	UUG	AUG	Methionine START
Serine	UCU	ACU	Threonine
Serine	UCC	ACC	Threonine
Serine	UCA	ACA	Threonine
Serine	UCG	ACG	Threonine
Tyrosine	UAU	AAU	Asparagine
Tyrosine	UAC	AAC	Asparagine
STOP	UAA	AAA	Lysine
STOP	UAG	AAG	Lysine
Cysteine	UGU	AGU	Serine
Cysteine	UGC	AGC	Serine
STOP	UGA	AGA	Arginine
Tryptophan	UGG	AGG	Arginine
Leucine	CUU	GUU	Valine
Leucine	CUC	GUC	Valine
Leucine	CUA	GUA	Valine
Leucine	CUG	GUG	Valine
Proline	CCU	GCU	Alanine
Proline	CCC	GCC	Alanine
Proline	CCA	GCA	Alanine
Proline	CCG	GCG	Alanine
Histidine	CAU	GAU	Aspartic Acid
Histidine	CAC	GAC	Aspartic Acid
Glutamine	CAA	GAA	Glutamine Acid
Glutamine	CAG	GAG	Glutamine Acid
Arginine	CGU	GGU	Glycine
Arginine	CGC	GGC	Glycine
Arginine	CGA	GGA	Glycine
Arginine	CGG	GGG	Glycine

The fours bases identified in DNA remain the same with the exception that Thymine is replaced by Uracil. Of the 64 codons that make up RNA, 37 contain Uracil while the remaining 27 do not.

Gnomons.

A gnomon is a figure that, added to a given figure, makes a larger figure of the same shape.

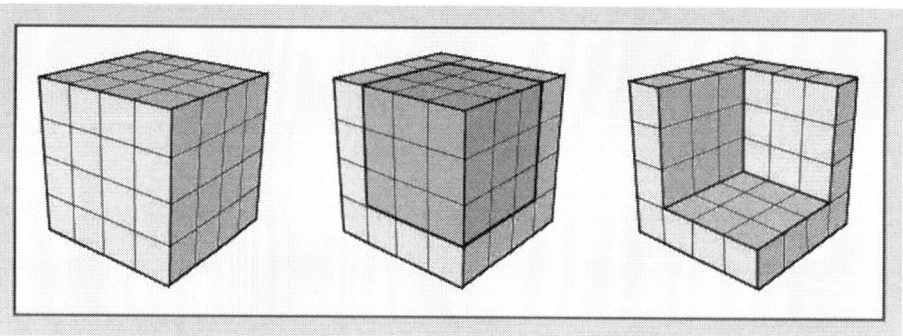

      4 x 4 x 4 Cube = 64   **3 x 3 x 3 Cube = 27**   **Shell = 37**

Above we see a 64 cube with a 27 cube removed to expose the 37 shell that allows the 27 cube to grow to a 64 cube.

We can see from a geometric viewpoint how 27 and 37 are in Gnomon relationship:-

$$4^3 - 3^3 = 37$$

$$64 - 27 = 37$$

. This is a general property of Centred Hexagons:-

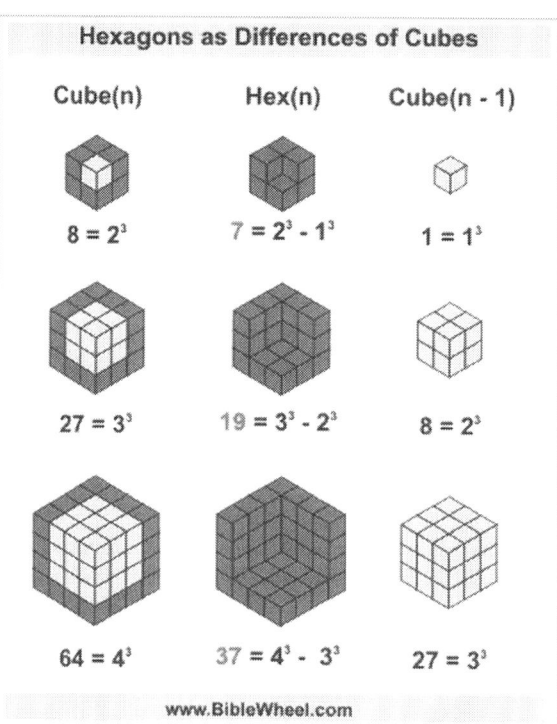

The reason for this is that Centred Hexagons are actually projections of three dimensional cubes onto a two dimensional plane.

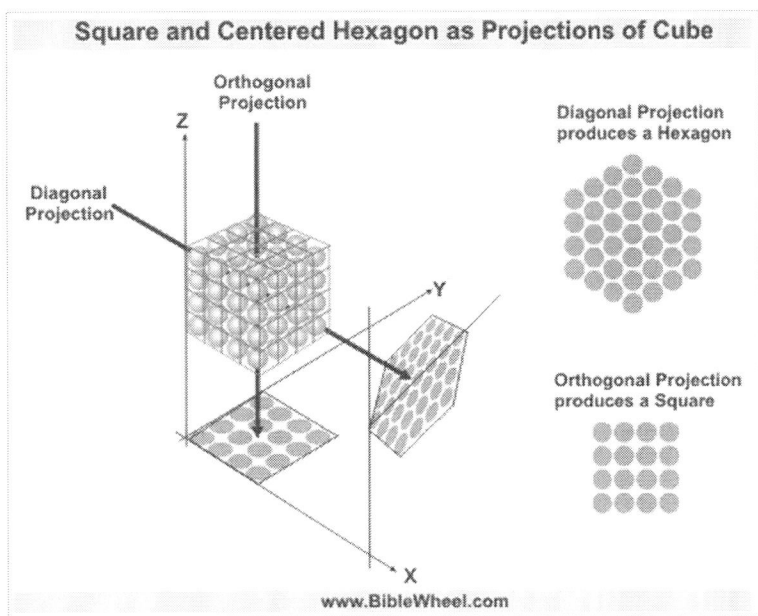

This means that there is a geometric relation between Hexagonal Numbers and Cubes.

Specifically, the figure above shows the geometric relation between the Number 37 and the Number 64.

Put another way, this means that 37-as-Hexagon is a diagonal projection of 64-as-Cube.

Chapter 7: Key Numbers - 55.

55 - The Fulcrum Point of Balance.

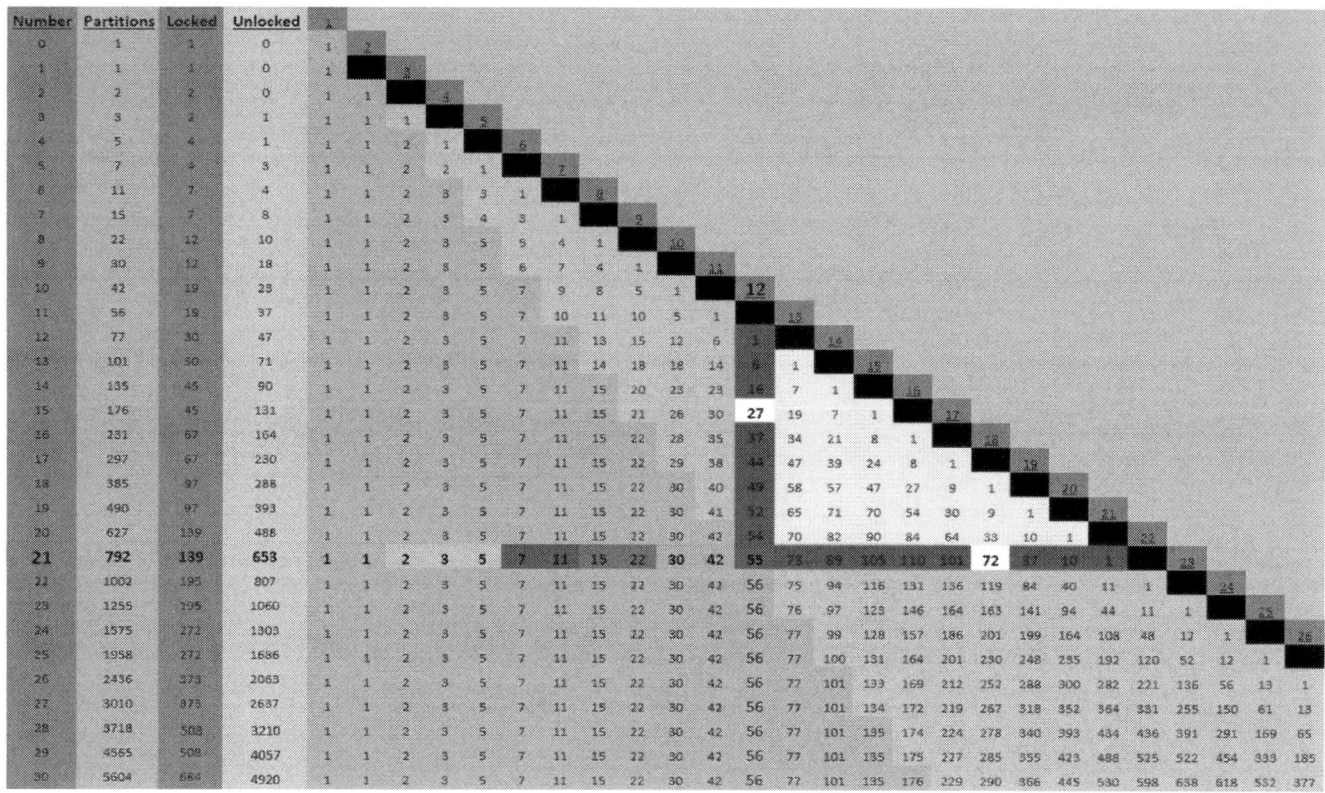

Looking along the Stage Values for the Number 21 we can see that the olive coloured number 55 occurs on the 12th Stage Value. 55 is the Fulcrum for the Hologram Projector about which the containing numbers, 37 and 73 are anchored.

In terms of the fundamental numbers 0 to 9 - the 55 represents the central axis of the Hologram Projector, the 0 and the 9.

55 is so beautifully connected to the 0 9 axis evident in the dimensions of number. Consider that:-

$$1+2+3...+10 = 55$$

$$1+2+3...+100 = 5050$$

$$1+2+3...+1000 = 500500 \text{ etc.}$$

With each dimension expansion of 10 we see how the Tree of Life grows.

This axis pervades all things and everything is formed about it.

The 37 and 73 are the 3 and 6 and together with the 0 9 central axis, contain all things which are electric in nature.

The theory goes that the integration of Spherical Sound Waves and Cubed Light Particles produces a Carbon based central axis to the universe, or Tree of Life, as represented by the number 55. Carbon's relation to the Number 55 is deduced from and inferred by the molecular structure of Chlorophyll and Hemoglobin which both contain 55 Carbon atoms.

This 'Tree of Life' is I believe none other than what mainstream Science have coined the aptly named 'Axis of Evil'.

Axis of Evil - Purposeful Universe.

More support for this theory of a Central Axis comes from the brilliant book 'Purposeful Universe' by Carl Johan Calleman.

'For some time now it has been realised that the study of cosmic background radiation, CMB for short, may serve to probe the relevance of the cosmological principle.

One of the predictions from the big bang theory was that the high temperatures generated at the sudden early expansion of the universe would give rise to an afterglow in the form of CMB at a few degrees above absolute zero. The detection and measurement of CMB exactly in accordance with this expectation was a very clear verification of the big bang theory.

The study of CMB also became the basis for projects designed to test the relevance of the assumptions of homogeneity and isotropy in the universe. If these assumptions were true, so it was argued, the temperature distribution would not display some clearly discernible pattern in the early universe some 15 billion years ago, and the cosmological principle could hence be considered as valid.

Data from the initial satellite launched in 1989 to measure the CMB indeed only discovered minor anisotropies (irregularities) in the background radiation, which were interpreted as indicative of randomly distributed early density fluctuations that later would give rise to galaxies. Hence no large scale structures were discovered in these initial studies and the CMB was still consistent with the cosmological principle.

In 2003, new satellite measurements recorder the CMB by means of the so called Wilkinson Microwave Anisotropy Probe (WMAP) which provided more accurate data from the afterglow. When this data set was analysed mathematically, an axis and polarised fields of temperature were discovered by Max Tegmark. These fields were organised somewhat like the panels of a basketball which defined the direction of an axis through the early universe. Kate Land and Joao Magueijo of Oxford University later verified these findings in an article entitled The Axis of Evil - the findings seem very disturbing to the foundations of established cosmology - they indicate that the universe has always contained a potentially organising structure.

Studies of the axis based on other datasets support the validity of its existence in dramatic ways. First it was found that the polarisation of light from quasars was influenced by the proximity to this central axis. An interesting twist of this particular study is that the polarisation corkscrewed around the axis and its authors suggested that a potential explanation of this effect is that the entire universe rotates around this central axis. This would then be consistent with the idea that the universe as a whole is a spinning vortex generated by the axis and that it emerged as such from the very beginning.

Michael Longo at Michigan University studied the handedness of spiral galaxies throughout the universe. He found that a line separating the preference for the two types of handedness approximately lined up with the axis previously discovered in the WMAP. The axis that our own particular galaxy revolves around as well as those of most other galaxies observed was also found to be aligned with the central axis. Longos study was based in a few thousand galaxies and he concluded that 'a well-defined axis for the universe on a scale of 170Mpc would mean a small but significant violation of the cosmological principle and of Lorentz symmetry and thus the underpinnings of special and general relativity.'

The Central Axis is the fundamental space time organiser of our universe.

If the handedness and spin axes of galaxies are directly related to the central axis this means that the formation of galaxies may not be a result of random fluctuations causing them to rotate. The spin of the galaxies would instead be related to that of the central axis and also possibly to one another in a form of entanglement at the largest possible scale. If the different galaxies of the universe are not independent but retain a connection to the overall polarised structure created by the central axis this would favour the idea that their evolution is connected and synchronised something that we will see later is crucial for the evolution of life"

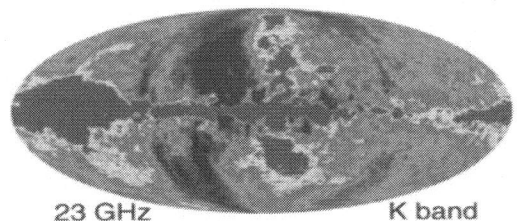

Image: Wikipedia.

"The discovery (Axis of Evil) casts doubts on all contemporary concepts of the nature and development of the universe. Even the Einstein theory of relativity seems obsolete now.

Until recently space and time were believed to have unfolded in a chaotic way after the Big Bang, and the universe was thought to be homogeneous and expanding continuously.

Now scientists will have to acquiesce to an ordered way of development of the universe as if it was born and develops in compliance with a scenario written beforehand." Leonid Speransky in Pravda an astrophysicist and professor with the Lomonosov State University in Moscow.' - http://www.pravdareport.com/science/tech/20-04-2006/79383-universe-0/

Yes! This scenario written beforehand is exactly what I am getting at, with the effect that the universe developes according to the universal template that is the canon of number - in the same way the seed knows how to grow into the oak; the universe is unfolding according to a specific design.

55 - Plato's Lambda.

As I mentioned earlier, everything appears to be a dance of the 2 and 3 and this resonates with the harmonic series of 2 and 3 as depicted in Plato's Lambda.

In his book Timaeus, Plato states that God created the Cosmic Soul using two mathematical strips

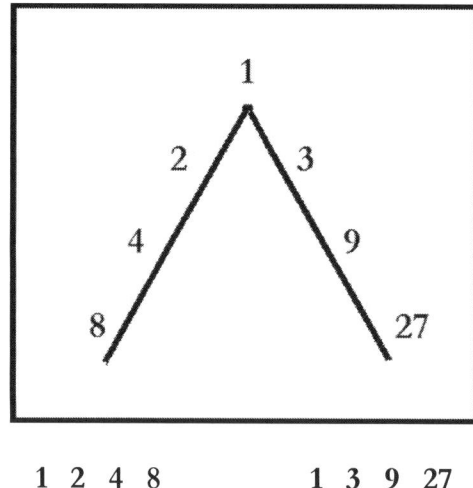

        1 2 4 8   1 3 9 27

These two strips follow the shape of an inverted "V" or the "Platonic Lambda" since it resembles the shape of the 11th letter of the Greek alphabet "Lambda". Apt, as 11 is appointed as the number of the Earth.

The only thing missing from the picture above, in my opinion, is an additional 1 at the peak of the triangle.

$$(2^0 + 2^1 + 2^2 + 2^3) + (3^0 + 3^1 + 3^2 + 3^3) = 55$$

$$(1 + 2 + 4 + 8) + (1 + 3 + 9 + 27) = 55$$

Interestingly, the Mod 9 total for this sequence is:-

$$1 + 2 + 4 + 8 + 1 + 3 + 9 + 9 = 37$$

Dante's 55 'Stelle'.

In the course of my research I came across Dante's work and his most famous work, a book called The Commedia. The book is in three parts named Inferno, Purgatorio and Paradiso or in English; Hell, Purgatory and Paradise, or Heaven.

I was interested to discover that Dante distributed 55 'Stelle', Italian for stars, across the three parts of his book and uses the Platonic Lambda's number 55 as a vehicle for the soul's ascent to the stars.

There has been much scholarly conjecture over just why that might be the case and I offer the following, perhaps original, understanding.

Dante distributes the 55 'Stelle' or Stars in the following way:-

11 Inferno   13 Purgatorio   31 Paradiso

Below we see a table of the distribution of these stars together with a Mod 9 analysis which threw up several items of interest unlikely to have been produced by chance.

Inferno Stelle (11)				Purgatorio Stelle (13)				Paradiso Stelle (31)			
Canto	Mod	Line	Mod	Canto	Mod	Line	Mod	Canto	Mod	Line	Mod
1	1	38	2	1	1	23	5	1	1	40	4
2	2	55	1	6	6	100	1	2	2	30	3
3	3	23	5	8	8	86	5	2	2	137	2
7	7	98	8	8	8	91	1	4	4	23	5
15	6	55	1	12	3	90	0	4	4	52	7
16	7	83	2	17	8	72	0	5	5	97	7
20	2	50	5	18	0	77	5	6	6	112	4
22	4	12	3	27	9	89	8	7	7	138	3
26	8	23	5	30	3	111	3	8	8	11	2
26	8	127	1	31	4	106	7	8	8	110	2
34	7	139	4	32	5	57	3	9	0	33	6
172	55	703	37	33	6	41	5	10	1	78	6
				33	6	145	1	12	3	29	2
				256	67	1088	44	13	4	4	4
								14	5	86	5
								15	6	16	7
								17	8	77	5
								18	0	68	5
								18	0	115	7
								22	4	112	4
								23	5	92	2
								24	6	147	3
								25	7	70	7
								28	1	19	1
								28	1	21	3
								28	1	87	6
								30	3	5	5
								31	4	28	1
								32	5	108	0
								33	6	145	1
								477	117	2090	119

11 Inferno   13 Purgatorio   31 Paradiso

11, the Number of the Earth, are placed in Inferno (Hell on Earth?). Plato attributes the cube to the Earth. When we cube the number 11 we see the other star proportions for Purgatorio and Paradiso.

$$11^3 = 1331$$

13 and 31 are mirrors of each other and are also mirrored in their squares which is very rare indeed, certainly for a 2 digit number.

$$13^2 = 169 \quad 31^2 = 961$$

$$961 - 169 = 792 \text{ Partitions for } 21$$

The only other two digit number that has this property is the number 12.

$$21^2 = 441 \quad 12^2 = 144$$

441 - 144 = 297 Partitions for 17

297 is the mirror of 792

Looking at the Stelle positioning in Inferno we see:-

Inferno Stelle (11)			
Chapter	Mod 9	Line	Mod 9
1	1	38	2
2	2	55	1
3	3	23	5
7	7	98	8
15	6	55	1
16	7	83	2
20	2	50	5
22	4	12	3
26	8	23	5
26	8	127	1
34	7	139	4
172	55	703	37

The total of the Chapter numbers where the stars are distributed comes to 172 = 73 Mod 99 where 73 is obviously one of the key numbers in Integer Partition Theory.

Then we see that the total of the Mod 9 values generated by the Canto (verse) numbers comes to 55, another key number.

Very interestingly, when I totalled the Occurrence values for the 11 stars in the Inferno, these came to 703.

I instantly recognised this number 703, again from Vernon Jenkins work, as the central triangle in the 73 based triangle as illustrated overleaf.

(3 x 666) + 703 = 2701

2701 = 37 x 73

703 = 19 x 37

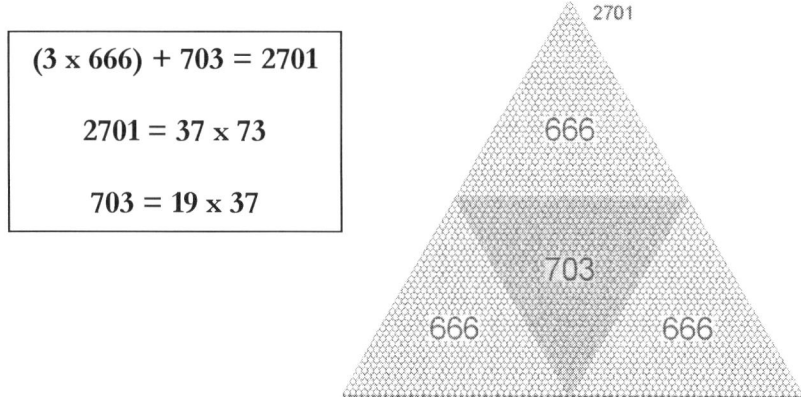

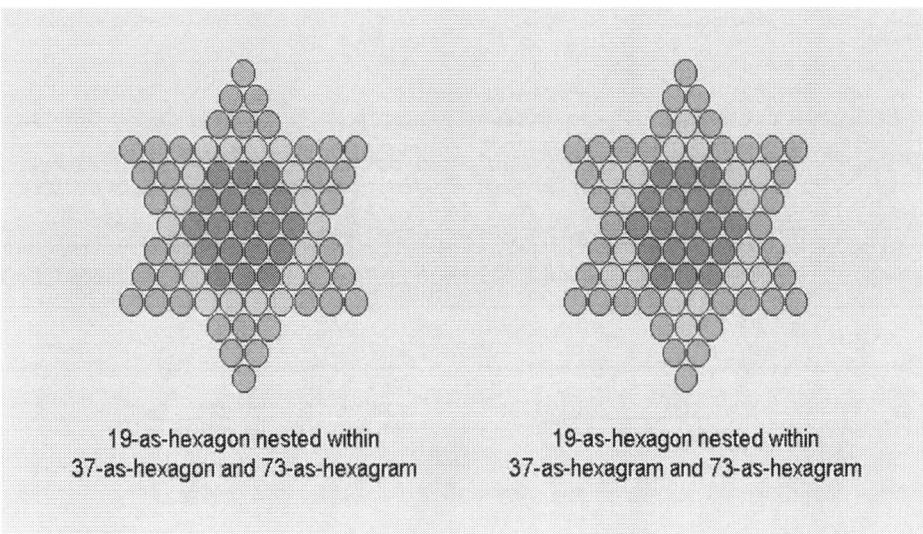

19-as-hexagon nested within
37-as-hexagon and 73-as-hexagram

19-as-hexagon nested within
37-as-hexagram and 73-as-hexagram

19, 37 and 73 are key geometrically related numbers in the Integer Partition Theory and further, 703 represents the 37th Triangle and 19th Hexagonal Number.

Last but not least, the Mod 9 total for the Occurrence values comes to our friend 37.

Canto Line Analysis.

Inferno			Purgatorio			Paradiso		
Canto	Lines	Mod 9	Canto	Lines	Mod 9	Canto	Lines	Mod 9
1	136	1	1	136	1	1	142	7
2	142	7	2	133	7	2	148	4
3	136	1	3	145	1	3	130	4
4	151	7	4	139	4	4	142	7
5	142	7	5	136	1	5	139	4
6	115	7	6	151	7	6	142	7
7	130	4	7	136	1	7	148	4
8	130	4	8	139	4	8	148	4
9	133	7	9	145	1	9	142	7
10	136	1	10	139	4	10	148	4
11	115	7	11	142	7	11	139	4
12	139	4	12	136	1	12	145	1
13	151	7	13	154	1	13	142	7
14	142	7	14	151	7	14	139	4
15	124	7	15	145	1	15	148	4
16	136	1	16	145	1	16	154	1
17	136	1	17	139	4	17	142	7
18	136	1	18	145	1	18	136	1
19	133	7	19	145	1	19	148	4
20	130	4	20	151	7	20	148	4
21	139	4	21	136	1	21	142	7
22	151	7	22	154	1	22	154	1
23	148	4	23	133	7	23	139	4
24	151	7	24	154	1	24	154	1
25	151	7	25	139	4	25	139	4
26	142	7	26	148	4	26	142	7
27	136	1	27	142	7	27	148	4
28	142	7	28	148	4	28	139	4
29	139	4	29	154	1	29	145	1
30	148	4	30	145	1	30	148	4
31	145	1	31	145	1	31	142	7
32	139	4	32	160	7	32	151	7
33	157	4	33	145	1	33	145	1
34	139	4						
Total	4720	157	Total	4755	102	Total	4758	141

There are a total of 100 Cantos distributed across the three parts of the Commedia and extraordinarily enough, when I looked at the line numbers as above using Mod 9 we can see something very unlikely to be by chance and which speaks to my work on the structure of Numbers themselves.

All line counts from each Canto return 1 4 or 7 Mod 9 exclusively.

Did Dante weave the secrets of universal mathematics throughout his Commedia? I rather think he may well have done. Living in Rome he may well have had access to the Vatican library where I strongly suspect all the knowledge of antiquity is held. The Romans were supposed to have burned the Great Library at Alexandria which was the repository for all this knowledge and I just don't buy that they would have torched that information. The Romans well understood that knowledge is power.

He may well have hidden the exposure of this numerical perfection so that it would go undetected by the Vatican who have been known, especially in his era, to get rid of people whose ideas are too close to the truth for comfort.

DNA & Amino Acids - 37 55 73.

I was very fortunate to find that there is a further direct connection between the Integer Partition Theory paper and my DNA Paper in great work done by a number of other researchers analysing the Group of 20 Amino Acids.

'Equations from Rakocevic's table:-

A comparison of the sums of the alternating cells with those for the columns from Rakocevic's original table reveals numerical integration of a high order.'

'Paardekooper has accordingly rewritten the three pairs of equations that result from aggregations of molar mass sums in Rakocevic:

1. $C(1) = 1405.63 = (2 \times 666) + (2 \times 37) - 0.37$
2. $C(2) = 1332.37 = (2 \times 703) - (2 \times 37) + 0.37$
3. $A(1) = 1406.55 = (2 \times 666) + (2 \times 37) + 0.55$
4. $A(2) = 1331.45 = (2 \times 703) - (2 \times 37) - 0.55$
5. $C(3) = 1478.73 = (2 \times 666) + (2 \times 73) + 0.73$
6. $C(4) = 1259.27 = (2 \times 703) - (2 \times 73) - 0.73$

The numerical integration of Paardekooper's three pairs of equations highlights the role of the integer pair [666, 703] in the numerical structuring of the molar mass sums of the amino acids, and the concomitant fine-tuning role played by the integer triplet [37, 55, 73].' - From www.craigdemo.co.uk/rakocevicpaper2.doc

From Vernon Jenkins - the original discoverer of the geometrical phenomenon - comes graphic realization of the geometrical integration of the integer triplet [37, 55, 73]:

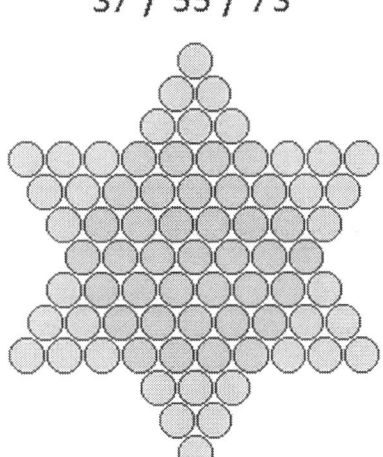

Depicting the related hexagon/hexagram pair 37/73
as the intersection/union of 55-as-triangle
with an inverted copy of itself

Chapter 8: Key Numbers - 37 & 73.

37 and 73 - The Boundary Markers for physical reality.

These two numbers are imbued with the properties of the 3 and 6, in 'fundamental', numerical terms.

The 3 and 6 Number Pair form the 'Containing' aspect of the magnetic infrastructure of all things and define the polarity or 'fatness' of a thing.

The 0 9 Number Pair represents the 'Forming' Central Axis or Fulcrum Point of Balance

To explain how this aspect of the Hologram Projector might work, I want to draw the reader's attention to the figuracy of these numbers and their numerical properties.

We have already stated that 37 and 73 are mirrors of each other numerically and further through their prime terms, being 12 and 21 respectively, are also mirrored.

It turns out that 37 & 73 are also geometrically related and this is where we really get into the 'geometric integration of numbers' aspect of the Integer Partition Theory, as inspired by the great work of Vernon Jenkins.

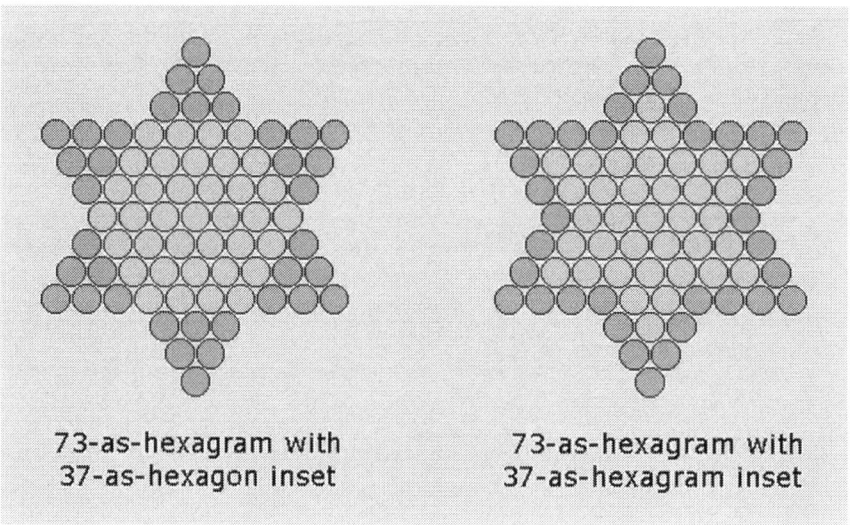

http://www.whatabeginning.com/ASPECTS/MEDIA_FN/ASM_37_73.gif

Vernon Jenkins work allowed me to see that these Star / Hexagrams, 37 and 73, may be formed by the conjoining of 2 triangles, which, in becoming inverted relative to each other, take the form of 6 pointed Stars, or Hexagrams. I believe that this interaction happens through some force of nature that I believe to be a natural property of electromagnetism.

I felt that the coming together of the particular triangles involved with these numbers could form Hexagrams and also then be thought to 'throw off' Centred Hexagons by way of redundancy.

Some interactions are clearly more special or meaningful than others, geometrically speaking, and the triangles that make these numbers, 13, 37 and 73, are very revealing indeed and in agreement with Plato's guidelines.

The Hologram of Physical Reality is proposed as a geometric integration of Triangles to form Hexagrams & Centred Hexagons.

37.

The 37 Hexagram can be produced as explained earlier, this time making use of a triangle with a base of 7, and again inverting one of them and laying it directly over the other to produce the Hexagram.

Each triangle would be made up as 7+6+5+4+3+2+1 = 28 counters per triangle. 2 Triangles of 28 x 2 = 56 counters reduced to a 37 Hexagram or Star.

This means that 19 of these counters will have become redundant or could be said to have been 'thrown off'.

$$56 - 19 = 37$$

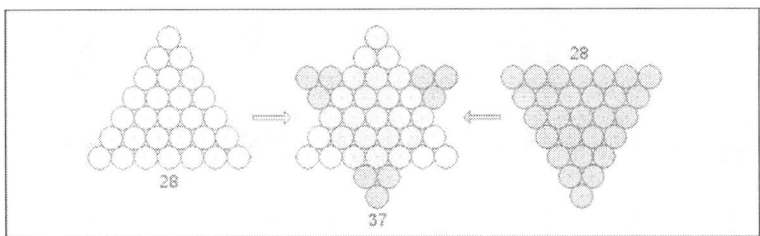

You will notice that 19 is the previous Hexagon number and the 'Special' one by virtue of its unique 'magicness' which we will come to shortly.

It is this that I am saying has been 'thrown off' as matter. A sort of stardust perhaps?

The above I feel reflects the essential duality of all things.

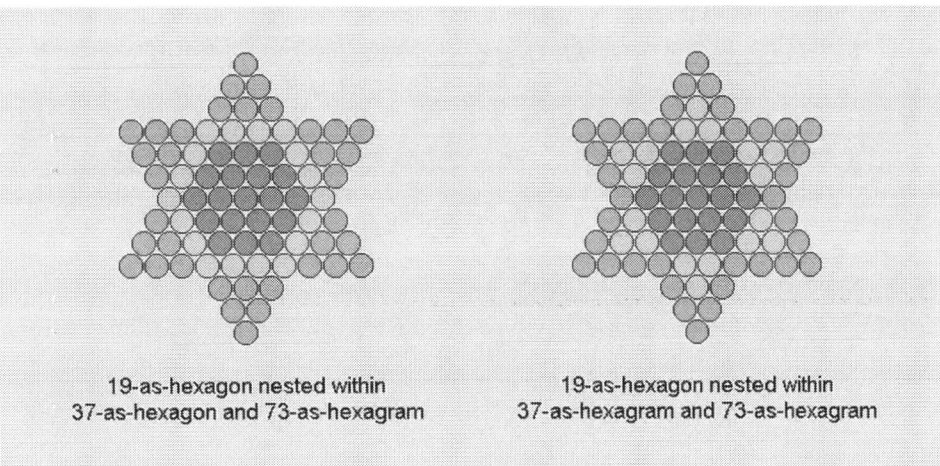

19-as-hexagon nested within 37-as-hexagon and 73-as-hexagram

19-as-hexagon nested within 37-as-hexagram and 73-as-hexagram

37 & Mitochondrial DNA.

From Wikipedia - Human mitochondrial DNA with the 37 genes on their respective H- and L-strands.

Mitochondrial DNA (mtDNA or mDNA) is the DNA located in mitochondria, cellular organelles within eukaryotic cells that convert chemical energy from food into a form that cells can use, adenosine triphosphate (ATP).

Mitochondrial DNA is only a small portion of the DNA in a eukaryotic cell; most of the DNA can be found in the cell nucleus and, in plants and algae, also in the plastids, like chloroplasts.

In humans, the 16,569 base pairs of mitochondrial DNA encode for only 37 genes. Human mitochondrial DNA was the first significant part of the human genome to be sequenced. In most species, including humans, mtDNA is inherited solely from the mother.

16569 Base Pairs - 37 Genes

16569 = 135 = 36 Mod 99

73.

In the same way as the 37 behaves, 'figuratively' speaking, the 73 Hexagram geometry may also be formed using 2 triangles, this time with a base of 10.

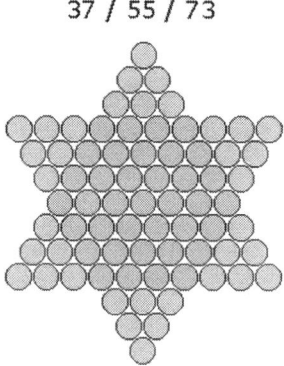

Depicting the related hexagon/hexagram pair 37/73
as the intersection/union of 55-as-triangle
with an inverted copy of itself

$$1+2+3+...+10 = 55 \times 2 = 110.$$

This time 37 counters effectively become redundant or 'thrown off'.

It is my conjecture that, in fact, it is these triangles, which interact / integrate to form, and subsequently 'throw off', these Centred Hexagons, are the incremental evolution of how manifestation of matter in reality all happens.

It became obvious to me that, owing to these particular number's respective, unique, geometric representations and relationships, that the Number 37 and its Mirror 73 are, numerically speaking, the component parts that form the 'containing', 'mechanistic' infrastructure for the production of the Hologram as evidenced in the Partition Table.

The Partition Table shows exactly the algorithm that the 'machine' uses to condition Light with Sound or Vibration to form the plane upon which what we think of as physical reality may play out.

The numerical unfolding must inevitably create the geometric integration of Sound and Light, which seem to me to be the only media required for creation out of the bounded infinite potential of the 'Holographic Plane'.

Everything that happens with numbers is inevitable and occurs as the algorithm simply unfolds.

This indicates to me that physical reality and man's consciousness is simply an unfolding, in the same way that a seed knows exactly how to become the great oak, its purpose and destiny is assured, as I feel is ours, if we can only get the world the right way up.

Evidence in Nature?

Figuracy of Chemical Bonds in Molecules.

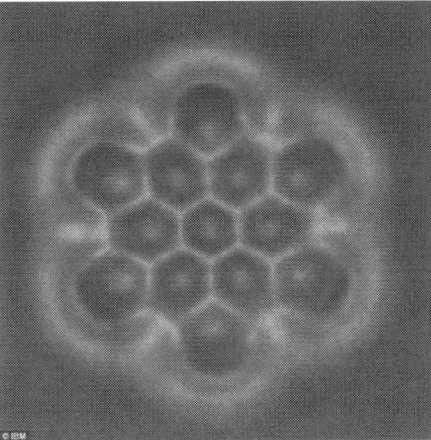

Above we see this incredible picture IBM scientists have taken, showing an atomic force microscope capable of seeing chemical bonds in individual molecules.

'This is a picture of a nanographene molecule exhibiting carbon - carbon bonds of different length and bond order imaged by non-contact atomic force microscopy using a carbon monoxide functionalised tip. The molecule was synthesised at the National Scientific Research Centre in Toulouse.' http://www.rdmag.com/news/2012/09/ibm-scientists-first-distinguish-individual-molecular-bonds

The above is the second order Hexagram of 13 which according to my theory would be the combination of two 4 based triangles, each of 10 counters coming together to produce the 13 Hexagram leaving a redundancy of 7 counters which form a 7 counter centred Hexagon.

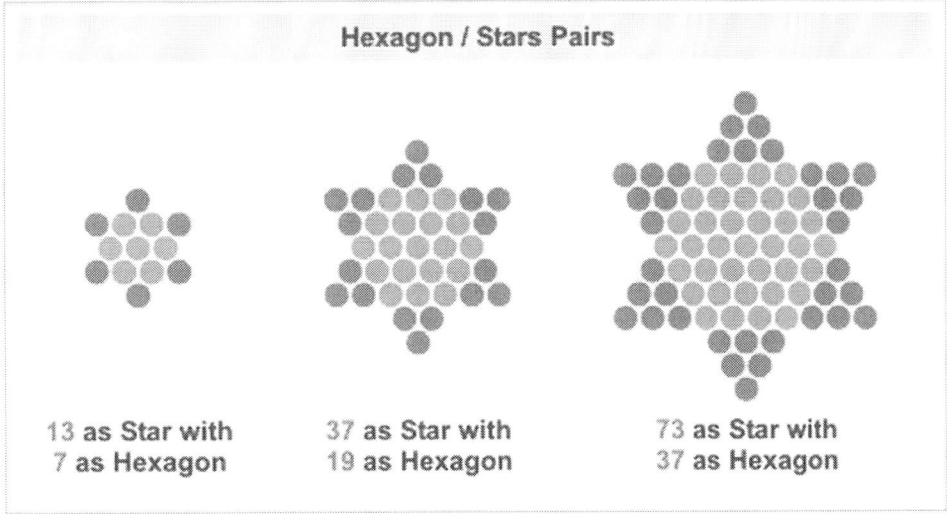

The 37 Hexagram would be formed by the combination of two 7 based triangles in the same way while the 73 Hexagram would be formed by the combination of two 10 based triangles.

37 x 73 in Genesis.

Vernon Jenkins has done some truly fantastic work on the numerical and geometric work on the Bible and I highly recommend you take a look at his work at www.whatabeginning.com, it is remarkable.

Vernon discovered that when we look at the opening 7 words of the Hebrew version of the Bible that constitutes the first verse in Genesis the total numerical value of the words used comes to exactly 2701.

$$2701 = 37 \times 73$$

As we saw in the last chapterurther, 2701 is the sum of the first 73 numbers and is therefore a triangular number as depicted below and which can be divided into 3 triangles of 666 and one of 703.

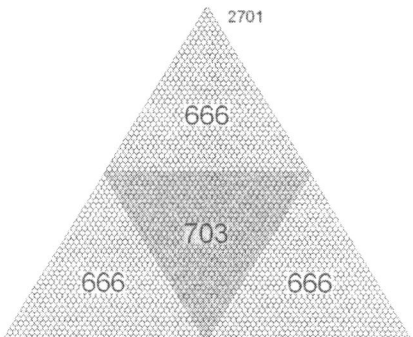

666 is the sum of the first 36 numbers while 703 is the sum of the first 37 numbers.

Below left we see that 2701 can also be depicted as 73 x 37 Hexagrams and on the right we can see this 703 depicted as a Centred Hexagon of Hexagrams with the illusion of the 3 x 3 x 3 cube which I propose represents Light.

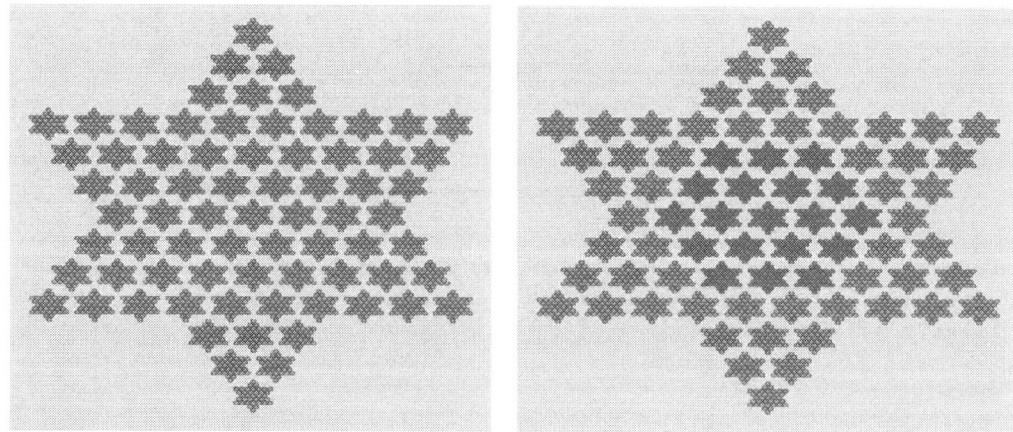

I would also remind the reader that 19 x 37 = 703.

Chapter 9: Key Numbers 19 & 91.

19 - The Magic Hexagon.

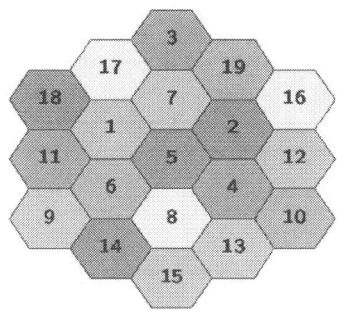

This particular Hexagon is unique in that it is the only Hexagon where the numbers it can carry can be arranged so as to create what is called a magic constant, that is, whichever way you add it up, it comes to the same sum. In the case of the Magic Hexagon, the Magic Constant is 38. A key number to bear in mind especially as it is related to that all important number 137 which is 38 Mod 99.

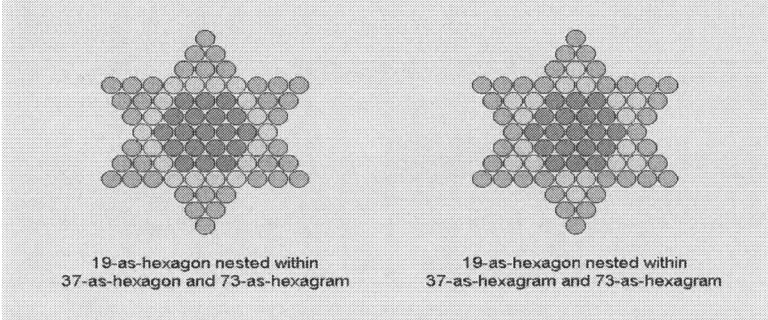

The Magic Hexagon is embedded within both the 37 Hexagram / Centred Hexagon and is my idea of the origin of duality.

19 is linked to 37 & 73 via the dimensions of the Earth.

$$19^2 = 361 \qquad 91^2 = 8281 \qquad 8281 - 361 = 7920$$

Mean Diameter of the Earth in Miles = 7920

Sum of the first 4 Platonic Solids = 7920   Dimensions of New Jerusalem = 7920

Smallest Sporadic Group M11 Matthieu Group = 7920

Total Number of Protons and Neutrons used in the Game of Life = 7920

7920 is also found in the Great Pyramid, Chartres, Wells, Stonehenge and Old English Metrology.

All just a coincidence?

Consider also that:-

$$37^2 = 1369 \qquad 73^2 = 5329$$

5329 – 1369 = 3960 Radius of the Earth in Miles….

91.

91 is obviously the reflection of 19 but you may remember it is a very special figurate number because, other than 37, it is the only trifigurate number.

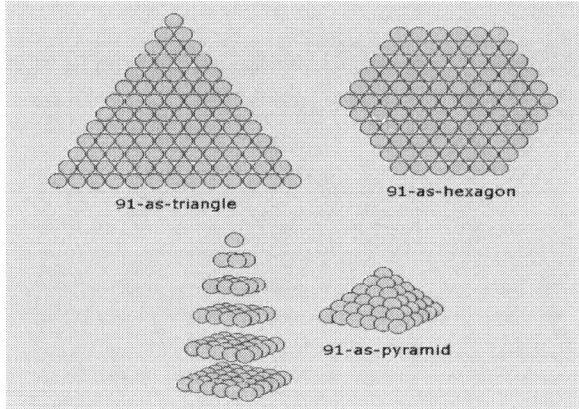

You will remember from the Prime Number section that I pointed to the square pyramid structure that I believe to be inherent to Prime Numbers.

Similarly, if we start with a dot representing the 0 dimension and say that the dimensions are formed as emanations from this dot in triangular form. The first thing to evolve is the Tetractys - a 4 based triangle of 10. This represents the first dimension.

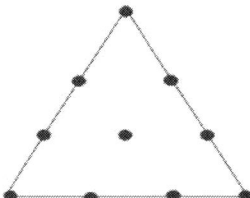

Then for the next dimension up you would need wrap a layer of 18 to get around that and you would have a 7 based triangle of 28 for 2d and Area. You wrap a further layer of 27 around to get a 10 based triangle of 55 for 3d and Volume. Finally you wrap around a layer of 36 to get a 13 based triangle of 91 for 4d and Time.

91 features again as the difference between the cubes of 5 and 6:-

$$6^3 - 5^3 = 216 - 125 = 91$$

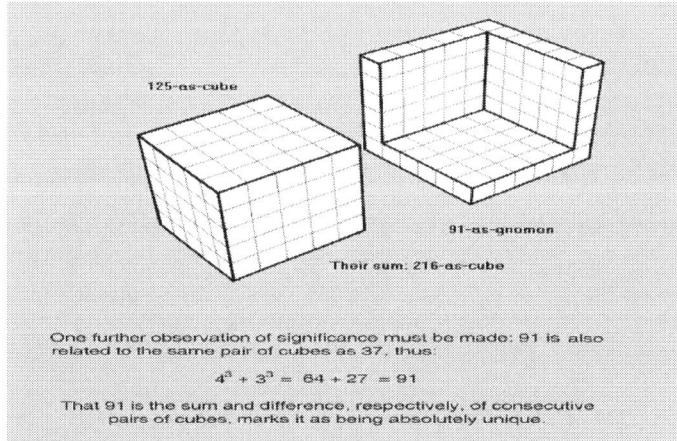

http://www.otherbiblecode.com/Symb.htm

Chapter 10: Key Numbers - 27 & 72 - Cause and Effect.

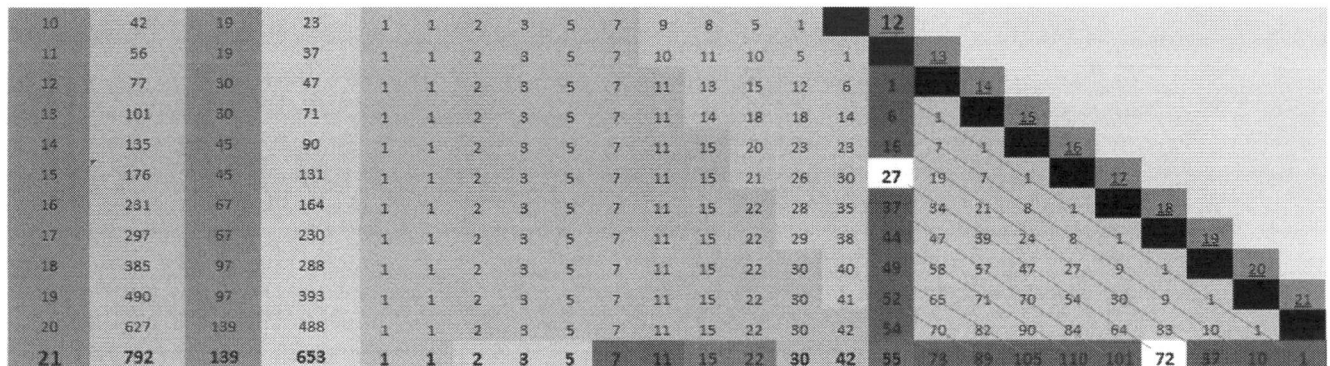

As we look vertically down the 12th Stage values for Numbers we see the Number 27 occurring on the Number 15.

15 x 12 = 180 which puts one in mind of a semi-circle and of course 15 + 12 = 27.

Why might 15 be significant?

My feeling is that it has to do with the fact that 15 is the Magic Constant for the Lo Shu 3 x 3 Magic Square, the smallest order of Magic Square.

The number 27 in this location is rotated through 90 degrees to produce the Number 72 which lies on the 18th Stage Value for the Number 21.

I think of 27 as the Light Source (Cubes of Light Particles 3^3) for the Hologram Projector, the Mirror of Physical Reality - Metaphysical Reality.

Light and Sound are geometrically integrated and focused via the 37 - 55 - 73 'lens / filter' to produce Physical Reality, as represented by the Number 72, the Numerical Mirror of 27.

Prime Numbers - Connection to 27 & 72.

Assuming I am correct about my re-classification of Prime Numbers then the first 6 Prime Numbers in Physical Reality, remarkably, total 72.

$$5 + 7 + 11 + 13 + 17 + 19 = 72...$$

And perhaps even more remarkably when you look at the above numbers in Mod 9 Digit Sum adds to its mirror, or reflection, 27.

$$5 + 7 + 2 + 4 + 8 + 1 = 27....$$

Cubes and Hexagons.

27 is the Third Cube and is also the sum of the first 3 centred hexagonal numbers and this is true for any 3 centred hexagons.

$$1 + 7 + 19 = 27$$

11	56	19	37	1	1	2	3	5	7	10	11	10	5	1		13		
12	77	30	47	1	1	2	3	5	7	11	13	15	12	6	1		14	
13	101	30	71	1	1	2	3	5	7	11	14	18	18	14	6	1		15
14	135	45	90	1	1	2	3	5	7	11	15	20	23	23	16	7	1	16
15	176	45	131	1	1	2	3	5	7	11	15	21	26	30	27	19	7	1

In light of that it is interesting to see above that by reading horizontally for the 13th 14th and 15th Stage Values of the number 15 and vertically for the 13th Stage Value of numbers 13, 14 and 15, we see 19 7 1 at right angles to each other. Each of these numbers represents a centred hexagon in their figuracy.

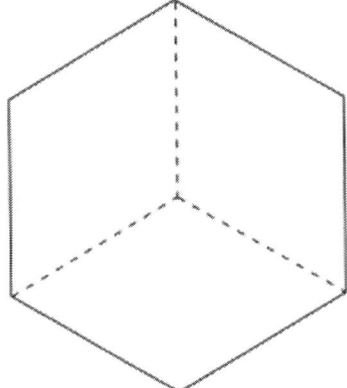

Interestingly, when you look at the centred hexagon above, it can also be perceived as a cube.

It is important to start to see the geometries that combine to make other geometries and the illusions, or different perceptions, one can have about the same object, both of which are true.

The above image was used to explain quantum physics. This demonstrates how a single three dimensional object can be two things at the same time in two dimensions.

Order	Pentagon	Mod 9
1	1	1
2	5	5
3	12	3
4	22	4
5	35	8
6	51	6
7	70	7
8	92	2
9	117	0
10	145	1
11	176	5
12	210	3
13	247	4
14	287	8
15	330	6
16	376	7
17	425	2
18	477	0
19	532	1
20	590	5
21	651	3
22	715	4
23	782	8
24	852	6
25	925	7
26	1001	2
27	1080	0

Pentagonal Numbers

Here we see a 9 digit repeating pattern of the Mod results

1 5 3 4 8 6 7 2 0

1 5 3
4 8 6
7 2 0

The 27th Order of Penatagonal Numbers is 1080 which is of interest as the mean radius of the Moon in miles.

72 - The Effect.

The 72 is the Projected Effect - The Hologram of Physical Reality. Why 72?

I had identified a 72 Mod 9 number sequence that repeats every 6 Octaves in my research into Music.

6 x 12 = 72 Notes.

72 Degrees is the Angle at which the Helix of our DNA rises.

72 years is the time taken by the Sun to move 1 degree in the Precession of the Equinoxes - the path of the Sun against the backdrop of the fixed stars.

The whole precession takes 360 x 72 = 25920 years.

25920 = 81 Mod 99 - 81 stable chemical elements.

The Solar Factor "72" and Pythagorean 3, 4, 5 Triangle.

From collaborator Michael Joyce – 'Circa-500 BC, the Greek mathematician and philosopher, Pythagoras, "found" the connection between the lengths of sides in triangles which contained a right angle (90 degrees), a 3, 4, 5 one being the first of an infinite series.

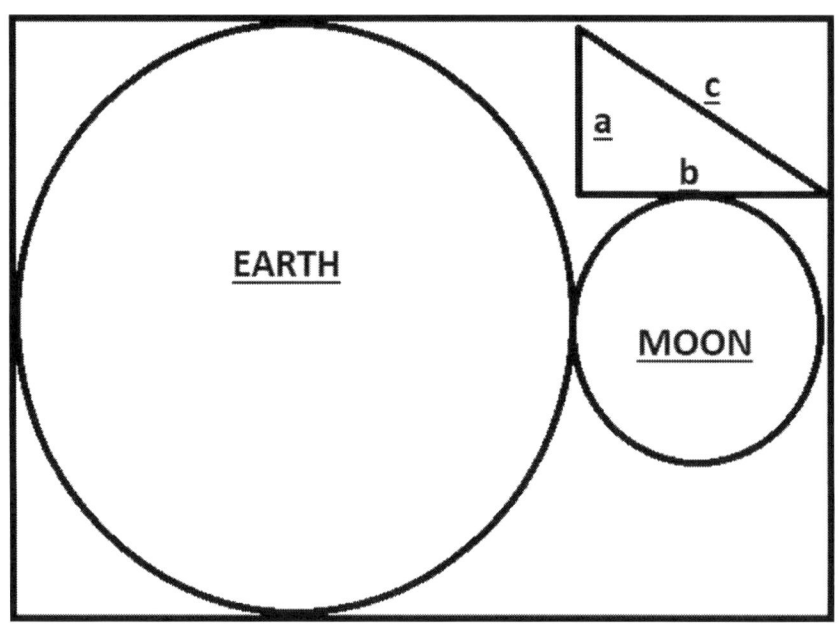

Earth Diameter = 7920 Miles
Moon Diameter = 2160 Miles

In the triangle above the Moon

$$a = ((2 \times 3960) - (2 \times 1080)) / 2 = 2880 = 72 \times 40$$

$$b = 2160 = 72 \times 30$$

Using Pythagoras' theorem $a^2 + b^2 = c^2$ for right angled triangles we get:-

$$c = \sqrt{(2880^2 + 2160^2)} = 3600 = 72 \times 50$$

Our Creator has imprinted the knowledge of 72 and 3, 4, 5, and more, in the heavens.

Thus contained in the Earth and Moon perceived dimensions there exist a 30, 40, 50 or 3, 4, 5, (Pythagorean Triangle) and the very important factor 72.'

Partitions of 21 - Mod 9.

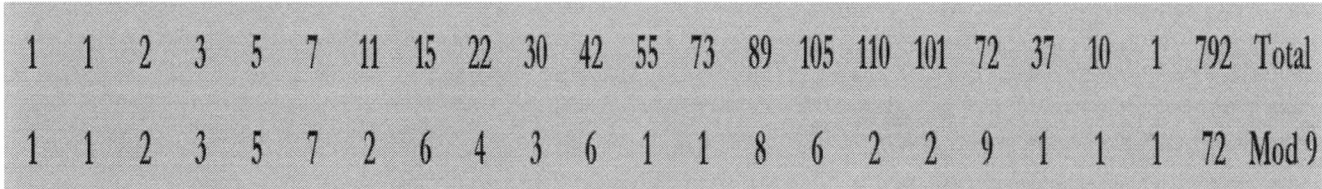

When one looks at the Mod 9 results for each of the Stage Values of the 21 we see that the total is 72.

Chapter 11: Preceding Fulcra Numbers - 37 & 91.

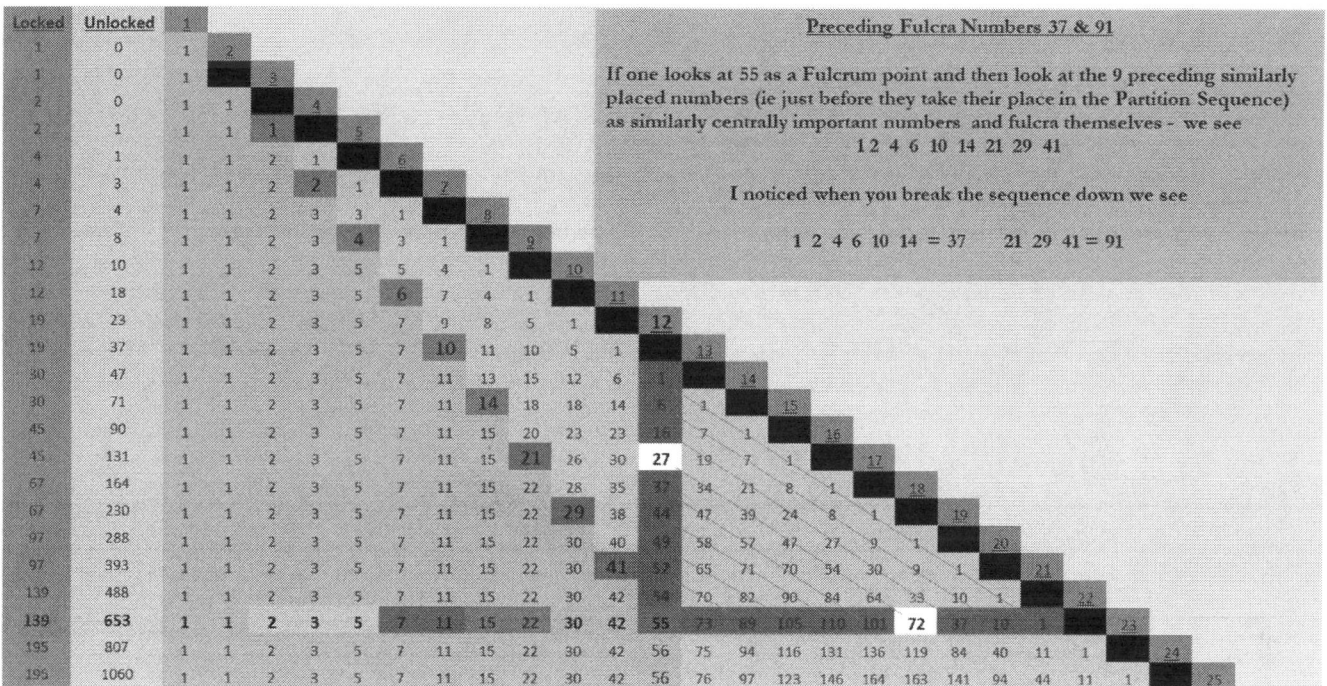

Very interestingly, if one looks at 55 as a Fulcrum point and then looks at the 9 preceding similarly placed numbers that I have highlighted in Green and Blue, they are:-

$$41\ 29\ 21\ 14\ 10\ 6\ 4\ 2\ 1$$

I noticed when you break the sequence down we see:-

$$1\ \ 2\ \ 4\ \ 6\ \ 10\ \ 14\ = 37$$

$$21\ \ 29\ \ 41\ = 91$$

As we have seen, 37 and 91 are the Only Trifigurate Numbers.

$$37 + 91 = 128 = 2^7$$

Another clue to the workings of the Integer Partition Theory and its geometric understanding.

Both 37 and 91 are deeply embedded in the Hologram Projector Apparatus as we have seen, one overtly and the other less obviously.

37, as a principal component of the Hologram Projector and 91 connected to 37 as the mirror of the number 19, itself a special centred hexagon, embedded in the Hologram Projector apparatus.

Interestingly 37, 19 and 91 are also related via Cubes:

$4^3 - 3^3 = 64 - 27 = 37$ $4^3 + 3^3 = 64 + 27 = 91$ $3^3 - 2^3 = 27 - 8 = 19$

#2 A study of the foregoing reveals that 37 is the difference between the cubes of 3 and 4, thus:

$$4^3 - 3^3 = 64 - 27 = 37$$

This outcome is represented pictorially, as follows:

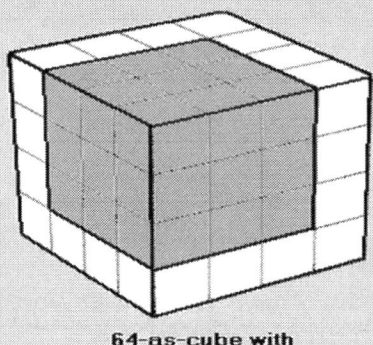
64-as-cube with 27-as-cube inset

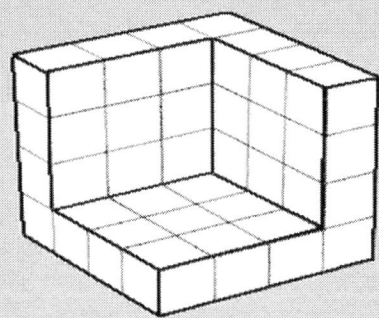
Their difference: 37-as-gnomon

The figure on the right comprises 37 unit cubes; it is referred to as a solid gnomon and is an allotrope of the hexagon depicted earlier.

Similarly, 91 is the difference between the cubes of 5 and 6, thus:

$$6^3 - 5^3 = 216 - 125 = 91$$

This operation is shown below:

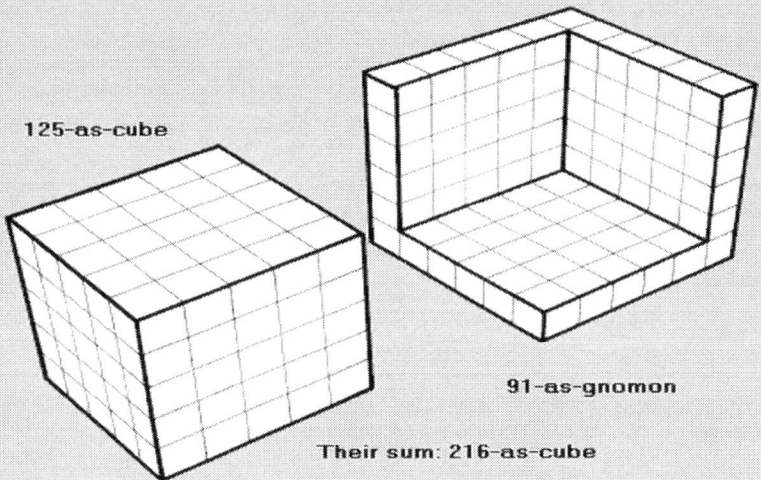
125-as-cube

91-as-gnomon

Their sum: 216-as-cube

One further observation of significance must be made: 91 is also related to the same pair of cubes as 37, thus:

$$4^3 + 3^3 = 64 + 27 = 91$$

That 91 is the sum and difference, respectively, of consecutive pairs of cubes, marks it as being absolutely unique.

http://www.otherbiblecode.com/Symb.htm

Chapter 12: Number 11 - 19 v 37 - Unlocked v Locked Potentials.

Very interestingly, the partitions for the Number 11, the proposed Number of the Earth, also total 56, and coincidentally just happen to be in the ratio 19 : 37 when the locked versus the unlocked potentials are compared as in the table below.

Number	Partitions	Locked	Unlocked
0	1	1	0
1	1	1	0
2	2	2	0
3	3	2	1
4	5	4	1
5	7	4	3
6	11	7	4
7	15	7	8
8	22	12	10
9	30	12	18
10	42	19	23
11	56	19	37
12	77	30	47
13	101	30	71
14	135	45	90
15	127	45	131
16	231	67	164
17	297	67	230
18	385	97	288
19	490	97	393
20	627	139	488
21	792	139	653

11, the Number of the Earth has 56 Partitions

19 for the Locked Potentials   1 1 2 3 5 7 = 19   37 for the Unlocked Potentials   10 11 10 5 1 = 37

I see the 56 Partitions for the Number 11 as 2 x 28 or 2 x 7 based triangles that as they conjoin forming the 37 Hexagram and the 19 Magic Hexagon.

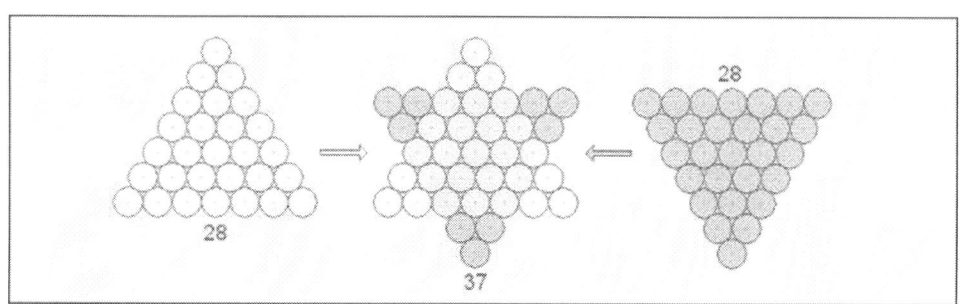

Remember 37 is one of only 2 trifigurate numbers, that is, a number that may take 3 different forms geometrically.

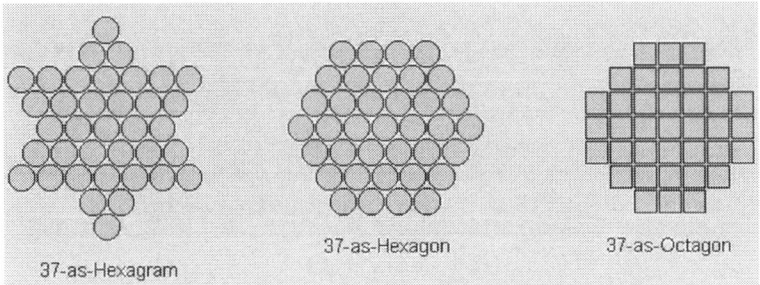

37-as-Hexagram 37-as-Hexagon 37-as-Octagon

The 37 Octagon using squares is of great interest as it is created using a 7 x 7 Grid and chopping off 3 from each corner.

7 x 7 = 49 connects deeply to the Hexagram sequence and to Light.

Term	Hexagram	Mod 9	Mod 99 1st Iteration	Mod 99 2nd Iteration
1	1	1	1	
2	13	4	13	
3	37	1	37	
4	73	1	73	
5	121	4	22	
6	181	1	82	
7	253	1	55	
8	337	4	40	
9	433	1	37	
10	541	1	46	
11	661	4	67	
12	793	1	100	1
13	937	1	46	
14	1093	4	103	4
15	1261	1	73	
16	1441	1	55	
17	1633	4	49	
18	1837	1	55	
19	2053	1	73	
20	2281	4	103	4
21	2521	1	46	
22	2773	1	100	1
23	3037	4	67	
24	3313	1	46	
25	3601	1	37	
26	3901	4	40	
27	4213	1	55	
28	4537	1	82	
29	4873	4	121	22
30	5221	1	73	
31	5581	1	136	37
32	5953	4	112	13
33	6337	1	100	1
34	6733	1	100	1
35	7141	4	112	13
36	7561	1	136	37
37	7993	1	172	73
38	8437	4	121	22
39	8893	1	181	82
40	9361	1	154	55
41	9841	4	139	40
42	10333	1	37	37

On the left we see the Hexagram sequence of numbers.

We see the repeating Mod 9 sequence in the third column showing 1 4 1 every three terms

Then we have the first iteration of applying the Mod 99 technique and where the total is more than 99 the second iteration shows the resulting number giving the final Mod 99 result.

For example, on the 12th Term of the sequence 793, the first iteration gives a total of 93 + 7 = 100 and then applying the rule again gives us 00 + 1 = 1

The analysis shows a clear 33 number palindromic sequence that repeats indefinitely. Palindromic meaning that it can be read backwards or frowards with the same result.

The 33 number repeating sequence is centred on the 17th term, number 49.

1 13 37 73 22 82 55 40 37 46 67 1 46 4 73 55

49

55 73 4 46 1 67 46 37 40 55 82 22 73 37 13 1

1 13 37 73 22 82 55 40 37 46 67 1 46 4 73 55 49 55 73 4 46 1 67 46 37 40 55 82 22 73 37 13 1

The total of the original numbers is 2145 which itself is 45 + 21 = 66 Mod 99

Applying the second iteration the total becomes 1353 = 53 + 13 = 66 Mod 99.

Extraordinarily, the difference between the total of the original numbers and the second iteration numbers is

2145 - 1353 = 792 - the Number of Partitions for the Number 21.

The Mod 9 Totals for each side 5 x 4 and 11 x 1 = 31 x 2 sides = 62 + 4 for centre = 66

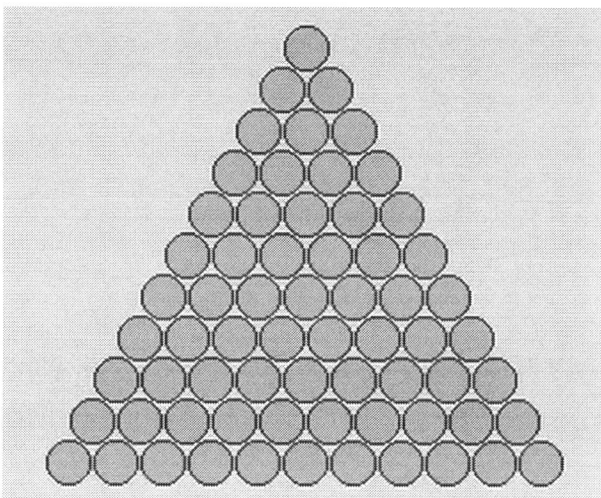

A lot of 66 going on here! - 66 is the number of the 11th Triangle or sum of the first 11 numbers.

Chapter 13: Inside the Integer Partition Table.

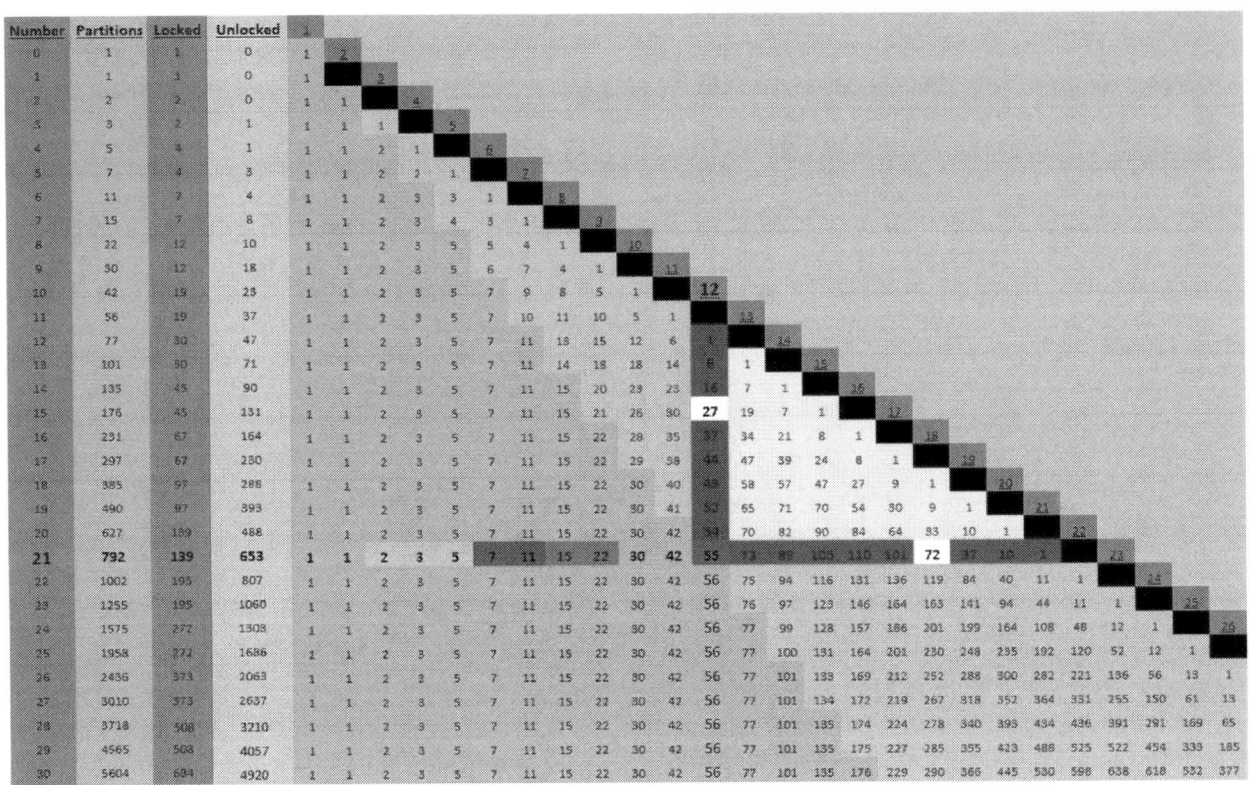

Orthogonal Relationships.

In noticing the orthogonal / right angled relationship between the 27 and 72, I was on the lookout for this and discovered it when looking vertically at the stages values.

If one goes down to the last unlocked potential in a Stage column and reads the sequence of numbers backwards, so upwards in effect, noting the difference between each term; that differential sequence produces the same sequence for all the partition stages of the previous number!

For example – let's take the 12th Stage column. Looking down we see 55 is the last 'unlocked potential' for the number 21. Vertically, reading upwards the sequence goes:

55 54 52 49 44 37 27 16 6 1

Looking at the difference between each number in the above sequence in turn reveals the sequence of:

1 2 3 5 7 10 11 10 5 1

which is in fact exactly the same sequence we see for the Stage Values for the number 11.

1 2 3 5 7 10 11 10 5 1

This works for every number and stage. The unlocked vertical stage numbers produce the previous number's partitions as described above.

This indicates that there is an orthogonal, or a right angled, 90 degree relationship evident and inherent throughout the Integer Partition Table.

The Holographic Plane.

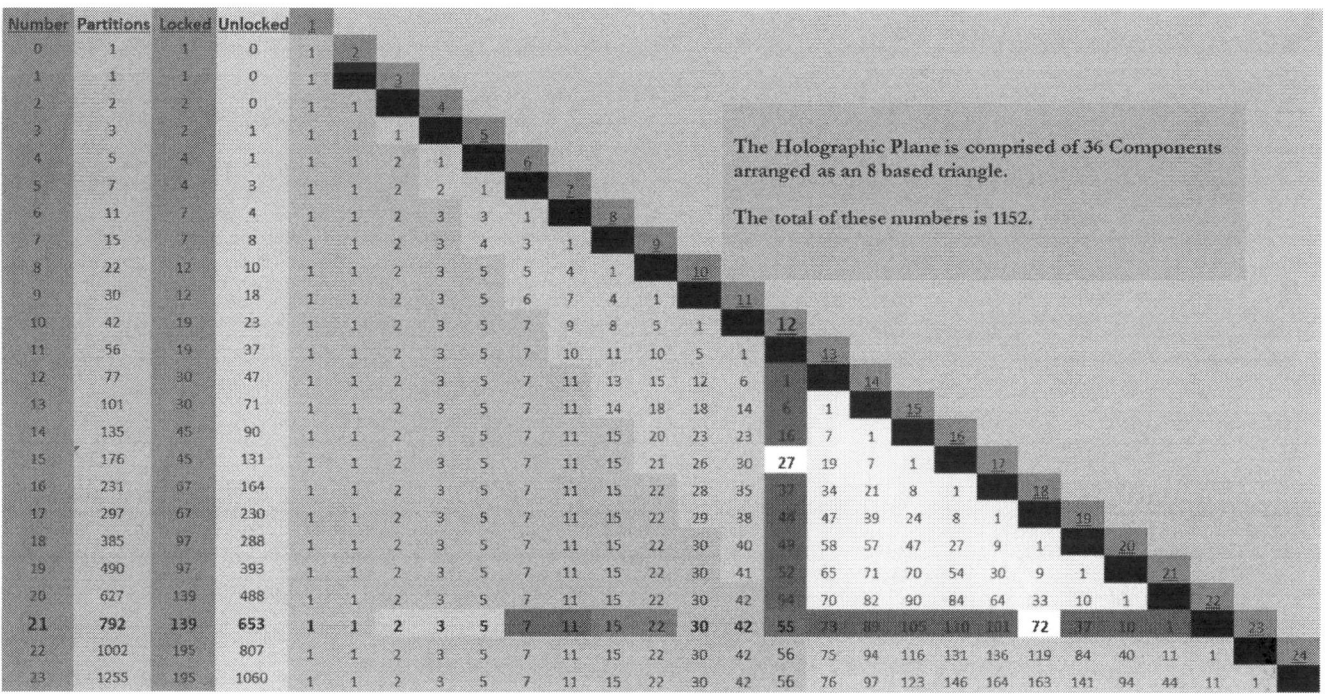

There are 19 components to the Hologram Projector Infrastructure (red and anchored on the olive coloured number 55) and 36 components to the Holographic Plane (the yellow numbers enclosed by the Hologram Projector Infrastructure), for a total of 55 components.

The Universe is Bounded by the Numbers 37 and 73

73 - 37 = 36

36 is a very special number in terms of its figuracy as the only number that is both Square (6th) and Triangular (8th).

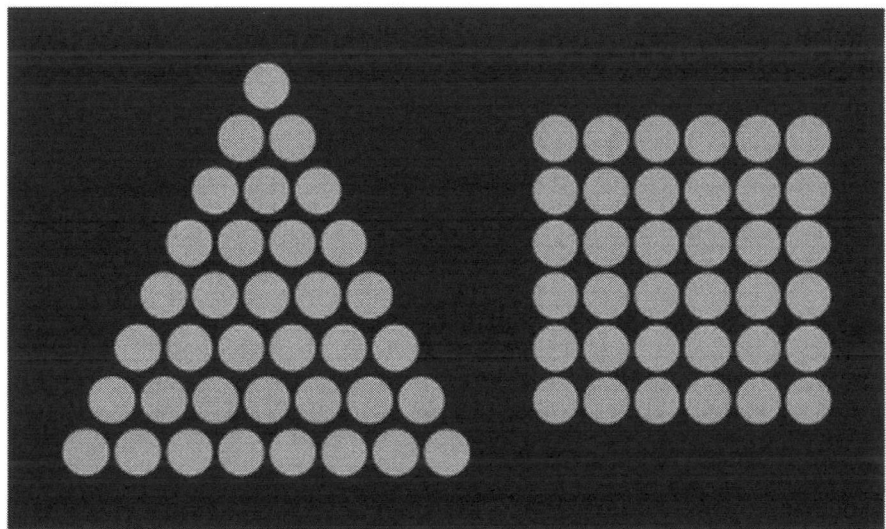

This is what I am theorising and naming the Holographic Plane.

Row Totals.

Interestingly, when we read across the row totals we see:-

1   8   27   64   119   199   300   434

The first four of which are the first four cube numbers, being 1^3, 2^3, 3^3 and 4^3.

Total								
1	1							
8	7	1						
27	19	7	1					
64	34	21	8	1				
119	47	39	24	8	1			
199	58	57	47	27	9	1		
300	65	71	70	54	30	9	1	
434	70	82	90	84	64	33	10	1
1152	301	278	240	174	104	43	11	1

In looking to see if these row totals add to some important number I was rewarded with:-

1 + 8 + 27 + 64 + 119 + 199 + 300 + 434 = 1152

1152.

Back to ancient metrology again we see that 1 Egyptian Foot is 1.152 British Feet. Even more interesting than that I stumbled across some rather astonishing information about this number from Martin Doutre who writes:-

"This particular number is of such immense importance that it will be found encoded, somewhere, within the larger standing stone circles or communal structures of antiquity. The number had a very significant purpose and universal application, which was:

The Heliacal or Sothic rise of the binary star, Sirius, occurs at a measurable interval of every 365.25 days and this event almost perfectly describes the duration of the true solar year. The Heliacal rise of Sirius, therefore, represented a gauge, by ancient astronomers, for a close approximation coding of the count of days in a year.

They did, however, know that the figure was slightly incorrect and so introduced a number, which could be used to correct the Sirius year. The amount of time required to deduct from the Sirius year, by their system, was 11.52 minutes, rendering the solar year as 365.2420 days...the exact figure occurring on the Mayan calendar.

It will be demonstrated, as we proceed, that this number, 11.52, is integral to the Menkaure Pyramid, but for the moment, let's extract it from the dimensions of the Great Pyramid.

1. The coded lengths up the side of the diagonal faces of the Great Pyramid were 576 feet...this is 1/2 of 1152.
2. Just as the sum of 51.84 was 1/500th of the 25920 year duration of the Precession of the Equinoxes, if the sum of 576 feet of diagonal side length is reduced to 1/500th of its value, the total is 1.152 feet.
3. It has been shown, mathematically, that the true square footage value for a face of the Great Pyramid was 230400 sq. ft. This is 115200 x 2.

The numbers 1.152, 11.52, 115.2, 1152 or larger expressions, will be seen to recur prolifically from Egypt to Stonehenge, then into the Pacific. This is a truly dynamic astronomical number, related to 288 and 5184. It is an important calibration increment coded into the Sarsen Circle at Stonehenge, where the total circumference (345.6 feet) is 11.52 feet per lintel (30 lintels) or 360 (degrees) x 11.52 inches per degree of arc." - http://www.celticnz.co.nz/US4.html

Cubes.

Total								
1	1							
8	7	1						
27	19	7	1					
64	34	21	8	1				
119	47	39	24	8	1			
199	58	57	47	27	9	1		
300	65	71	70	54	30	9	1	
434	70	82	90	84	64	33	10	1
1152	301	278	240	174	104	43	11	1

Note the first 4 terms of the Row Totals are:-

$$1^3 = 1 \qquad 2^3 = 8 \qquad 3^3 = 27 \qquad 4^3 = 64$$

Interestingly the sequence stops obliging with cubes at 5, where we would hope for:-

$5^3 = 125$ we see only 119 'losing' 6   **$6^3 = 216$, we see only 199 'losing' 17**

$7^3 = 343$, we see only 300 'losing' 43   **$8^3 = 512$, we see only 434 'losing' 78**

These 'lost amounts' total $6 + 17 + 43 + 78 = 144 = 12^2$

144 is not an insignificant number in the Bible as it is mentioned three times in the Book of Revelation:- 'Do not harm the earth or the sea or the trees, until after we have sealed the servants of God on their foreheads. And I heard the number of the sealed, a hundred and forty-four thousand, sealed from every tribe of the sons of Israel.' - 'Then I looked, and behold, on Mount Zion stood the Lamb, and with him 144,000 who had his name and his Father's name written on their foreheads.' - 'No one could learn that song except the 144,000 who had been redeemed from the earth.'

If we include the missing 144, the total would come to $144 + 1152 = 1296$. 1296 is a lovely number - it is 36×36 which is obviously apt and can also be expressed as $2^4 \times 3^4$ or 6^4 implying perhaps a fourth dimensional cube known as a hypercube.

Holographic Plane and 153.

Total								
1	1							
8	7	1						
9	1	7	1					
19	7	3	8	1				
20	2	3	6	8	1			
28	4	3	2	9	9	1		
39	2	8	7	9	3	9	1	
29	7	1	9	3	1	6	1	1
153	31	26	33	30	14	16	2	1

Looking at Holographic Plane Mod 9 revealed a very interesting number – 153, the sum of the first 17 numbers, the 9th Hexagon number and most famously the number of fish caught in the Bible John 21:11.

There are 153 contained regions when we draw an 8 based triangle of circles.

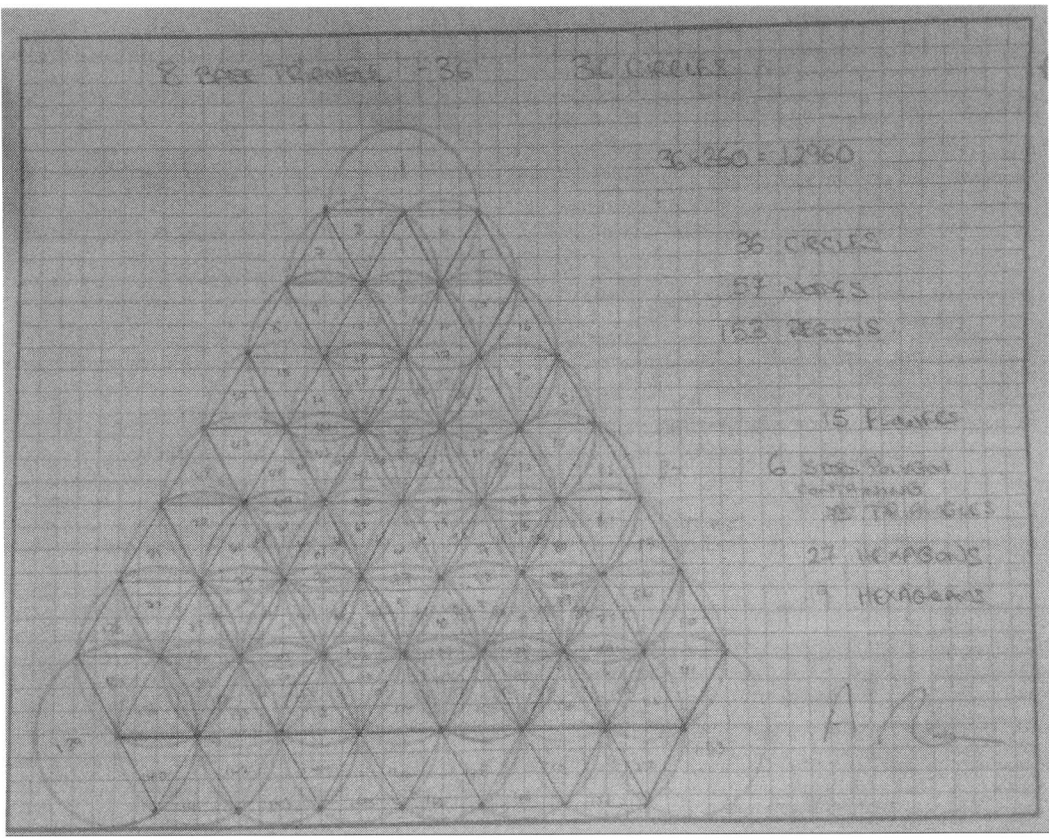

The precision of the number of fish as 153 has long been considered, and various writers have argued that the number 153 has some deeper significance, with many conflicting theories having been offered.

As seen below, if we draw two circles intersecting each other at their centres we form what is called the vesical piscis which looks like a rugby ball.

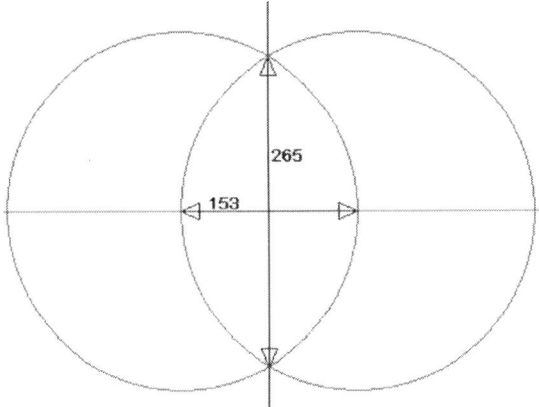

The mathematical ratio of the height of the vesica piscis to the width across its centre is the square root of 3, or 1.7320508.

The ratio 265:153 = 1.7320261 is one of a series of approximations to this value.

153 - Difference between Mirrored Squares.

$351^2 - 153^2 = 123201 - 23409 = 99792$ which is rather elegant

The reverse of 153 is 351, which is also happens to be the 26th triangular number since $1 + 2 + 3 + ... + 24 + 25 + 26 = 351$.

153 and the Moon.

There are 365.25 days in a year on average, and each lunar cycle (days between full moons) is 29.53 on average. Thus, the average number of full moons per year is 365.25/29.53, or approximately 12.369. Oddly enough, the square root of 153 is approximately 12.369 as well.

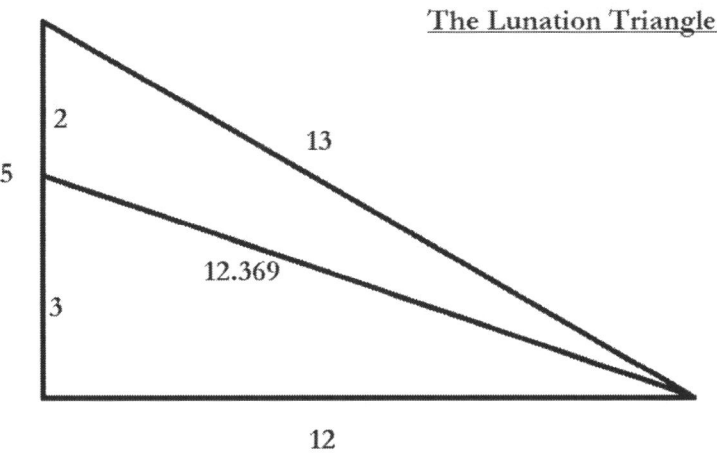

The Lunation Triangle.

I particularly like how the 5 is divided 3 to 2 which puts me in mind of the Perfect Fifth we saw in Music and in DNA where the Hydrogen count in the Side Chain for Thymine led codons, 72 and its pair Adenine, 108. 108/72 = 3/2.

153 and Stonehenge.

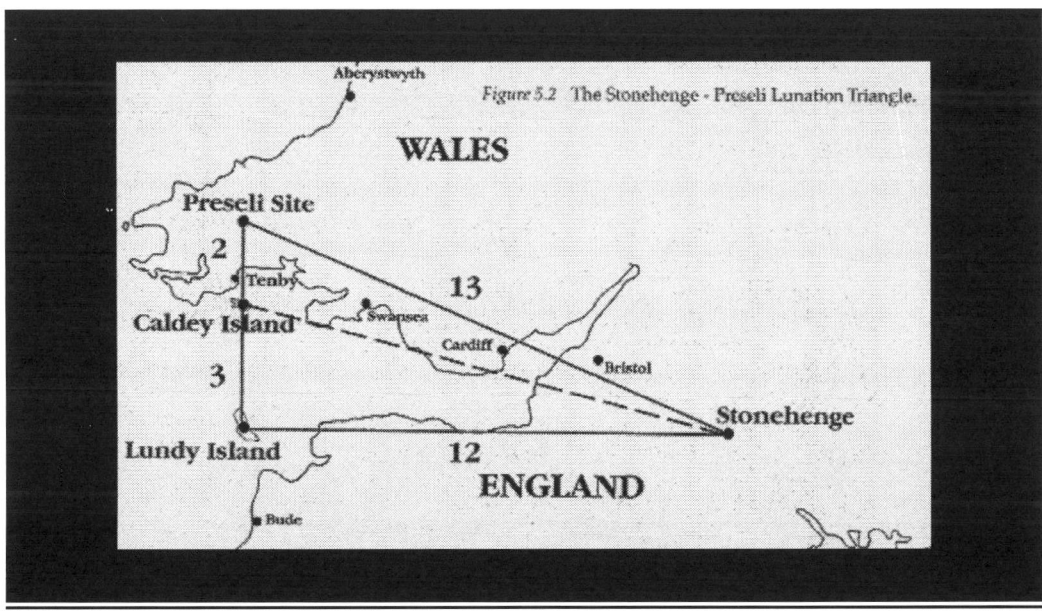

Author Robin Heath theorises about the lunation triangle that links the bluestone quarry in Wales, Lundy Island and Stonehenge.

Mod 9 Count with 0's.

Total								
1	1							
8	7	1						
9	1	7	1					
19	7	3	8	1				
20	2	3	6	8	1			
10	4	3	2	0	0	1		
21	2	8	7	0	3	0	1	
20	7	1	0	3	1	6	1	1
108	31	26	24	12	5	7	2	1

In this instance we find that the Mod 9 total is now 108 instead of 153 - a difference of 45. 108 is associated with so many things throughout my work, most notably the Moon, whose mean radius is 1080 miles.

The importance of the number 108 can be seen repetitively in astronomy and astrology:-

In Vedic Astrology there are 12 Solar Houses and 9 Lunar Houses - 9 x 12 = 108.

The Mean Diameter of the Sun is 108 times Mean Diameter of Earth.

Mean Distance between the Earth and the Sun is 108 times the Sun's Mean Diameter.

Mean Distance from Earth to the Moon is 238,800 miles, 108 times the Moon's Mean Diameter.

This is why the moon appears the same size as the sun during eclipses.

Holographic Plane Numbers - Range and Mean.

Total								
1	1							
8	7	1						
27	19	7	1					
64	34	21	8	1				
119	47	39	24	8	1			
199	58	57	47	27	9	1		
300	65	71	70	54	30	9	1	
434	70	82	90	84	64	33	10	1
1152	301	278	240	174	104	43	11	1

1 1 1 1 1 1 1 7 7 8 8 9 9 10 19 21 24 27 30 33 34 39 47 47 54 57 58 64 65 70 70 71 82 84 90

The Range is 1 – 90 which makes one immediately think of right angled triangles.

$$\text{Mean} = 1152 / 36 = 32 = 2^5$$

Chapter 14: The 11 Octaves of Physical Reality.

The Holographic Plane - Vertical & Horizontal Geometric Integration of Light & Sound.

The Universe is made up of the eight dimensions of the Holographic Plane with two more dimensions for the containing Hologram Infrastructure and a further dimension to accommodate the Fine Structure Constant, the so called dimensionless dimension, for a total of 11 dimensions.

1										Vertical	55	54	52	49	44	37	27	16	6	1
6	1									Horizontal	55	73	89	105	110	101	72	37	10	1
16	7	1								Difference	0	19	37	56	66	64	45	21	4	0
27	19	7	1						
37	34	21	8	1					
44	47	39	24	8	1				
49	58	57	47	27	9	1			
52	65	71	70	54	30	9	1		
54	70	82	90	84	64	33	10	1	
55	73	89	105	110	101	72	37	10	1

Earlier, I mentioned having spotted that the number 27 mapped perfectly to the reflected 72 through an orthogonal relationship and I soon realised that each of the numbers in the Hologram Infrastructure were similarly connected to each other in this way.

The infrastructure of the 'Numerical Machine' that manifests and contains Physical Reality seems to operate through an orthogonal relationship of the vertical unlocked potentials for the 12th Stage, which reading upwards, the sequence goes:-

$$(55) \quad 54 \quad 52 \quad 49 \quad 44 \quad 37 \quad 27 \quad 16 \quad 6 \quad 1$$

and looking horizontally, at the stage values for 21 from the 12th term onwards, the sequence goes:-

$$(55) \quad 73 \quad 89 \quad 105 \quad 110 \quad 101 \quad 72 \quad 37 \quad 10 \quad 1$$

I saw that when you looked at the differences between the numbers in the Hologram Infrastructure, connected in this way, we see all the key numbers and geometric relationships emerging again:-

$$89 - 52 = 37 \qquad 105 - 49 = 56 \qquad 110 - 44 = 66 \qquad 101 - 37 = 64$$

$$72 - 27 = 45 \qquad 37 - 16 = 21 \qquad 10 - 6 = 4 \qquad 1 - 1 = 0$$

I believe that the Holographic Plane and Hologram Infrastructure emanates from the Central Axis represented by the number 55 and that it can be thought of as a system of nested Tori, the Torus being the geometric representation of an octave by virtue of the fact that the surface of a torus divides itself naturally into seven regions. Although there are 8 octaves inside the Holographic Plane, if we do not count the one that consists only of 1's, which we may be able to get away with, then we are left with 7.

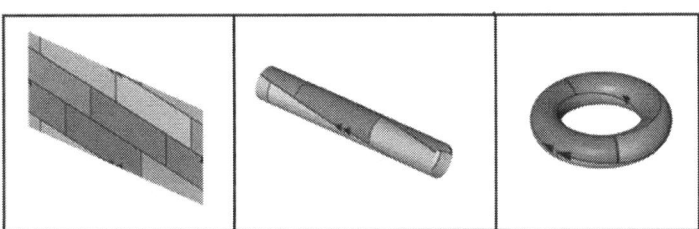

Author: Sebastian Kostal - Wikipedia.

This construction shows the torus divided into the maximum of seven regions, every one of which touches every other.

I also saw that if I we join the two connected numbers and included the Hologram Infrastructure only connection between 54 and 73 then we can essentially imagine 9 Strings of Numbers or what I like to think of currently as the 9 Octaves of Physical Reality, to mimic those proposed in Walter Russell's Octave Wave Theory.

STRING / OCTAVE 1 - 54 & 73.

2 Components 54 and 73.

Difference 73 - 54 = 19.

The figuracy of 19 produces the Magic Hexagon.

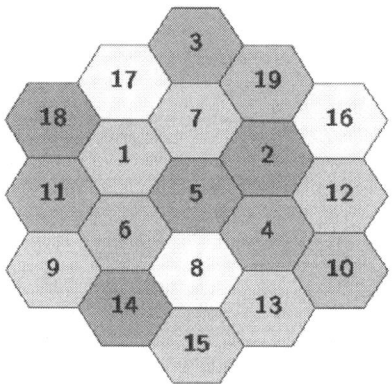

Sum 54 + 73 = 127 = 28 Mod 99 (7 base triangle).

Product 54 x 73 = 3942 = 81 Mod 99 - There are 81 stable chemical elements.

STRING / OCTAVE 2 - 52 & 89.

3 components 52, 70 and 89.

Difference is 89 - 52 = 37. (Centred Hexagon, Hexagram and Octagon).

Sum 52 + 89 = 141 = 42 Mod 99 (Two 6 based triangles of 21)

Total Sum 52 + 70 + 89 = 211.

Product 52 x 89 = 4628 = 74 Mod 99 (2 x 37)

					Vertical	55	54	52	49	44	37	27	16	6	1
1					Horizontal	55	73	89	105	110	101	72	37	10	1
8	1				Difference	0	19	37	56	66	64	45	21	4	0
16	7	1													
27	19	7	1												
37	34	21	8	1											
44	47	39	24	8	1										
49	58	57	47	27	9	1									
52	65	71	70	54	30	9	1								
54	70	82	90	84	64	33	10	1							
55	73	89	105	110	101	72	37	10	1						

STRING / OCTAVE 3 - 49 & 105.

4 Components 49 65 82 105.

Difference is 105 - 49 = 56.

56 is the number of partitions for Number 11.

Two 7 based Triangles of 28 = 56.

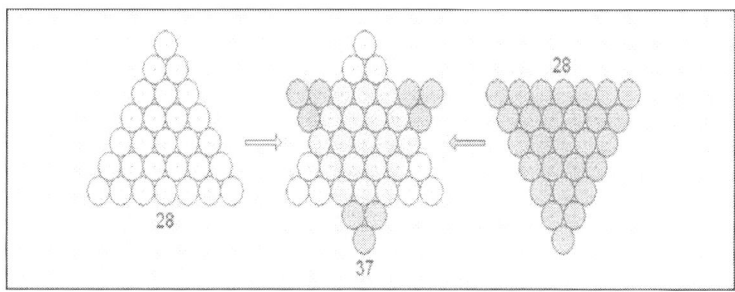

Further, it is the sum of previous two terms of this sequence - 19 + 37 = 56.

Sum is 105 + 49 = 154 = 55 Mod 99 (Central Axis and the 10 based Triangle).

Total Sum is 49 + 65 + 82 + 105 = 301.

Product 105 x 49 = 5145 = 96 Mod 99.

STRING / OCTAVE 4 - 44 & 110.

5 Components 44 58 71 90 110.

Difference is 110 - 44 = 66.

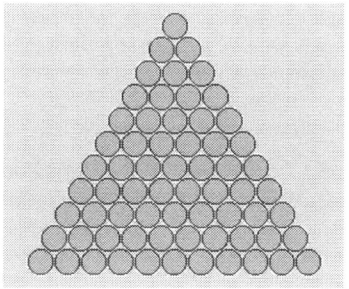

Sum is 110 + 44 = 154 = 55 Mod 99 (Central Axis and the 10 based Triangle).

Total Sum is 44 + 58 + 71 + 90 + 110 = 373 (373 Kelvin is the boiling temperature of water).

Product 110 x 44 = 4840 = 88 Mod 99.

STRING / OCTAVE 5 - 37 & 101.

6 Components 37 47 57 70 84 101.

Difference is 101 - 37 = 64 = 4^3 (Number of Codons of DNA).

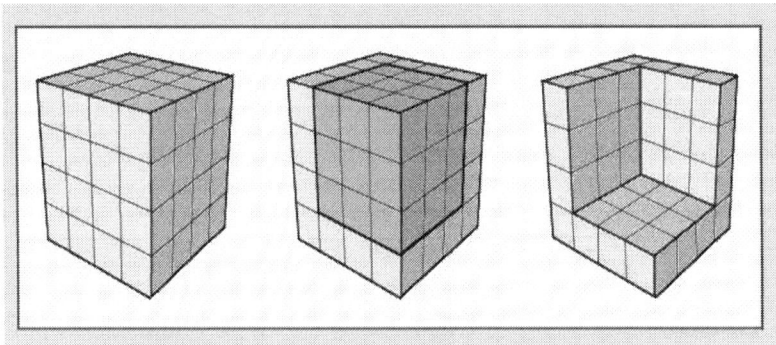

Sum is 101 + 37 = 138.

Total Sum 37 + 47 + 57 + 70 + 84 + 101 = 396.

Product 37 x 101 = 3737 = 74 Mod 99 (2 x 37).

STRING / OCTAVE 6 - 27 & 72.

7 Components 27 34 39 47 54 64 72.

Difference is 72 - 27 = 45 (9 base triangle).

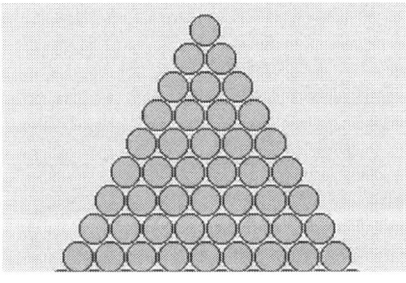

Sum is 72 + 27 = 99.

Total Sum 27 + 34 + 39 + 47 + 54 + 64 + 72 = 337.

Product 27 x 72 = 1944 = 63 Mod 99.

					Vertical	55	54	52	49	44	37	27	16	6	1
					Horizontal	55	73	89	105	110	101	72	37	10	1
					Difference	0	19	37	56	66	64	45	21	4	0
27	19	7	1												
37	34	21	8	1											
44	47	39	24	8	1										
49	58	57	47	27	9	1									
52	65	71	70	54	30	9	1								
54	70	82	90	84	64	33	10	1							
55	73	89	105	110	101	72	37	10	1						

STRING / OCTAVE 7 - 16 & 37.

8 Components 16 19 21 24 27 30 33 37.

Difference is 37 - 16 = 21 (6 base triangle and square).

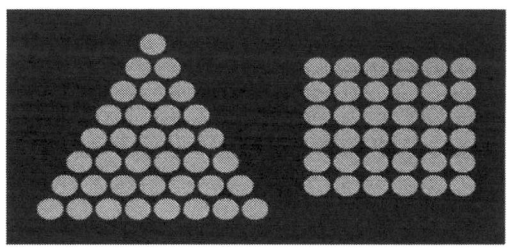

Sum is 37 + 16 = 53.

Total Sum 16 + 19 + 21 + 24 + 27 + 30 + 33 + 37 = 207 (3 x 69).

Product 37 x 16 = 592.

STRING / OCTAVE 8 - 6 & 10.

9 Components 6 7 7 8 8 9 9 10 10.

Difference is 10 - 6 = 4 (2 base square)

The 4 points of the Compass. 4 bases of our DNA etc.

Sum is 10 + 6 = 16.

Total Sum 6 + 7 + 7 + 8 + 8 + 9 + 9 + 10 + 10 = 74 (2x37).

STRING / OCTAVE 9 – 1 & 1.

10 Components 1 1 1 1 1 1 1 1 1 1.

1 - 1 = 0.

Sum is 1 + 1 = 2.

Total Sum = 1 x 10 = 10 (4 based triangle).

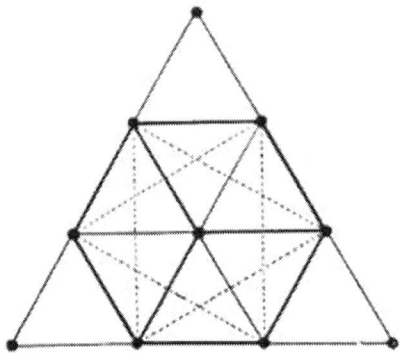

Product is 1 x 1 = 1.

Chapter 15: Remaining Hologram Projector Numbers for Number 21.

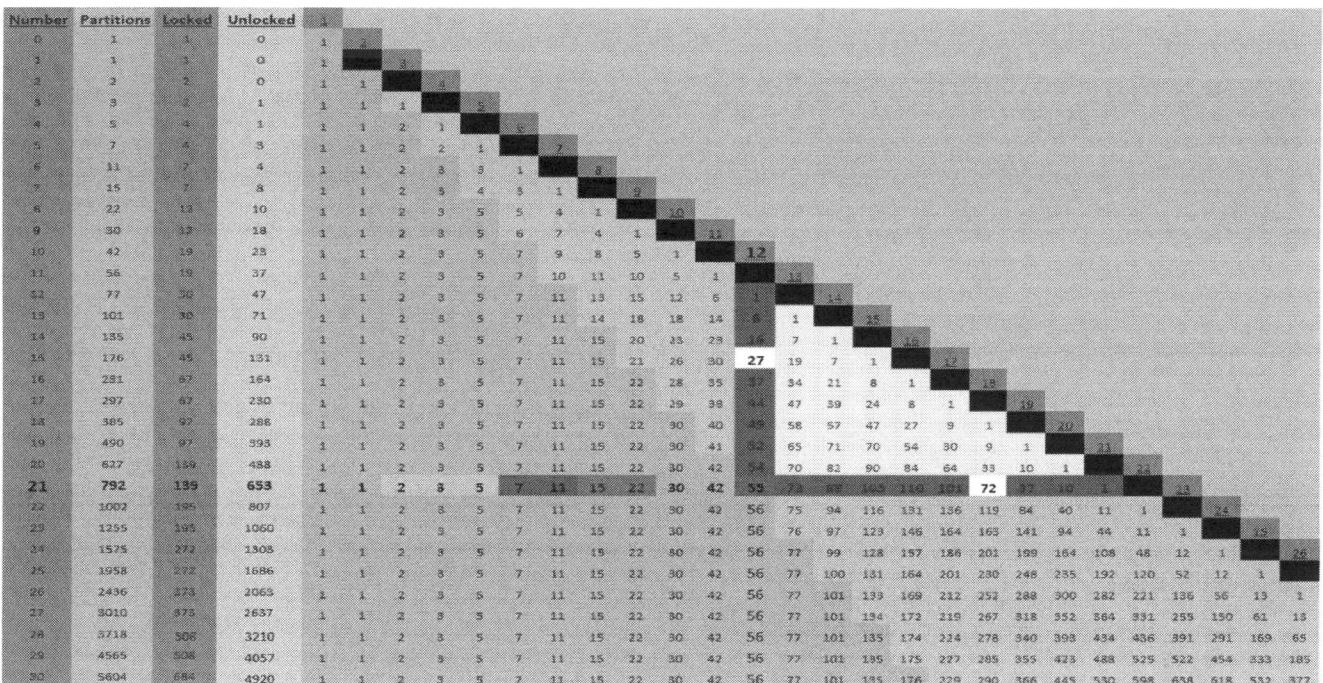

110.

$$73 + 37 = 110 \qquad 2 \times 55 = 110$$

110 is also the Total for all Projector Numbers for all the Number Pairs.

$$91 + 19 = 110 \quad 82 + 28 = 110 \quad 73 + 37 = 110 \quad 64 + 46 = 110$$

Projector	Vibrational Filter	Projected	Fine Structure Constant	
(0) 9	91 55 19	90	9+19+91 = 119	Here I show the other possible Fine Structure Constants using all of the Number Pairs.
18	82 55 28	81	18+28+82 = 128	I have generated this chart by looking at 27 as Cause and 72 as Effect and the boundaries created by adding 1 to the effect, being 73 and using its mirror 37. Then I add the three causal numbers 27 37 and 73 to find the fine structure constant at 137.
27	73 55 37	72	27+37+73 = 137	In this way I can theorise about the fine structure constant for the other number pairs.
36	64 55 46	63	36+46+64 = 146	I believe this may be interesting and have something to do with energy boundaries in physics because the fine structure constant at 1/137 is measured at normal lab energies but it seems to tend towards 1/128 at high energies which would be the fine structure constant generated by the 1 and 8 Number Pair.
45	55 55 55	54	45+55+55 = 155	Food for thought.

Boundaries 37 + 73 = 110

110 x 72 (The Solar Constant) = 7920 Diameter of the Earth etc.

110 x 1080 (Moon Radius in Miles) = 118800 Miles = Distance between Earth and Moon

The "perfect" circle coding of the Sarsen Circle at Stonehenge was 110 ft.

2 x 10 Base Triangles - 2 x 55 = 110

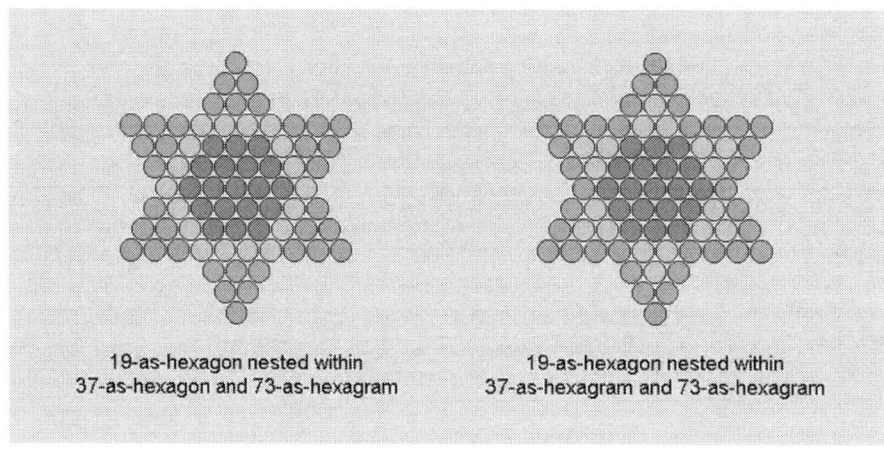

19-as-hexagon nested within 37-as-hexagon and 73-as-hexagram

19-as-hexagon nested within 37-as-hexagram and 73-as-hexagram

89.

89 is the 11th Fibonacci number. 11 being the number of the Earth

89 is the 24th Prime Number

Obviously there is the 24 hr clock but less obviously there is the 24 digit sequence Mod 9 that repeats in Fibonacci.

'The decimal expansion of 1/89 is just the Fibonacci series, added together in an appropriate fashion. Specifically, think of the Fibonacci series as being a sequence of decimal fractions, arranged so the right most digit of the nth Fibonacci number is in the $n+1$th decimal place. Then add:-

.01	1/89 =
.001	
.0002	0.01123595505617977528089887640449438202247191
.00003	
.000005	01123595505617977752808
.0000008	
.00000013	9887640449438202247191
.000000021	
.0000000034	9999999999999999999999
.00000000055	
.000000000089	Two Dimensional Numerical Symmetry
.0000000000144	
+ .	22 Number Pairs.

.01123595505...	

There are 89 Partitions of Number 14, where 14 is important as the mirror of 41, the centre of the 9 x 9 Magic Square of the Moon and the only root number to produce 5 digit recurring decimals, Pentagons, which clearly link to the Dodecahedron and Time.

37	78	29	70	21	62	13	54	5
6	38	79	30	71	22	63	14	46
47	7	39	80	31	72	23	55	15
16	48	8	40	81	32	64	24	56
57	17	49	9	41	73	33	65	25
26	58	18	50	1	42	74	34	66
67	27	59	10	51	2	43	75	35
36	68	19	60	11	52	3	44	76
77	28	69	20	61	12	53	4	45

89 is the total number of chapters in the four Gospels of the New Testament:-

Matthews 28 + Mark 16 + Luke 24 + John 21 = 89

105.

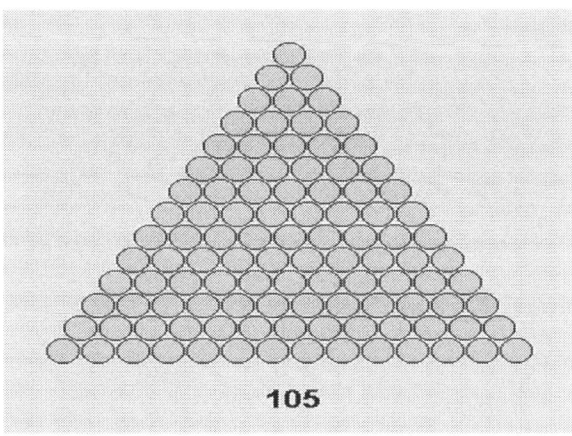
105

105 is the 5th Dodecagon and the 14th Triangle Number.

105 = 5 x 21 - 5 being the first Prime in physical reality and associated with Pentagons, Dodecahedrons and Time in conjunction with the 6 based triangle of 21, the key number for Effect in the theory.

This will also display the unique properties of 21, namely that the 6 digit recurring decimals it produces do not have two dimensional numerical symmetry.

101.

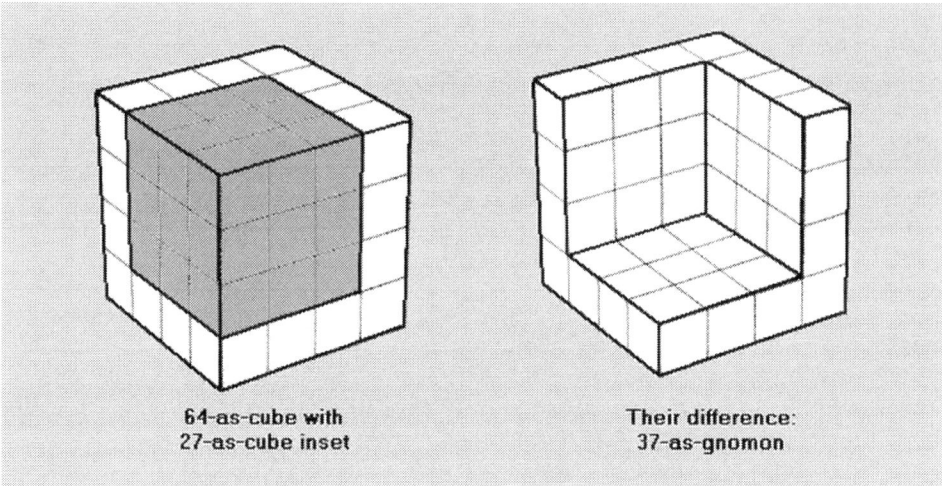

4^3 + (4^3 − 3^3) - Cube plus Gnomon = 101

64 + (64 − 27) = 64 + 37 = 101

101 is the 27th Prime Number reinforcing the idea that Light is Cubed by which I mean that the medium through which Light operates is cubic.

101 uniquely produces 4 digit recurring decimals as its effect on all other numbers - Squares.

1/101 = 0.0099 0099 R

10.

10 is the 4th Triangle – Pythagoras' Tetractys.

The Greek word Tetractys signifies, literally, the number four, and is therefore synonymous with the quaternion; but it has been peculiarly applied to a symbol of the Pythagoreans, which is composed of ten dots arranged in a triangular form of four rows.

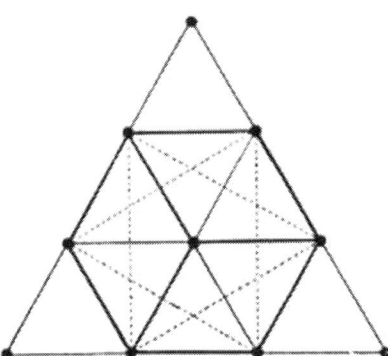

2 of these 4 based triangles produces a 13 Star Hexagram Inverting one of these triangles and overlaying one upon the other we can see that 13 Counters remain, meaning that 7 have become redundant or have been 'thrown off'.

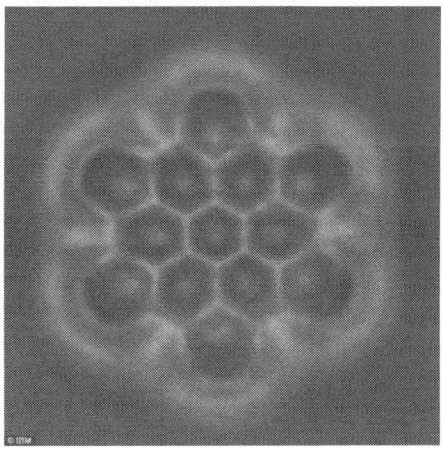

The number 10 is fundamental to the whole of my theory and when we look at a Tetractys of circles we arrive at the front cover picture below which gives many clues as to the nature of Nature.

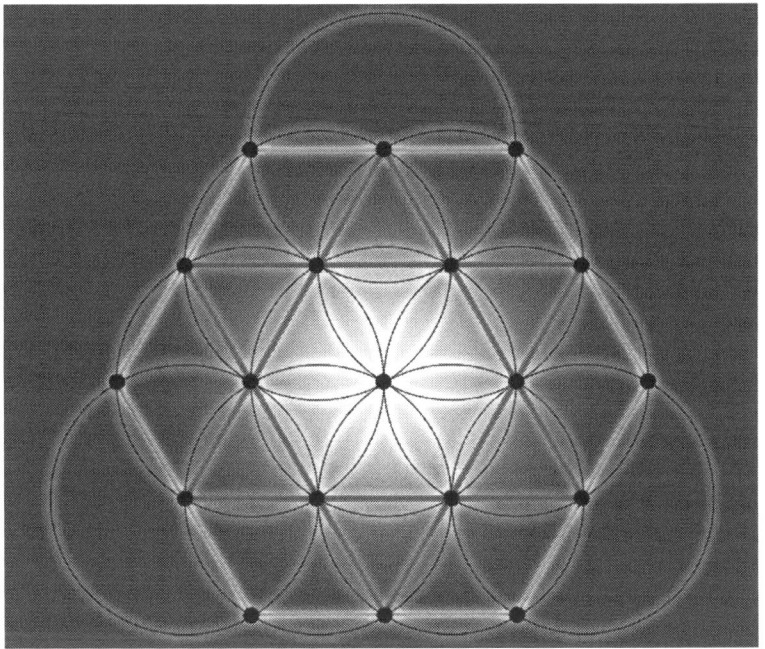

Above we can see the geometric qualities embedded within a Tetractys of circles.

10 Circles - Tetractys

19 Nodes - Magic Hexagon

36 Regions - Number of Holographic Plane Components

Order 13 Hexagram   Order 19 Magic Hexagon

Anthony Morris

Chapter 16: Remaining Hologram Projector Numbers for 12th Stage Value.

6.

Number 6 is the First 'Perfect' Number, defined by a number whose factors summed or multiplied are the same.

1 x 2 x 3 = 6 and 1 + 2 + 3 = 6

Cubes are obviously deeply connected to my theory and these have a total of 6 faces.

Dice, the plural of die, have been around since before recorded history and it is uncertain where or when they originated.

A die has 6 numbered sides which total the key number 21 with each opposite face equalling 7.

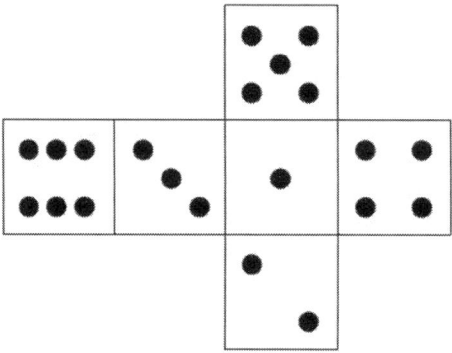

6 Points on the Star of David.

God took 6 days to create the world, on the 7th day he rested.

n	a	b	c																					
10	42	19	23	1	1	2	3	5	7	9	8	5	1		12									
11	56	19	37	1	1	2	3	5	7	10	11	10	5	1	13									
12	77	30	47	1	1	2	3	5	7	11	13	15	12	6	1	14								
13	101	30	71	1	1	2	3	5	7	11	14	18	18	14	6	1	15							
14	135	45	90	1	1	2	3	5	7	11	15	20	23	23	16	7	1	16						
15	176	45	131	1	1	2	3	5	7	11	15	21	26	30	27	19	7	1	17					
16	251	67	164	1	1	2	3	5	7	11	15	22	28	35	37	34	21	8	1	18				
17	297	67	230	1	1	2	3	5	7	11	15	22	29	38	44	47	39	24	8	1	19			
18	385	97	288	1	1	2	3	5	7	11	15	22	30	40	49	58	57	47	27	9	1	20		
19	490	97	393	1	1	2	3	5	7	11	15	22	30	41	52	65	71	70	54	30	9	1	21	
20	627	139	488	1	1	2	3	5	7	11	15	22	30	42	54	70	82	90	84	64	33	10	1	
21	792	139	653	1	1	2	3	5	7	11	15	22	30	42	55	73	89	105	110	101	72	37	10	1

Both 27 and 72, the numbers for cause and effect in the theory, are 6 places away from the 55 Central Axis.

In the Old Testament Book of Genesis; humankind was created on day 6.

In Islam there are Six articles of belief.

16.

16 is a square number, being $4^2 = 4 \times 4$.

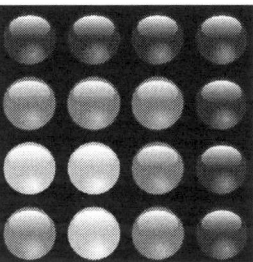

Because 16 is 2^4, the number 16 was used in weighing light objects in several cultures. The British have 16 ounces in one pound; the Chinese used to have 16 liangs in one jin. In old days, weighing was done with a beam balance to make equal splits. It would be easier to split a heap of grains into sixteen equal parts through successive divisions than to split into ten parts.

Chinese Taoists did finger computation on the trigrams and hexagrams by counting the finger tips and joints of the fingers with the tip of the thumb. Each hand can count up to 16 in such a manner.

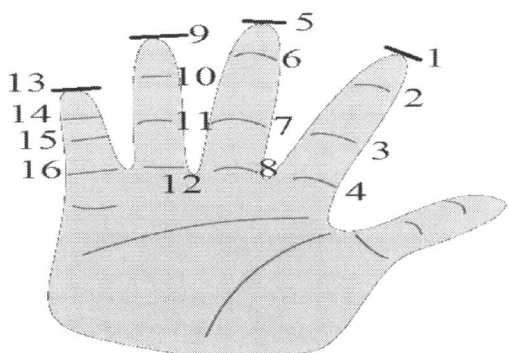

49.

49 we already saw as centring the Mod 99 sequence in the Hexagram numbers and the 49th Octave is the only Octave of the electro-magnetic scale that we can see.

Eight 3, 4, 5 Pythagorean Triangles surround a single empty space in a 7 x 7 square. I think this is very important to appreciate more deeply and particularly because of the fact that 8 x 3 4 5 triangles may be placed in a 7 x 7 square leaving the central space empty.

The number of partitions for the special number 19 total 490 for 10 of these perfect space containers.

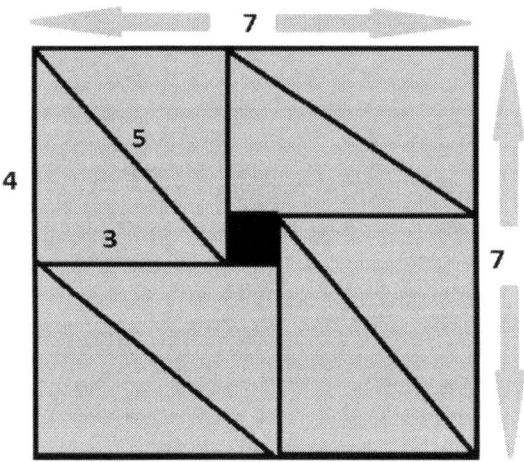

52.

4 x 13 = 52 Weeks in Year.

52 playing cards in a regular deck.

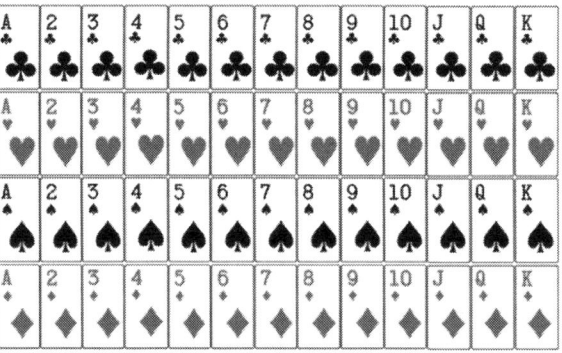

52 white keys on a piano

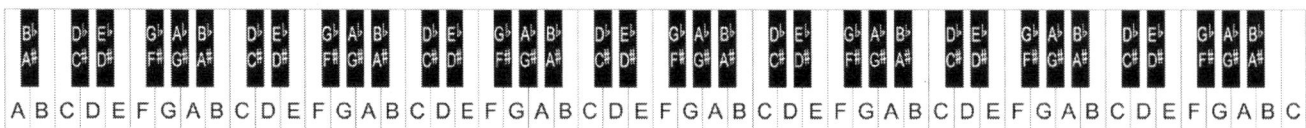

52 is a significant number in the Mayan Calendar - A Calendar Round date is a date that gives both the Tzolk'in and Haab'. This date will repeat after 52 Haab' years or 18,980 days, a Calendar Round.

For example, the current creation started on 4 Ahau 8 Kumk'u. When this date recurs it is known as a Calendar Round completion.

Arithmetically, the duration of the Calendar Round is the least common multiple of 260 and 365; 18,980 is 73 × 260 Tzolk'in days and 52 × 365 Haab' days. – Wikipedia.

54.

2 x 27 - The Number for Light.

2 cubes of 3.

Chapter 17: Conway's Game of Life.

The Game of Life - John Conway.

The Game of Life is not your typical computer game. It is a 'cellular automaton', and was invented by Cambridge mathematician John Conway. This game became widely known when it was mentioned in an article published by Scientific American in 1970. It consists of a collection of cells which, based on a few mathematical rules, can live, die or multiply. Depending on the initial conditions, the cells form various patterns throughout the course of the game. - From 'Life is a bit' by Edwin Martin.

The Rules.
For a space that is 'populated':-
1. **Each cell with one or no neighbours dies, as if by loneliness.**
2. **Each cell with four or more neighbours dies, as if by overpopulation.**
3. **Each cell with two or three neighbours survives.**

For a space that is 'empty' or 'unpopulated':-
1. **Each cell with three neighbours becomes populated.**

This is a fascinating game and one which, in my view, may tell us a great deal about Numbers 1 to 9. I started by simply placing a row of squares according to the number I wanted to investigate the effects of, with the following very interesting results: You can try these for yourself - http://www.bitstorm.org/gameoflife/

Numbers 1 – 9.

Here I examine each of the numbers 1 to 9 as starting points for the Game of Life - I was interested to see how each of them completes their journey.

Number 1: A single cell - just blinks out of existence after a single generation.

Number 2: Again, just blinks out of existence after a single generation.

Number 3:

Start.	1st Generation.

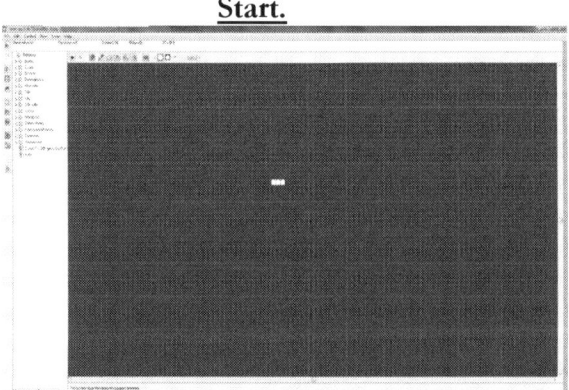 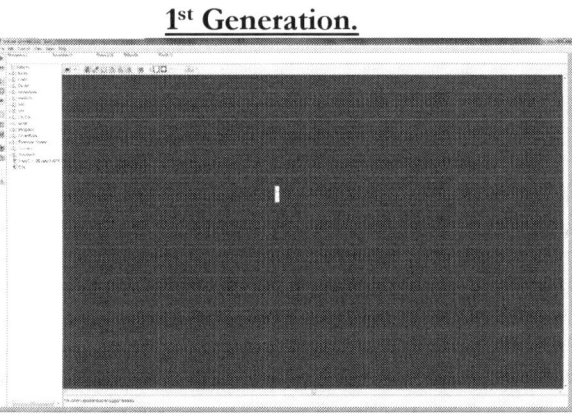

3 is immediately coherent in that it settles immediately into an infinitely repeating sequence of duality, at right angles, or orthogonal to each other. I take this to mean it's alive in some way. Both patterns taken together occupy a 3 x 3 square and in my view this may give some insight into the self-contained nature or completeness of the number 3. The 3 is alive and self-perpetuating.

Number 4.

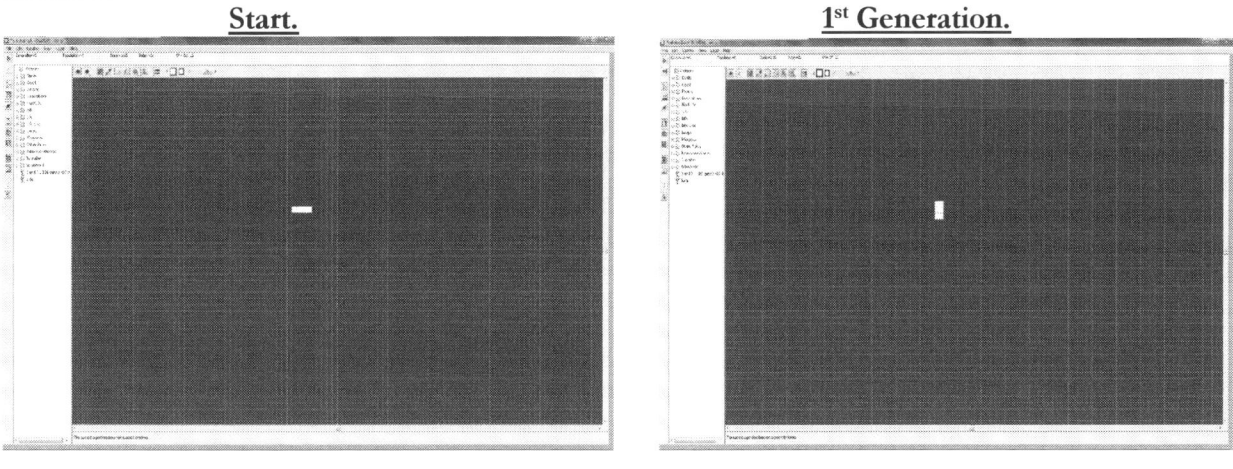

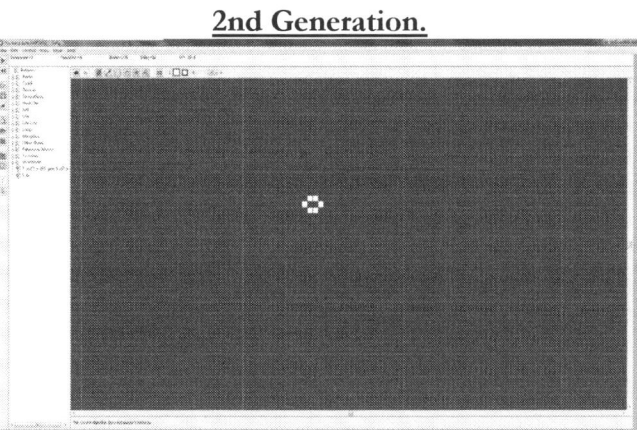

The journey ends with the static geometry displaying an enclosed area, a line. 4 can be understood as 2^2 which also introduces the dimension of Area. The pattern occupies a 4 x 3 Rectangle which when divided give 2 Pythagorean 3 4 5 triangles.

Number 5.

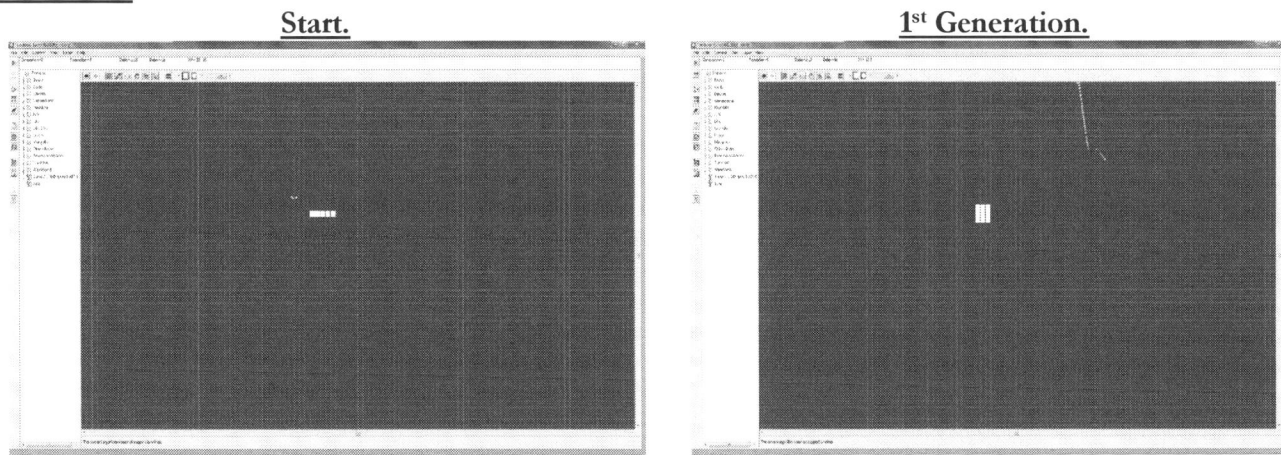

Number 5, the first prime in physical reality. We see the starting 5 squares in a row becomes a 3 x 3 block of 9 squares.

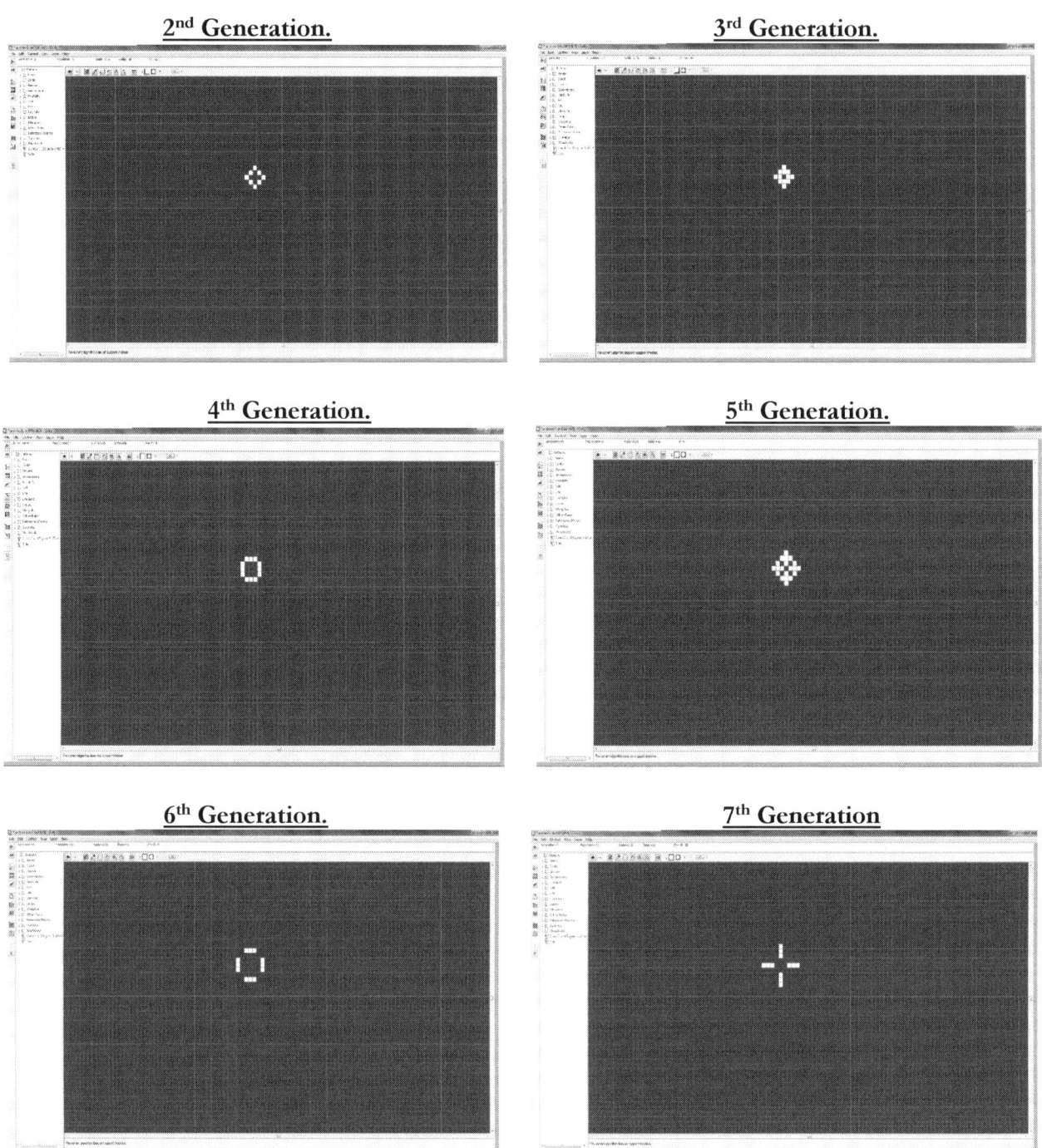

This sequence goes on to produce a repeating or oscillating 2 pattern sequence, starting from the 6th and 7th generation.

The 1st Oscillation of the duality at the 6th Generation makes use of a 7 x 7 Square Space - Area is 49 units. Remember the eight 3, 4, 5 Pythagorean Triangles surround a single empty space in a 7 x 7 square that we saw earlier. We also see a contained square 5 x 5 = 25 units.

The 2nd Oscillation at the 7th Generation makes use of a 9 x 9 Square Space. Area is 81 units. There are 81 stable chemical elements. We also see a contained space of a 3 x 3 box.

My feeling is that this stable, flickering, duality effect we see is supportive of Biology. In other words Life can be sustained.

Number 6.

Blinks out completely on the 12th Generation - no life there with 6 as a starting point.

Number 7.

14th Generation

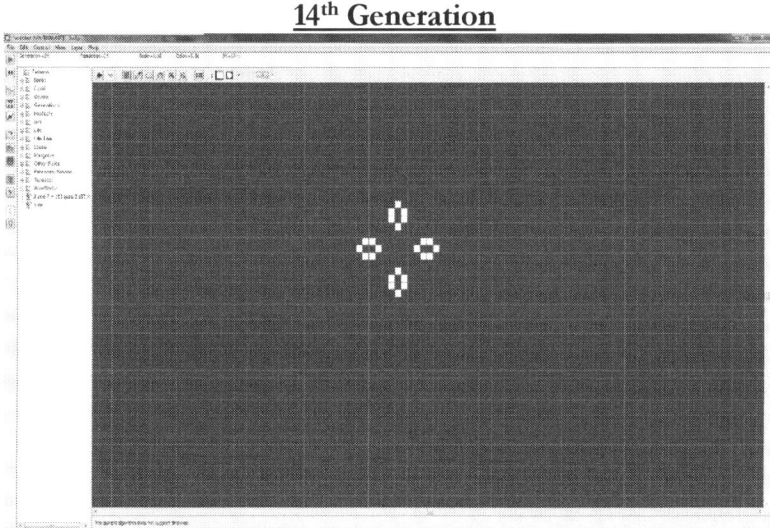

Really quite a beautiful journey, but again cannot go on and support Life as it quickly reaches a static status at the 14th generation. We can see 4 of the same final pattern displayed by the Number 4 which each contain 2 units of Space, at right angles to each other. It uses an overall 13 x 13 square of space for 169 Units.

Number 8.

48th Generation.

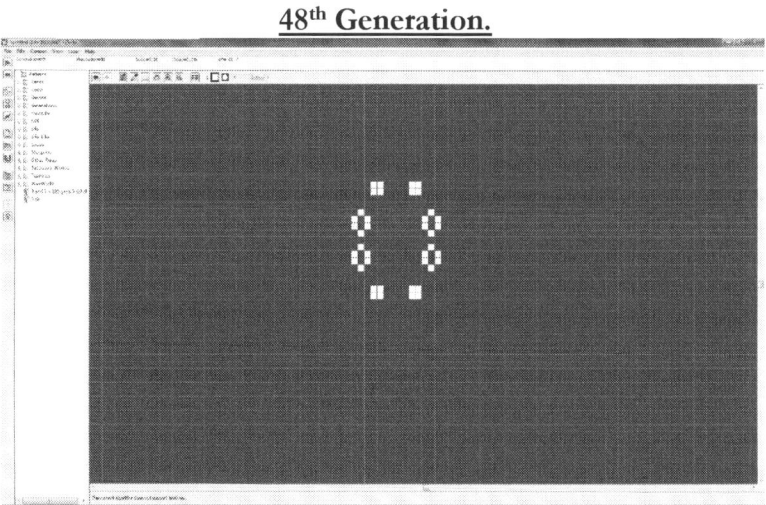

Again, very interesting journey until the 48th generation where it reaches a static state. We can see 4 of the pattern for the Number 4 together with four 2 x 2 squares, occupying an overall area of 14 x 17 = 238 units.

Taking each pattern and noting their dimensions also, we see a 14 x 9 = 126 (27 Mod 99) units for the Number 4 patterns and then for the 2x2 boxes it is 8 x 17 = 136 (37 Mod 99).

Number 9.

Start.

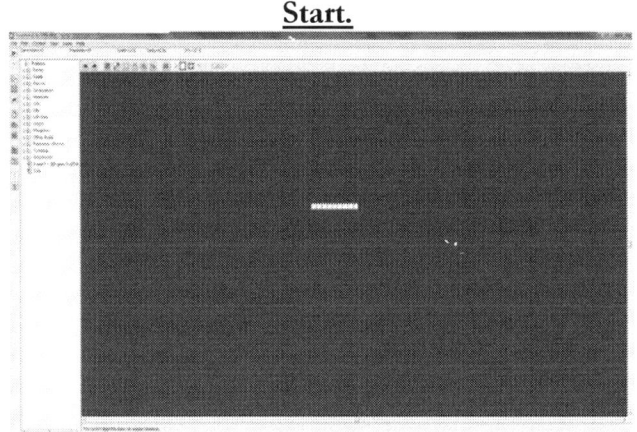

Area used 1 x 9 = 9 Units.

1st Generation.

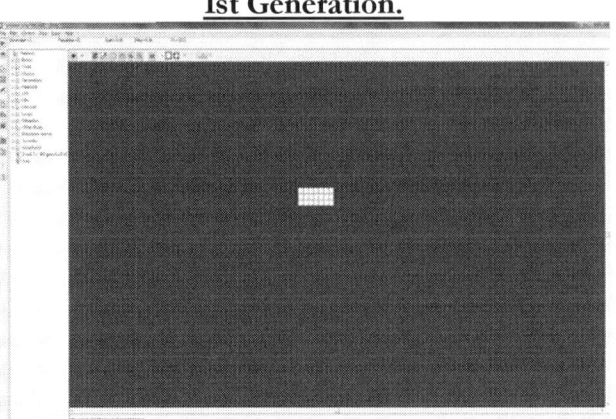

Area used 3 x 7 = 21 units.

2nd Generation.

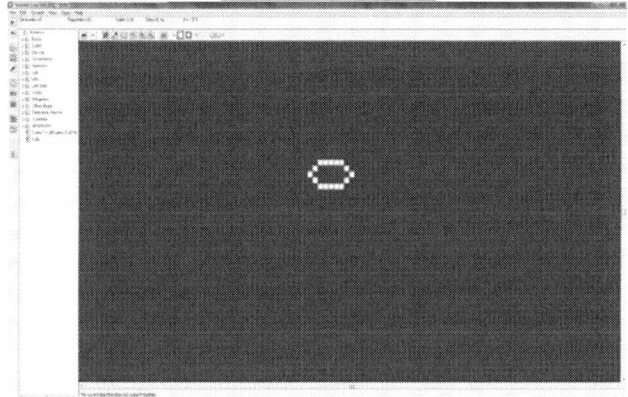

Area used 5 x 9 = 45 units. Contained Space = 17 units

3rd Generation.

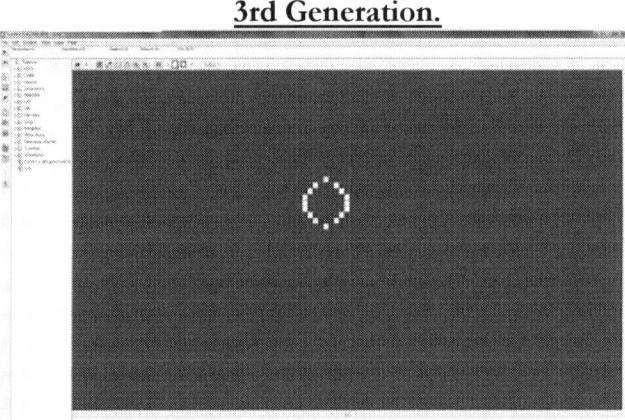

Area used 9 x 9 = 81 units Contained Space = 37 units

4th Generation.

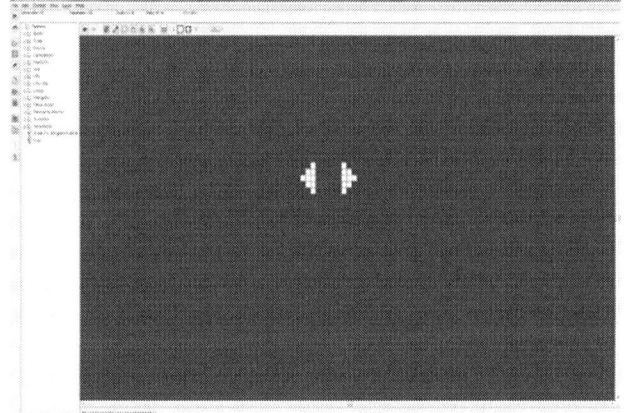

Area used 5 x 11 = 55 units. Contained Space = 6 units.

5th Generation.

Area used 5 x 11 = 55 units. Contained Space = 6 units.

6th Generation.

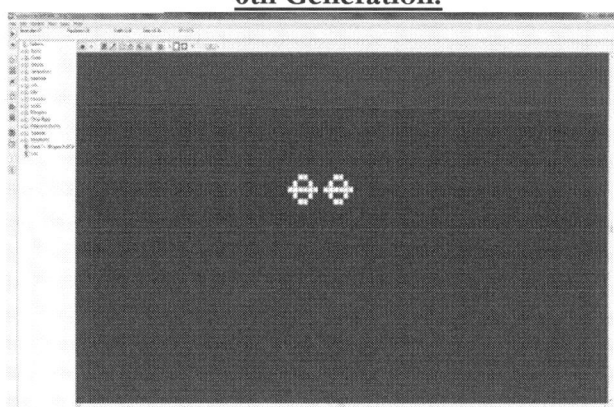

Area used 5 x 13 = 65 units. Contained Space = 8 units.

7th Generation.

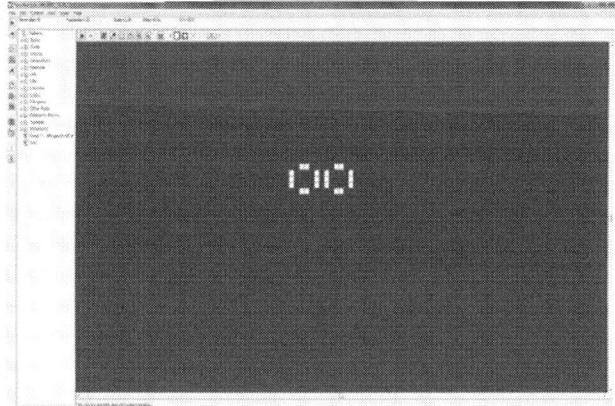

Area used 5 x 13 = 65 units. Contained Space = 12 units.

8th Generation.

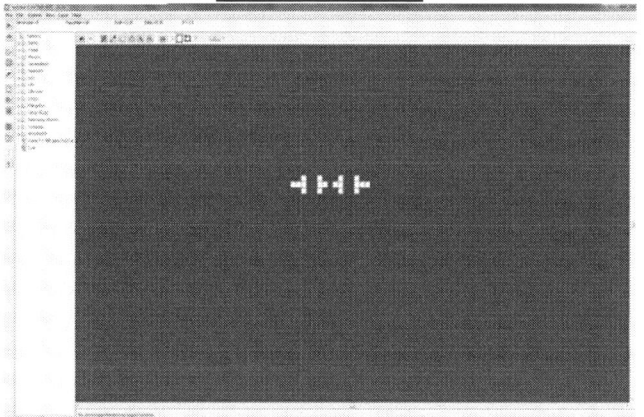

Area used 3 x 15 = 45 units.

9th Generation.

Area used 3 x 11 = 33 units

10th Generation.

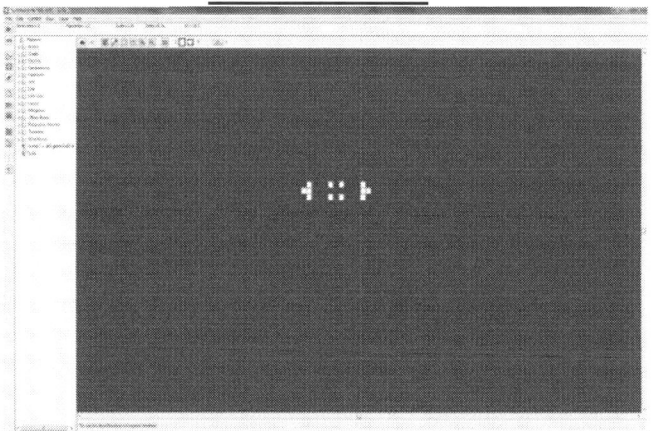

Area used 3 x 13 = 39 units.

11th Generation.

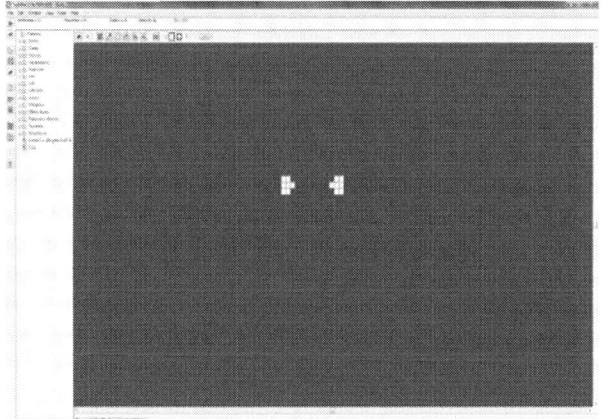

Area used 3 x 13 = 39 units

12th Generation.

Area used 3 x 15 = 45 units.

13th Generation.

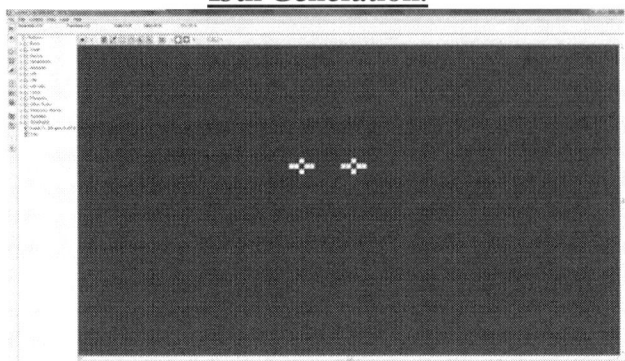

Area used 3 x 15 = 45 units

2 separate entities, each containing a single enclosed space. Each entity using a 3 x 5 = 15 unit space.

14th Generation.

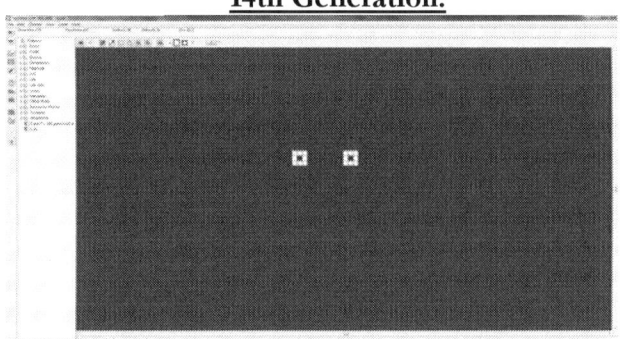

Area used 3 x 13 = 39 units.

15th Generation.

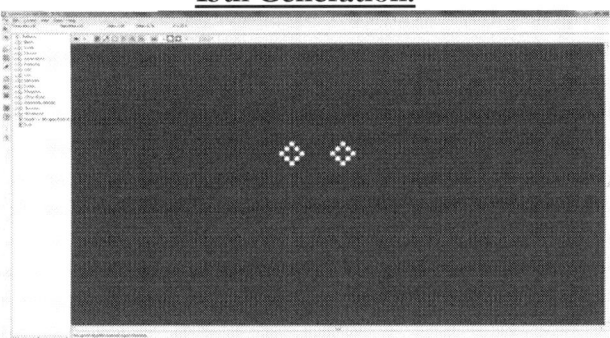

Area used 5 x 15 = 75 units.

14th Generation we see 2 separate entities, each containing a single enclosed space and occupying 3 x 3 = 9 unit space.
15th Generation we see 2 separate entities, each containing a 5 unit enclosed space and occupying 5 x 5 = 25 unit space.

16th Generation.

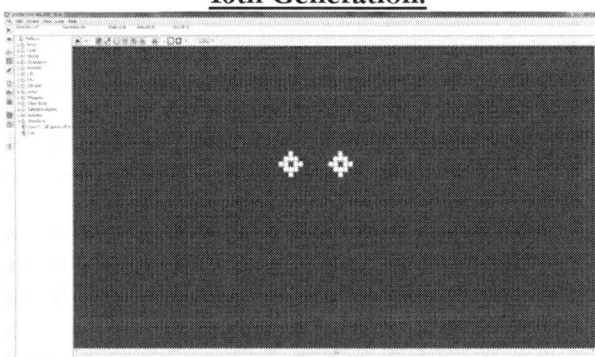

Area used 5 x 15 = 75 units.

17th Generation.

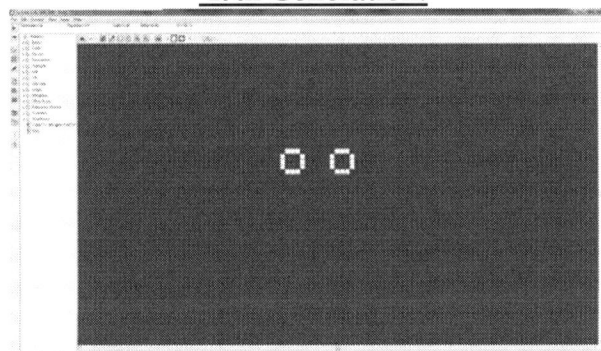

Area used 5 x 15 = 75 units.

16th Generation - 2 separate entities, each containing a single enclosed space. Each entity occupying 5 x 5 = 25 units.
17th Generation - 2 separate entities, each containing a 3 x 3 square of 3 units. Each entity occupying 5 x 5 = 25 units.

18th Generation.

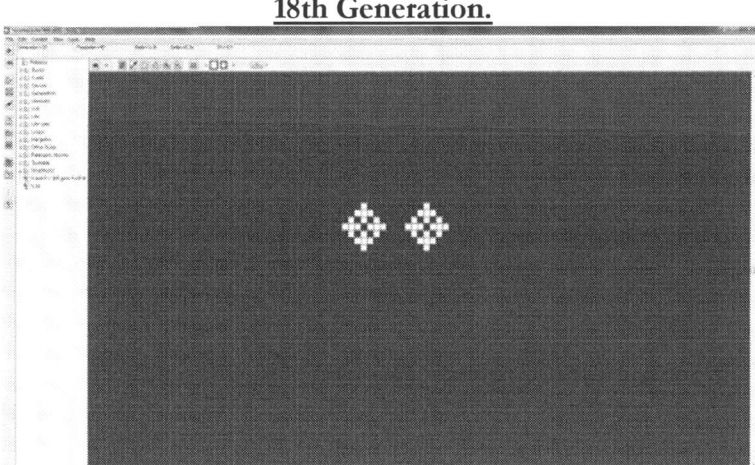

Area used 7 x 17 = 119 units.

2 separate entities, each containing a centred pentagon, 5 spaces arranged as the 5 on a die. Each entity occupying 7 x 7 = 49 units.

19th Generation - The First Oscillation Pattern.   20th Generation - The Second Oscillation Pattern.

 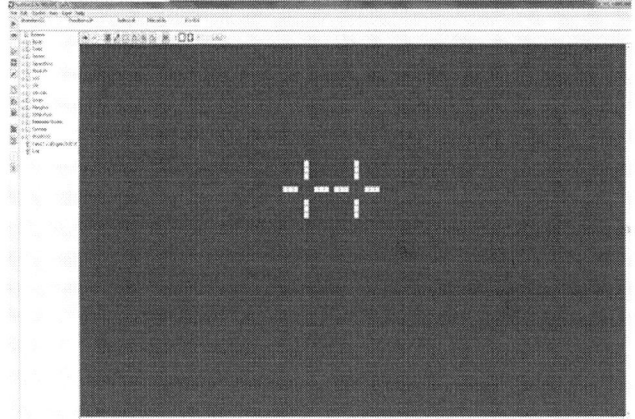

Area used 7 x 17 = 119 units. (20 Mod 99)   Area used 9 x 19 = 171 units. (72 Mod 99)

19th Generation - 2 separate entities, each containing a 5 x 5 square of 25 units. Each entity occupying 7 x 7 = 49 units.
20th Generation - 2 separate entities, each containing a 3 x 3 square of 9 units. Each entity occupying 9 x 9 = 81 units.

Very interesting to see that the sequence finishes at the 20th generation, and a stable duality is perpetuated with both patterns simultaneously flashing -

20 Amino Acids - 20 Sporadic Groups contained within the Monster Group.

Conclusions.

Only the 3, 5 and 9 produce a sustainable duality. I believe Conway's rules might just be the same rules that the numerical algorithm of life employs to create a coherent and congruent system and I believe we may be able to map these rules to the Amino Acids exactly.

The 3 and 7.

Keeping things super simple in my thinking I wondered what the effect of two numbers separated by a single space would have and my first go to was the 3 and 7 because they are shown to be the reciprocal of each other in the section Octaves of Perception.

The combination of a line of 3 squares, a space, and then a line of 7 squares produces a fascinating journey to symmetrical oscillatory balance which is finally attained on the 185th generation. Note 185 = 5 x 37.

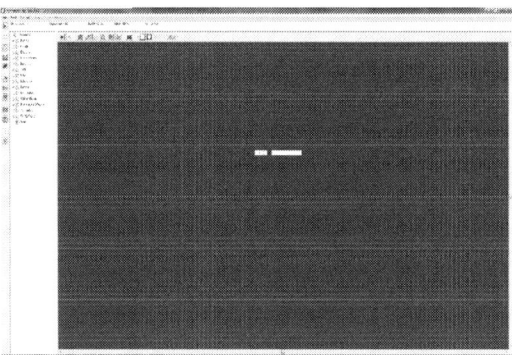

Following The Conway Game of Life rules the above simple line becomes the oscillating patterns below.

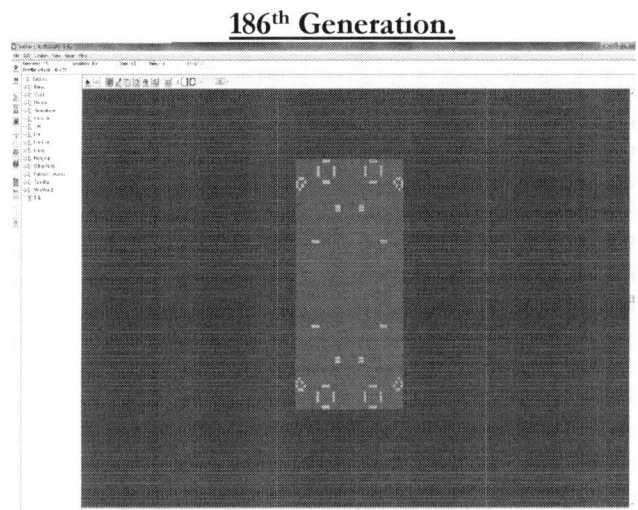

Very interestingly, the space required for this unfolding reaches a maximum rectangle size at 40 x 72 = 2880 units for the first oscillation at the 185th Generation while the second oscillation makes use of a rectangular space 40 x 70 = 2800 units.

$$2880 - 2800 = 80$$

$$80/2880 = 0.02\,777777\,R \qquad 80/2800 = 0.0\,285714\,R$$

$$2800/2880 = 0.97222222R \qquad 2880/2800 = 1.0\,285714\,R$$

Please also notice that

3960 Mean Radius of Earth - 1080 Mean Radius of Moon = 2880.

288(0) in New Jerusalem.

More from John Michell - 'In constructing the New Jerusalem, the geometer begins by imitating the first act described in the Old Testament, 'In the beginning God created the heaven and the earth.' The corresponding geometric operation is to draw the circle of the heavens together with the earthly square and to harmonize them by giving them both an equal perimeter.

This is achieved by the means of the Pythagorean 3 4 5 triangle. The sum of these 3 numbers 3 + 4 + 5 = 12; their product, 3 x 4 x 5 = 60; and the numbers 12 and 60 are at the root of the numerical code that supplies all the dimensions of the New Jerusalem. Added together 12 + 60 = 72 and multiplied together 12 x 60 = 720.

This is the number of Truth and it reveals the truth of things in providing the multiplier that raises the dimensions of the squared - circle figure, created by the 3 4 5 triangle, to those of the New Jerusalem. Multiplied by 720, the three sides of the 3 4 5 triangle become 2160 2880 3600; the side of the New Jerusalem square becomes 7920 and the radius of its circle, 5040; and both the perimeter of the square and the circumference of the circle measure 31680.

The graphic below shows the rings of the New Jerusalem and their dimensions. The most convenient way of calculating the area of an individual ring is to subtract the square on its inner radius from the square on its outer radius, multiplying the remainder by 22/7. Thus the areas of the separate rings are:-

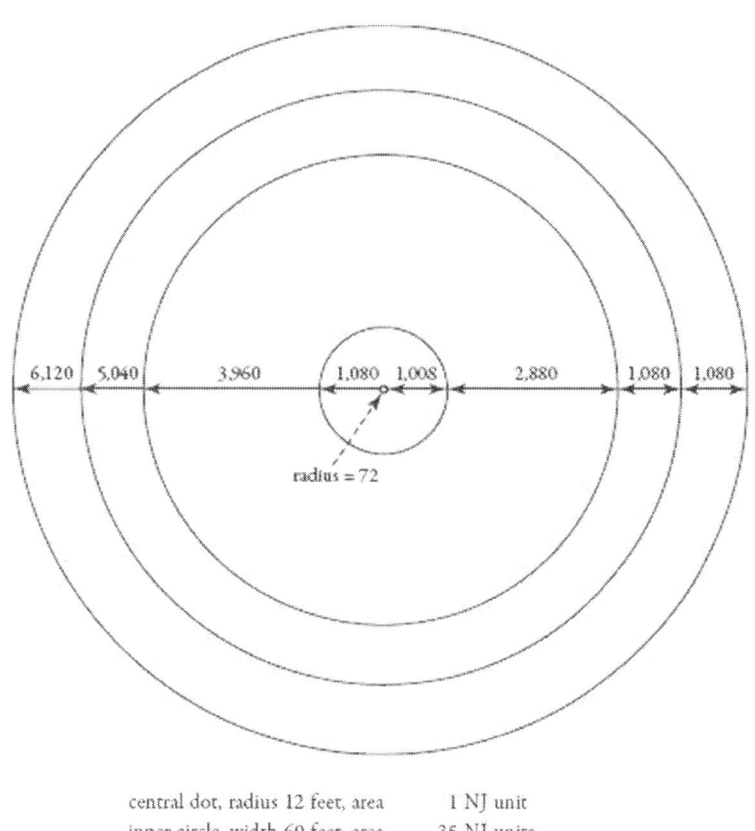

```
        central dot, radius 12 feet, area         1 NJ unit
        inner circle, width 60 feet, area        35 NJ units
        first ring, width 1,008 feet, area     8,064 NJ units   = 504 × 16
        second ring, width 2,880 feet, area  100,800 NJ units   = 504 × 200
        third ring, width 1,080 feet, area    67,500 NJ units ⎫
        outer ring, width 1,080 feet, area    83,700 NJ units ⎭ = 504 × 300
        total, radius 6,120 feet, area       260,100 NJ units
```

The total area, deducting the 8,100 units contained by the first ring, radius 1,080 feet, is 252,000 = 5,040 × 50 NJ units. St. John's New Jerusalem is an image of the sublunary world, all that lies within the influence of the moon. Its various parts represent:

Inner circle of radius 72, the pole of the Earth and Universe.
Circle of radius 1,080, the Underworld.
Circle of radius 3,960, the Earth.
Outer rings of width 2,160, the heavens below the Moon.

The ratio between the numbers 216 and 144 is 3:2 (a musical fifth), and this is also the ratio between the areas of two parts of the New Jerusalem, the ring of width 2,880 feet in cross-section and the outer pair of rings with combined width of 2,160 feet Their respective areas are 144 × 316,800 square feet and 216 × 316,800 square feet. Numerically, therefore, these two parts of the diagram represent the Inferior Principle of earth and the Superior Principle of the heavens.

As an unlimited source of mathematical harmonies and felicities, the New Jerusalem has obvious appeal to those who delight in such things. Yet its prestige in ancient times was far greater than would be attributed to mere curiosities of number. To the extent that its functions corresponded to those of the Chinese River Map, it provided a standard of proportion for the right conduct of human affairs and a means of attracting the benevolent powers of the cosmos. It is stated as follows by an ancient commentator on the Book of Diagrams:

'By means of the doctrine of numbers, virtuous conduct is brought into contact with invisible beings. The diagrams are useful in manifesting the right course of things to men and in bringing the virtuous conduct of men into contact with invisible beings. It follows, therefore, that when men harbour any doubts which they cannot decide, the diagrams are then useful in the visible world, in the intercourse of men and in elucidating their doubts.'

Similar claims are made by Plato on behalf of the number system that he weaves through the diagrams of his ideal cities and cosmologies: that its study refines and elevates the mind and thus leads to individual enlightenment and to the best possible forms of society. Plato's geometric and musical expressions of the cosmos, examined later, are all developments of the New Jerusalem scheme and its underlying pattern of number.' *The Dimensions of Paradise* by John Michell published by Inner Traditions International and Bear & Company, ©2008. All rights reserved. http://www.Innertraditions.com Reprinted with permission of publisher.

288 Connections.

Fibonacci.

Earlier we saw that the number 144 was important and this too is a Fibonacci Number, the 12th.

-144 89 -55 34 -21 13 -8 5 -3 2 -1 1 0 1 1 2 3 5 8 13 21 34 55 89 144

To the left of zero, the total of the terms:-

$$-144 \quad 89 \quad -55 \quad 34 \quad -21 \quad 13 \quad -8 \quad 5 \quad -3 \quad 2 \quad -1 \quad 1 \quad = \quad -88$$

To the right of zero, the total or the terms:-

$$1 \quad 1 \quad 2 \quad 3 \quad 5 \quad 8 \quad 13 \quad 21 \quad 34 \quad 55 \quad 89 \quad 144 = 376$$

$$-88 + 376 = 288$$

288 - Physical Correlations.

'The diameter of the Aubrey Circle at Stonehenge was 288 feet.

The total number of gifts given to Moses in Numbers chapter 7 of the Bible was 288.

The design length for the diagonal face of the Menkaure Pyramid was intended to be 288 feet.

The number of grid squares for the Giza Plateau, into which all structures were geometrically positioned, was 288.

The pyramid acre was 28800 sq. feet. The geomancer's mile of Great Britain was 28800 feet x 2.

The diagonal face length of the Great Pyramid was 288 feet x 2.

The overall north-south diameter of the Waitapu standing stone circle in New Zealand (two centrally overlapping circles of 192 feet each, culminating in 3 segments of 96 feet each) was 288 feet.' Source: Michael Doutre http://www.celticnz.co.nz/US6.html

Chapter 18: The Numerical World Soul.

Metaphysical v Physical Numbers.

Armed with the understanding that numbers in pairs operate specifically and for example where the key numbers 27 and 37 in my theory come together, they mean a combination of the 'quality' or 'energy' of the 2 with that of the 7 etc.

2 and 3 are part of what I call the 'a priori', causal or metaphysical numbers - 1 2 3 and 4, with 5 being the first Prime / Atom (that which is indivisible) in physical reality joining 6, 7 and 8 as the physical numbers.

As 27 and 37 are central to my theory, 17 and 47 complete this particular grouping and its causal interaction with the number 7.

17 27 37 47.

Adding these numbers together we see that:-

$$17 + 27 + 37 + 47 = 128 \text{ or } 2^7 \text{ or two } 4 \times 4 \times 4 \text{ cubes.}$$

Multiplying these numbers together we see that:-

$$17 \times 27 \times 37 \times 47 = 798201 \ (162 = 63 \text{ Mod } 99)$$

798201 is a two dimensional numerically symmetrical number:-

$$798$$
$$201$$

$$999$$

57 67 77 87.

Equally important will be the interaction with the numbers that describe effect, with the number 7.

Adding these numbers together:-

$$57 + 67 + 77 + 87 = 288$$

Our friend 288 again.

Multiplying these numbers together:-

$$57 \times 67 \times 77 \times 87 = 25583481 \ (= 198 = 99 \text{ or } 0 \text{ Mod } 99)$$

17 27 37 47 v 57 67 77 87.

When we look at the two groups of numbers in relation to each other we see that:-

$$128 / 288 = 4/11 \text{ or } 0.4444444 \text{ R}$$

$$288 / 128 = 9/4 \text{ or } 2.25$$

Then as quotients of the whole, being 128 + 288 = 416 we see that:-

$$128/416 = 0.307692 \text{ R}$$

$$307$$

$$692$$

$$999$$

Displaying the Number Pairs in the 2 dimensional numerical symmetry.

$$288/416 = 0.692307 \text{ R}$$

$$692$$

$$307$$

$$999$$

Also displaying the Number Pairs in the 2 dimensional numerical symmetry where it is also very interesting to see the recurring decimals these two numbers throw up being the inverse of each other.

The Numerical World Soul.

Thinking about numbers grouped in this way and seeing these familiar numbers, lead me to the rather beautiful discovery below which links the decimal system of number to the dimensions of the Earth and Moon, to Gematria and the Bible.

110										270	
	11	21	31	41	104	51	61	71	81	264	368
	12	22	32	42	108	52	62	72	82	268	376
	13	23	33	43	112	53	63	73	83	272	384
	14	24	34	44	116	54	64	74	84	276	392
	50	90	130	170	440	210	250	290	330	1080	1520
	15	25	35	45	120	55	65	75	85	280	400
	16	26	36	46	124	56	66	76	86	284	408
	17	27	37	47	128	57	67	77	87	288	416
	18	28	38	48	132	58	68	78	88	292	424
126	66	106	146	186	504	226	266	306	346	286	1144	1648
	116	196	276	356	944	436	516	596	676	2224	3168

We saw 3168 appearing repeatedly in the DNA section where we were focusing on information about 7920 from John Michell's seminal book - Dimensions of Paradise where he shows that the numerical qualities of this number 3168 are profound and numerous.

'The circle with circumference 31680 and radius 5040 consists of two semicircles each with an area of 11!

The perimeter of the square New Jerusalem is 4 x 12 furlongs, or 31680 feet.

The perimeter of the square 12 hides of Glastonbury is 31680 feet.

The mean circumference of the Stonehenge sarsen circle is 316.8 feet, or a hundredth part of 6 miles.

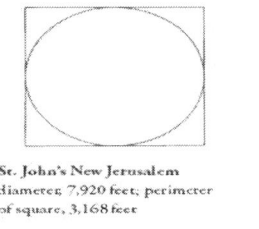

St. John's New Jerusalem
diameter, 7,920 feet; perimeter of square, 3,168 feet

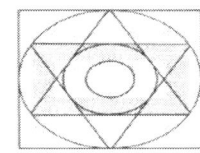

St. Mary's Chapel, Glastonbury
diameter, 79.2 feet; perimeter of square, 316.8 feet; diameter of inner circle, 21.6 feet

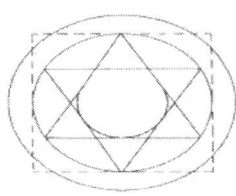

Stonehenge
diameter of (bluestone) circle within square, 79.2 feet; perimeters of outer (sarsen) circle and square, 316.8 feet; diameter of inner horseshoe, 39.6 feet

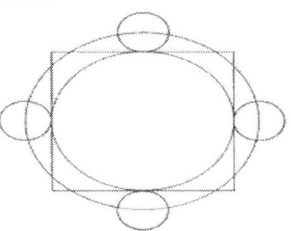

Earth and moon
diameter of inner (earth) circle, 7,920 miles; perimeters of outer circle and square, 31,680 miles; diameter of small (moon) circles, 2,160 miles

A square containing the circle of the Earth, average diameter 7920 miles, has a perimeter of 31680 miles, and if the moon, diameter 2160 miles, is drawn tangent to the Earth, a circle struck from the centre of the Earth circle to pass through the center of the Moon has a circumference of 31680 miles.

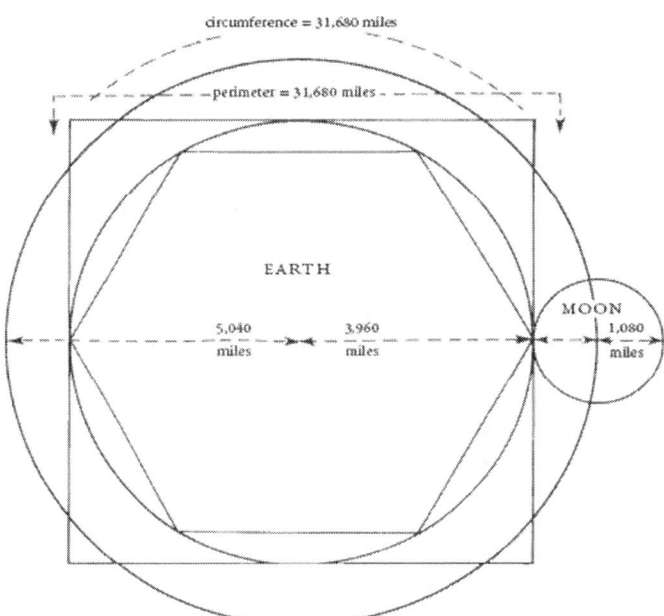

In his Natural History, at the end of the second book, Pliny implies that the world circumference is 3168000 miles.

The number 3168 has been preserved in our traditional system of metrology from at least as early as the building of Stonehenge.

For example,

31680 x 2 inches = 1 mile

31680 feet = 6 miles

31680 furlongs = 3960 miles - mean radius of the Earth.

The longer Greek furlong or stade of 625 feet, which appears in the dimensions of the Parthenon and Stonehenge, is equal to 633.6 or twice 316.8 feet. This Greek furlong and the English furlong of 660 feet come together in the acre.

For example, a rectangle with sides measuring 1 English by 5 Greek furlongs (660 x 3168 feet) contains 48 acres exactly, and the 144 acres of the New Jerusalem square are also contained by a rectangle of 3 English furlongs by 5 Greek furlongs.

The general character of the number 3168, as conveyed by its position in ancient cosmological diagrams and the phases associated with it through Gematria, is that it represents the spirit which passes through and encircles the universe, Plato's world soul.'

The Dimensions of Paradise by John Michell published by Inner Traditions International and Bear & Company, ©2008. All rights reserved. http://www.Innertraditions.com Reprinted with permission of publisher.

Extraordinarily, we find that Bethlehem, the claimed birthplace of Jesus, sits on 31.68 N latitude.

Even more extraordinarily - 3168 is the Greek Gematria for the words that translate to Lord Jesus Christ!

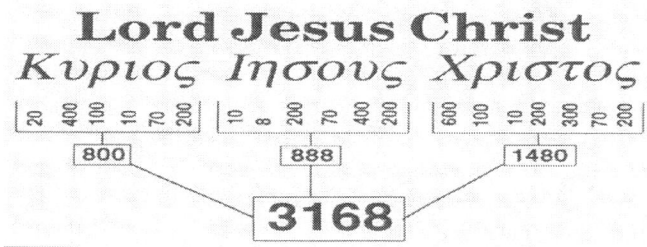

The 12 Tribes of Israel.

If you take the Hebrew names for the 12 tribes of Israel and look at the total sum using Gematria:

Asher 501 Benjamin 152 Dan 55 Gad 8 Issachar 830 Joseph 156

Judah 30 Levi 46 Naphtali 570 Rueben 259 Simeon 466 Zebulon 95

The 12 sons of Jacob total to 3,168.

Stonehenge.

The Station Stone rectangle precisely traces the risings and settings of the Sun and Moon. The long side is 3168 inches, bearing the signature of the Designer. The perimeter is precisely the length of one side of the Great Pyramid at its base that rests on the natural bedrock, namely .144 of a mile. - Bonnie Gaunt - Beginnings - The Sacred Design

Factors of 3168.

There are 36 factors excluding 1 and itself where the sum of its factors excluding 1 and 3168 is 6660 (= 126 = 27 Mod 99) 6660 = 10 x 666 - The Number of the Beast.

Diagonals.

110									270		
	11	21	31	41	104	51	61	71	81	264	368
	12	22	32	42	108	52	62	72	82	268	376
	13	23	33	43	112	53	63	73	83	272	384
	14	24	34	44	116	54	64	74	84	276	392
	50	90	130	170	440	210	250	290	330	1080	1520
	15	25	35	45	120	55	65	75	85	280	400
	16	26	36	46	124	56	66	76	86	284	408
	17	27	37	47	128	57	67	77	87	288	416
	18	28	38	48	132	58	68	78	88	292	424
126	66	106	146	186	504	226	266	306	346	286	1144	1648
	116	196	276	356	944	436	516	596	676		2224	3168

The Diagonals both Total 396 = 792, the Number of Partitions for 21.

Looking at the four quadrants we see some interesting totals:-

51 61 71 81 52 62 72 82 53 63 73 83 54 64 74 84 = 1080 Radius Moon

11 21 31 41 12 22 32 42 13 23 33 43 14 24 34 44 = 440 Standard Tuning for A on the Middle Octave

15 25 35 45 16 26 36 46 17 27 37 47 18 28 38 48 = 504 = 7 x 6 x 5 x 4 x 3 x 2 x 1 / 10

55 65 75 85 56 66 76 86 57 67 77 87 58 68 78 88 = 1144 = 55 Mod 99

3168 - Number Pairs.

Organised into their respective number pairs this same information can be represented as:-

NUMBER PAIRS									
11	81	21	71	31	61	41	51	368
12	82	22	72	32	62	42	52	376
13	83	23	73	33	63	43	53	384
14	84	24	74	34	64	44	54	392
50	330	90	290	130	250	170	210	1520
15	85	25	75	35	65	45	55	400
16	86	26	76	36	66	46	56	408
17	87	27	77	37	67	47	57	416
18	88	28	78	38	68	48	58	424
66	346	106	306	146	266	186	226	1648
Sum	792	Sum	792	Sum	792	Sum	792	3168

4 lots of 792 (Partitions for Number 21) = 3168.

3168 - The Great Design.

Collaborator James Heyworth has done some extraordinary work and has graciously allowed me to reproduce a great deal of his work here. It's absolutely stunning and profoundly supportive of The Numerical Universe in general. To see all of his great work in situ, please visit his website at www.thegreatdesign.com

'The unveiling of the Great Design reveals that the earth, the moon and the sun are factors in a great design. Other factors of the Great Design are the units of measure from various systems that are used to reveal the message embedded in the Great Design.

The time span of a day is dictated by the earth's rotation speed which is a finely tuned Great Design factor as are the units used by the whole world to divide the day and measure time called the second, the minute and the hour.

The whole world divides circles into 360 degrees with each degree containing 60 minutes that are subdivided into 60 seconds. This system was first used in the Great Design. Music's laws of harmony are Great Design factors, the numbers and number sequences found on the Musical Scale Chart are very thoroughly woven into the Great Design.

The Greek and Hebrew languages, their alphabets along with their embedded systems of numbers, were used in the Great Design before man even existed, thus demonstrating that the source of languages, alphabets and writing is neither mortal man nor evolution. The British system of latitudes and longitudes are used by the whole world but they were first used in the Great Design before man existed.

The purpose of the Stonehenge group of monuments and the Great Pyramid was to reveal the Great Design in detail and they have succeeded. These monuments are geometric time capsules containing a message from the past to the present. The seals are broken and the message can be read.

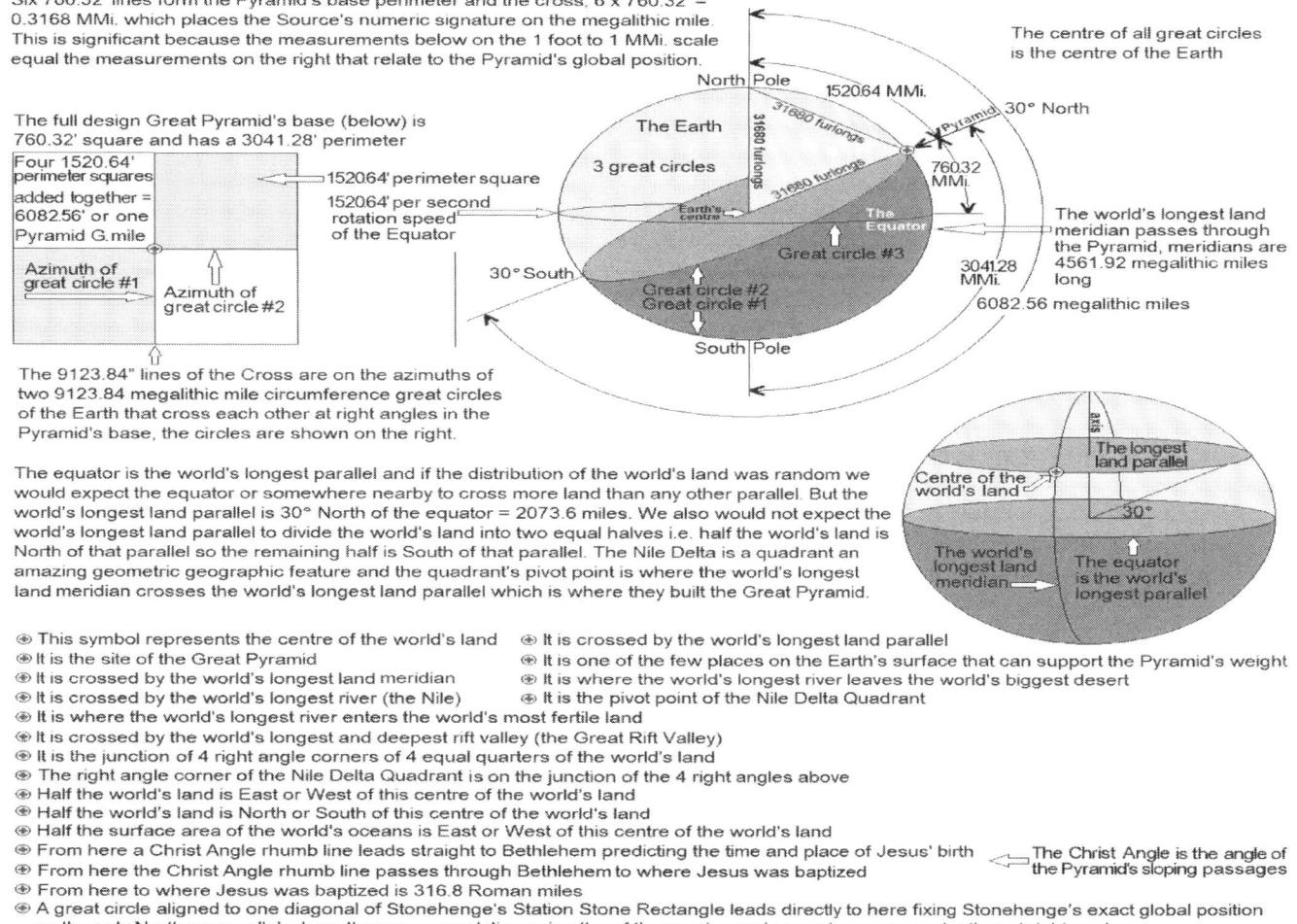

The New Testament was written using the Greek alphanumeric alphabet where every letter was also a numeric symbol and therefore every word has a numeric value called Gematria. Lord Jesus Christ has a Gematria value of 3168 in the Bible's Greek text.

Below are examples of the first use in our solar system of numbers and various units of measure, some Greek words are used for the first time via a code based on Biblical Gematria. The numeric system embedded in the Greek alphabet identifies the source by the number of his name, the Lord Jesus Christ, aka the Creator, has signed his work using the number of His Name which is 3168.

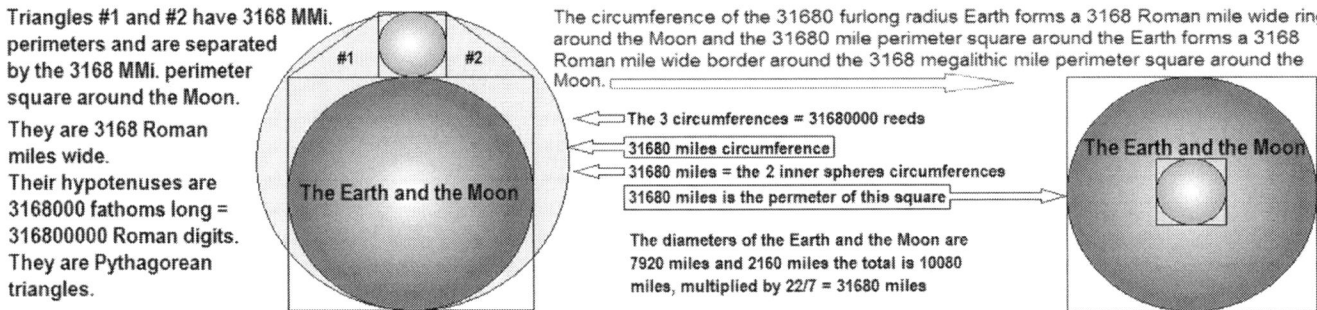

Triangles #1 and #2 have 3168 MMi. perimeters and are separated by the 3168 MMi. perimeter square around the Moon.
They are 3168 Roman miles wide.
Their hypotenuses are 3168000 fathoms long = 316800000 Roman digits.
They are Pythagorean triangles.

The circumference of the 31680 furlong radius Earth forms a 3168 Roman mile wide ring around the Moon and the 31680 mile perimeter square around the Earth forms a 3168 Roman mile wide border around the 3168 megalithic mile perimeter square around the Moon.

The 3 circumferences = 31680000 reeds
31680 miles circumference
31680 miles = the 2 inner spheres circumferences
31680 miles is the perimeter of this square

The diameters of the Earth and the Moon are 7920 miles and 2160 miles the total is 10080 miles, multiplied by 22/7 = 31680 miles

Each year the Earth completes an elliptical orbit of the Sun and 4 times per year the Sun's apparent angular diameter is 31.68 minutes of arc. The Moon's elliptical orbit of the Earth takes about a month and 4 times per orbit the Moon's apparent angular diameter is also 31.68 minutes of arc. A 3168 megalithic mile perimeter square fits tangent around the Moon, when a total solar eclipse occurs the square around the Moon would also contain the blacked out image of the 316800 megalithic mile diameter Sun (radius 316800 Roman leagues).

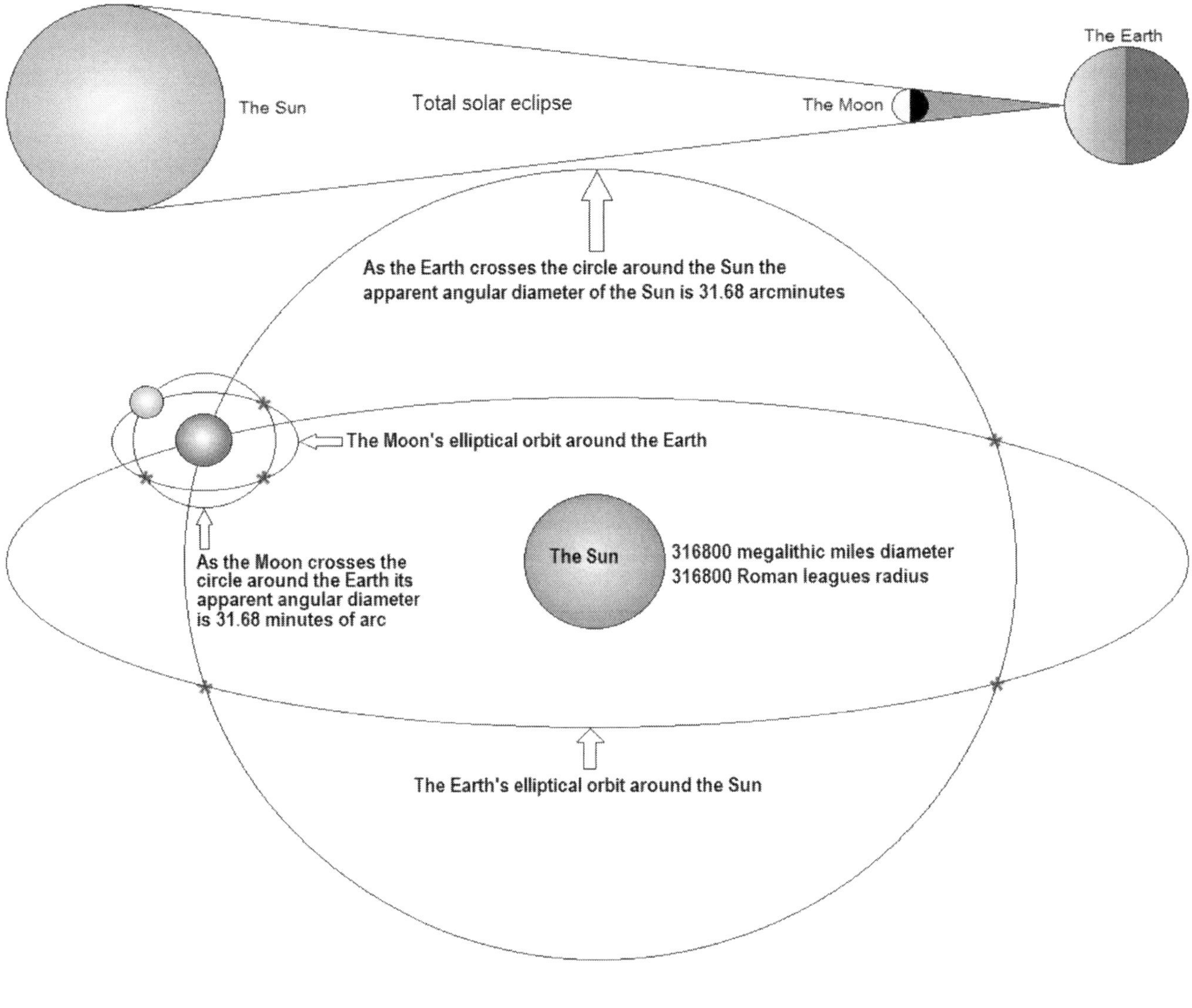

The Numerical Universe

Placing circles inside squares is suggested by the simple but profound geometric relationship between the earth and the moon, their circumferences via 22/7 = the perimeter of a square placed tangent around the earth.

When we add basic dimensions using appropriate units of measure the numbers produced are a code that reveal the identity of the Designer.

Further investigation reveals that the designs of the units of measure and our solar system have a common Source.

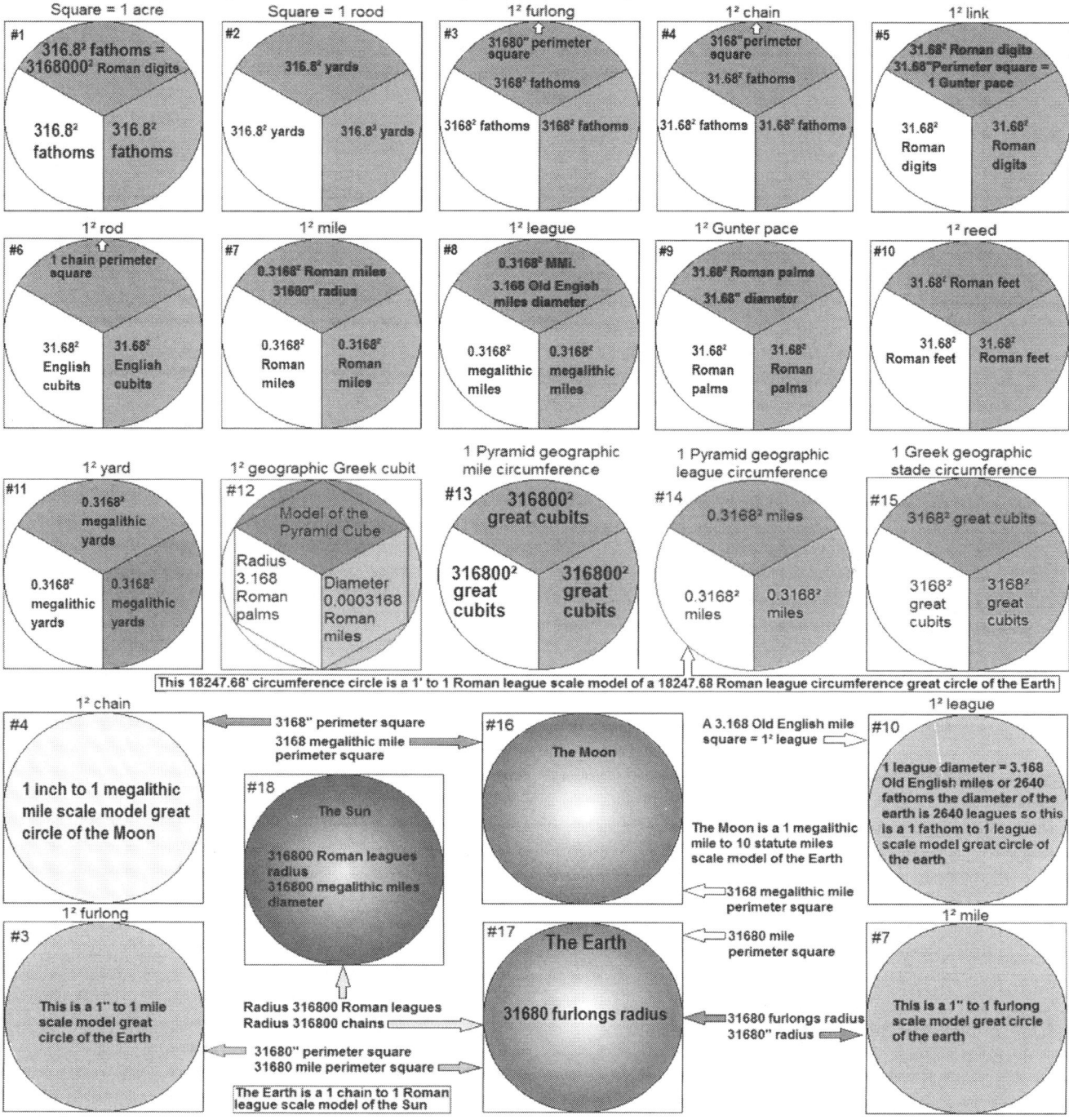

Simple geometric relationships between relevant units of measure reveal the 3168 signature everywhere showing that the units of measure and the solar system design have one Source and the first known use of those units of measure was in the cosmic design with the 3168 signatures before man existed. For example the earth's 31680 furlong radius equator turned 0.3168 Roman miles per second while orbiting the 316800 MMi. diameter sun = 316800 Roman leagues radius.

The Greek geographic foot equalled 1.01376 Imperial feet, if you type Greek foot 1.01376 into your internet browser you will get enough hits to demonstrate that the above geographic version of the Greek foot existed, the geographic factor separates this version of the Greek foot from the other versions that are not relevant to this investigation. The Parthenon's stylobate (floor) is 100 Greek geographic feet wide which is equal to 1 arc second of a great circle of the earth, some say this is evidence that the Greeks inherited their geographic units of measure from a more advanced prehistoric civilisation involved in global commerce such as Atlantis for example.

The Greek geographic cubit = 1½ Greek geographic feet and 1½ geographic cubits = 1 Greek geographic great cubit, the Parthenon's stylobate (floor) is 100 Greek great cubits long. So the Parthenon's floor is 100 Greek geographic feet x 100 Greek geographic great cubits and used to be called Hekatompendom which means 100 new feet. As we said 100 Greek feet = 1 arc second of the earth's circumference and 1000 Greek geographic cubits = the pre bulge equatorial rotation speed per second (0.3168 Roman miles). The Greek stade of 600 Greek geographic feet = 633.6 Roman feet and is 1/10 of a Pyramid geographic mile therefore 10 Greek geographic stades are equal to 1 arcminute of a 24883.2 mile circumference great circle of the earth; (the earth's 7920 mile diameter x 864/275 = 24883.2 miles circumference), note that the 864/275 method for pi is a factor in the formulation of the Greek geographic units of linear measure. The Greek geographic cubit = 0.3168 Roman paces or 0.0003168 Roman miles and the Greek geographic great cubit = 0.0003168 Roman leagues which indicates the real source of the Greek geographic units of measure.

The Roman mile = 5000 Roman feet or 1000 Roman paces, the Roman league = 1.5 Roman miles, the Roman foot = 16 Roman digits or 12 Roman inches, the Roman palm = 4 Roman digits (fingers). Noah Webster's 1828 Dictionary defined the Roman mile as 1600 yards which enables us to define the other units of measure, the Romans used other versions of the above units of measure during their long history but as they are not factors of the Great Design we don't need to know their details.

The megalithic mile = 14400 Imperial feet or 2.727272727 statute miles, the megalithic rod = 2.5 megalithic yards and 5280 megalithic yards = 1 megalithic mile. Professor Alexander Thom did decades of research on megalithic sites and rediscovered the megalithic units of linear measure he said the megalithic yard was 2.72 feet which is correct as far as he went but it has to be taken further to be compatible with the megalithic mile and the designs of the Stonehenge group of monuments. The megalithic yard has to be 2.727272727 feet long making 5280 megalithic yards to 1 megalithic mile instead of 5294.117647 megalithic yards of 2.72 Imperial feet to the megalithic mile, so in this investigation we use the megalithic yard of 2.727272727 Imperial feet and use 2.5 of these as the megalithic rod. These 3 units of measure are now compatible with each other as well as the designs of the monuments and the numbers on the Musical Scale Chart.

The mile i.e. the statute mile is 5280 feet, 3 miles = 1 league or 3.168 Old English miles, the Old English mile = 5000 feet, the chain = 66 feet or 100 links, the furlong is 660 feet, the rod = 16.5 feet, the English cubit = 1.5 feet, the yard = 3 feet, the fathom = 6 feet, 12 inches = 1 foot and the Gunter pace = 31.68 inches. The acre = 4840 square yards and 4 roods = 1 acre.

The Pyramid geographic mile = 6082.56 feet and is 1 arc minute of the earth's circumference i.e. calculated via the 864/275 method for pi and the earth's 7920 mile diameter, 3 Pyramid geographic miles = 1 Pyramid geographic league. The reed = 10.56 feet, the great cubit = 1.76 feet and 6 great cubits = 1 reed; 3000 great cubits = 1 statute mile which is 500 reeds.

In 1 second the original perfect equator turned 0.3168 Roman miles which is the circle's diameter, in 4 seconds it turned 1 Pyramid geographic mile this is the square's perimeter and equals 1 arc-minute of the equator's circumference. That is an interesting geometric relationship between 1 second of time on the equator and 1 arc minute of its original circumference (calculated via 864/275). The 0.3168 megalithic mile perimeter regular hexagon drawn tangent inside the circle is transformed into a three dimensional view of a transparent cube by the addition of the radial lines. Each side of the cube is the same size as the full design Pyramid's base therefore it is the smallest true cube that can contain the Great Pyramid's full design; we call it the Pyramid Cube.

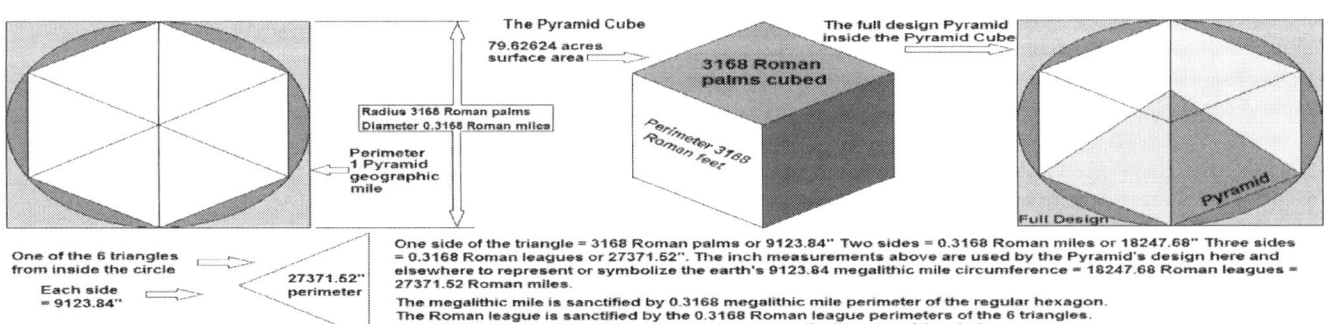

Every equation in the table below begins with 1 unit x 0.3168 then it is multiplied by a designated number of inches and the number of inches is always equal to the number of the selected units in the earth's circumference. Every equation produces 79.62624 acres which is the surface area of the Pyramid Cube.

	Linear Units of measure	Every equation produces the Pyramid Cube's surface area	The Earth's circumference
#1	Roman mile	0.3168 Roman miles x 27371.52" = 79.62624 acres	27371.52 Roman miles
#2	megalithic mile	0.3168 megalithic miles x 9123.84" = 79.62624 acres	9123.84 megalithic miles
#3	Roman league	0.3168 Roman leagues x 18247.68" = 79.62624 acres	18247.68 Roman leagues
#4	Statute mile	0.3168 statute miles x 24883.2" = 79.62624 acres	24883.2 miles
#5	Old English mile	0.3168 Old English miles x 26276.2592" = 79.62624 acres	26276.2592 Old English miles
#6	League (Imperial)	0.3168 leagues x 82924.4" = 79.62624 acres	8294.4 leagues
#7	Pyramid geographic mile	0.3168 Pyramid geographic miles x 21600" = 79.62624 acres	21600 Pyramid geographic miles
#8	Pyramid geographic league	0.3168 Pyramid geographic leagues x 7200" = 79.62624 acres	7200 Pyramid geographic leagues
#9	Furlong	0.3168 furlongs x 199065.6" = 79.62624 acres	199065.6 furlongs
#10	Rod	0.3168 rods x 796264" = 79.62624 acres	7962624 rods
#11	Reed	0.3168 reeds x 12441600" = 79.62624 acres	12441600 reeds
#12	Fathom	0.3168 fathoms x 21897.216" = 79.62624 acres	21897.216 fathoms
#13	Yard	0.3168 yards x 43794432" = 79.62624 acres	43794432 yards
#14	English cubit	0.3168 English cubits x 87588864" = 79.62624 acres	87588864 English cubits
#15	Foot (Imperial)	0.3168 feet x 131383296" = 79.62624 acres	131383296 feet
#16	Great cubit	0.3168 great cubits x 74649600" = 79.62624 acres	74649600 great cubits
#17	Roman foot	0.3168 Roman feet x 136857600" = 79.62624 acres	136857600 Roman feet
#18	Roman palm	0.3168 Roman palms x 547430400" = 79.62624 acres	547430400 Roman palms
#19	Roman inch	0.3168 Roman inches x 1654291200" = 79.62624 acres	1654291200 Roman inches
#20	Inch (Imperial)	0.3168 inches x 1576599552" = 79.62624 acres	1576599552 inches

The full design Pyramid's 0.0006336^3 mile volume focuses attention on the fact that 63360' = 1 mile, this is telling us to take note of the role of the inch in the Pyramid's design.

The inch is the star performer involved with every one of the equations above and it is very appropriate that 0.3168 inches multiplied by the earth's circumference produces the surface area of the Pyramid Cube; for example 24883.2 miles or 1576599552 inches x 0.3168 inches = the surface area of the Pyramid Cube.

	Units of measure	Surface area of the Pyramid Cube = 79.62624 acres	The Earth's circumference
#1	Roman mile	0.3168 Roman miles x 27371.52" = 79.62624 acres	27371.52 Roman miles
#2	Megalithic mile	0.3168 megalithic miles x 9123.84" = 79.62624 acres	9123.84 megalithic miles
#3	Roman league	0.3168 Roman leagues x 18247.68" = 79.62624 acres	18247.68 Roman leagues

The first 3 equations from the previous table

The rectangles below illustrate the three equations above which are from the previous list of 20 equations, each rectangle is formed by six 3168 Roman feet perimeter squares which are equal to the 6 sides of the Pyramid Cube, notice that the first rectangle represents both #1 and #3 equations.

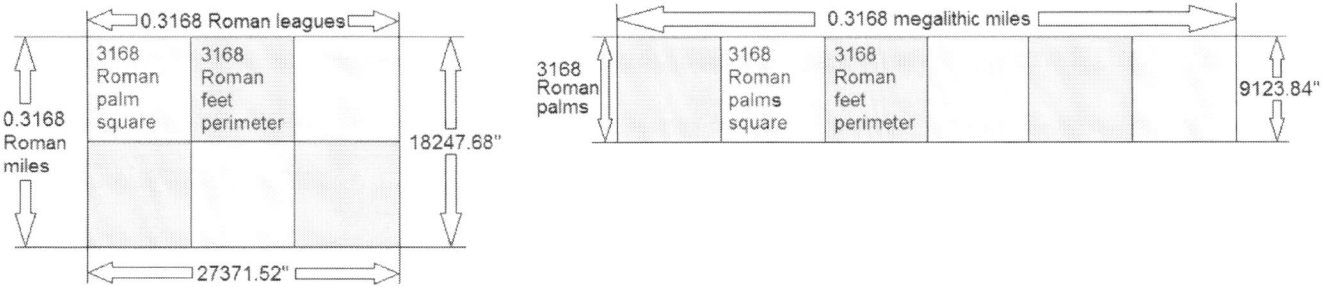

Weights and Measures and the number 3168, note that 3.168 million tons, 31.68 million hundredweight, 31.68 million quarters and 31.68 million stones are also features of the Pyramid's full design (shown opposite).

The circles, hemispheres, rings and quadrants below have been selected for attention by factors of the Pyramid's full design, we are also looking at 1/3 sectors of circles as suggested by the design of Silbury Hill, there are two series of circles below which overlap and interact.

The first series begins with Circle A and it's 31.68 English cubit circumference x 2 = 31.68 yards which is Circle B's circumference and we keep doubling the circumferences to produce 10 circles.

The second series of circles start with the 1 Pyramid geographic mile circumference circle and each following circle's circumference is increased by 1 Pyramid geographic mile until 10 circles are produced.

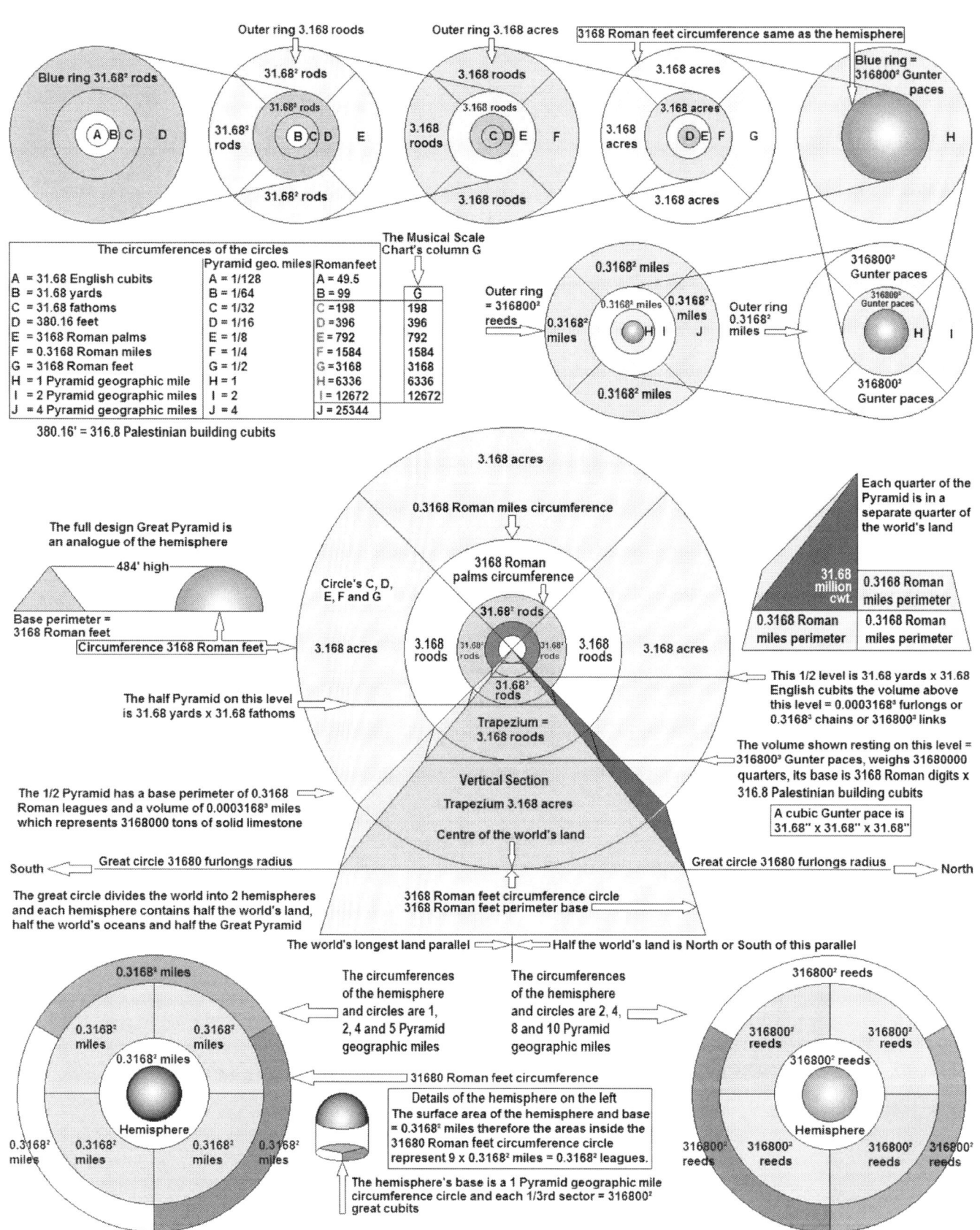

Silbury Hill's full cone design is 0.03168 Old English miles high, it is a model of a 1 league or 3.168 Old English miles high cone, the 1 league high cone is useful for observing how the longer units of linear measure relate to a tan 0.576 cone.

Below we examine how the 1, 2 and 3 mile diameter circles and the 1, 2 and 3 Pyramid geographic league circumference circles above relate to the design of the 1 league high cone.

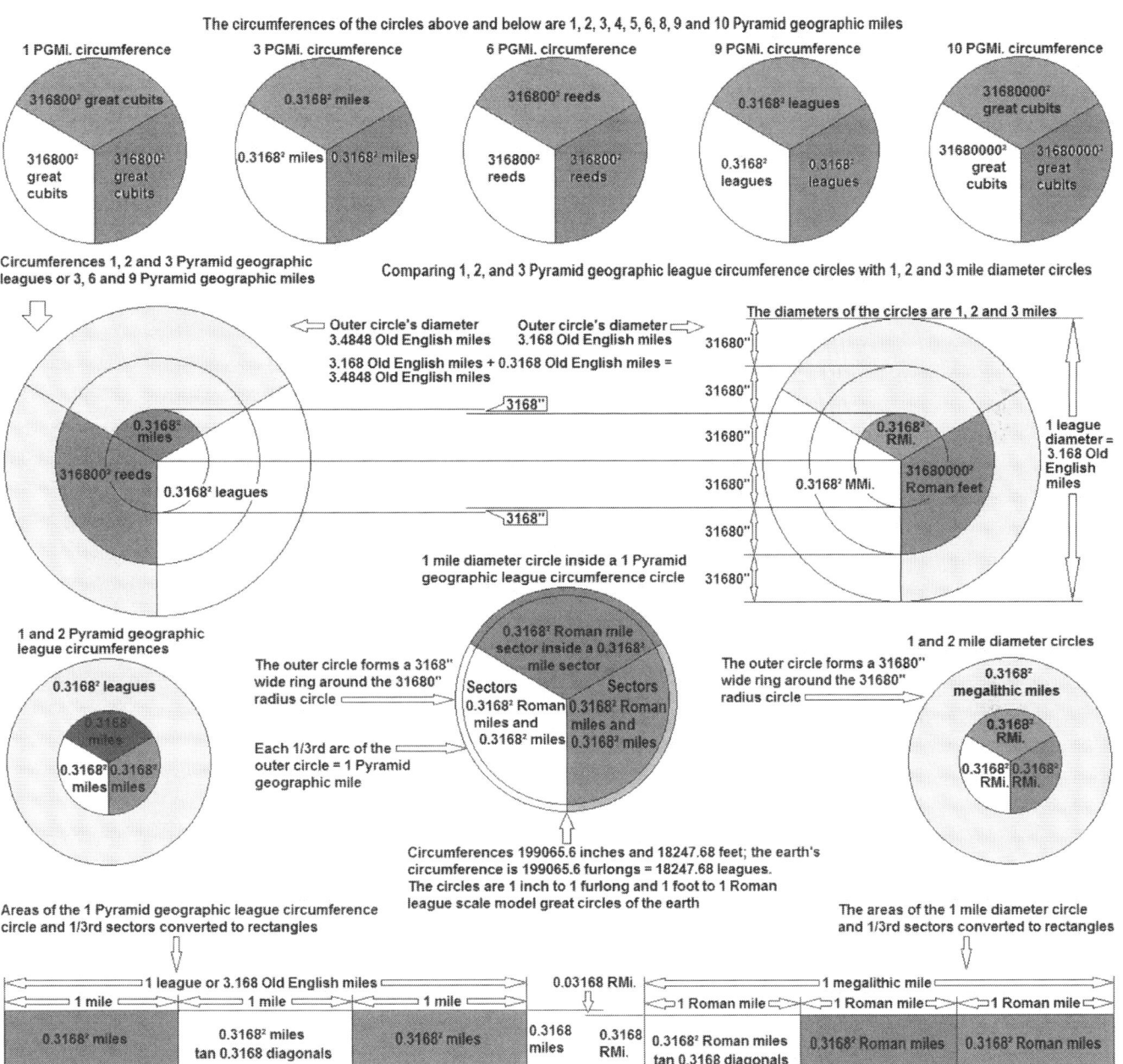

Silbury Hill is a model of a 1 league or 3.168 Old English mile high cone and details of the upper portion of the full cone full scale design are shown below. The illustration below focuses attention on the Pyramid geographic mile being the product of 4 x 0.3168 Roman miles or 2 x 3168 Roman feet or 0.3168 Roman miles + 0.3168 megalithic miles and also focuses attention on the fact that 31680 inches + 31680 inches = 1 mile. Those factors of the Pyramid geographic mile and the mile are also emphasized elsewhere by Silbury Hill's design and also by the Great Pyramid's full design.

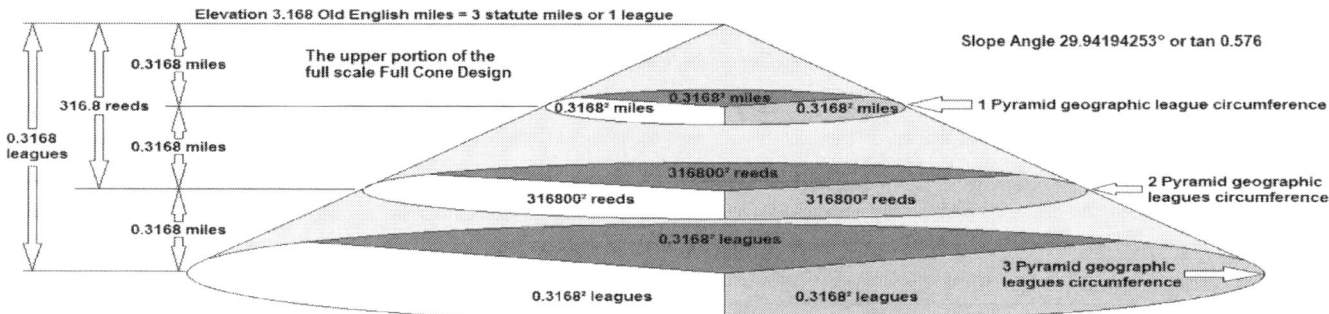

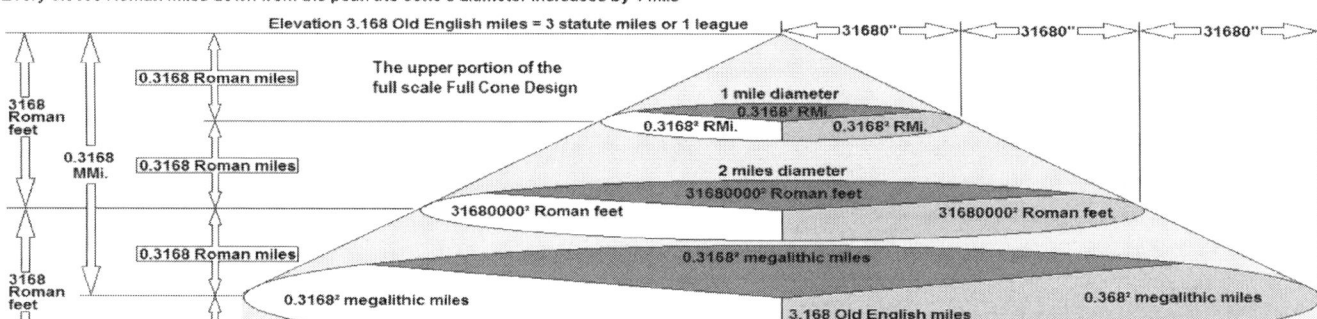

Converting the areas of the 1 mile diameter circle and 1/3 sectors into rectangles is in harmony with Silbury Hill's design in fact this is a designed function of the cone's design to demonstrate the geometric connections of various units measure from different systems to reveal that they have the same Source and the identity of the Source is also revealed.

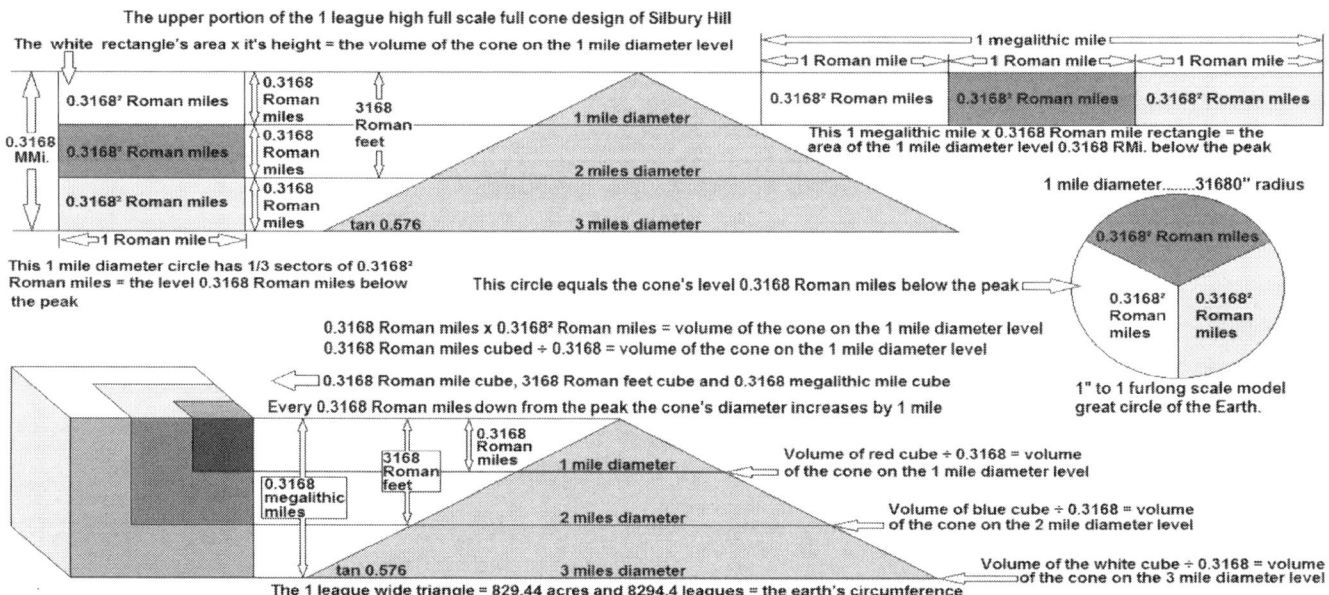

0.3168 RMi. x 0.3168 RMi. x 0.3168 RMi. = the red cube's volume ÷ 0.3168 = the cone's volume on the 1 mile diameter level 0.3168 Roman miles below the peak
3168 Roman feet x 3168 Roman feet x 3168 Roman feet = the blue cube's volume ÷ 0.3168 = the volume on the 2 mile diameter level 3168 Roman feet below the peak
0.3168 MMi. x 0.3168 MMi. x 0.3168 MMi. = the white cube's volume ÷ 0.3168 = the cone's volume on the 3 mile diameter level 0.3168 MMi. below the peak.

Geometric Geography.

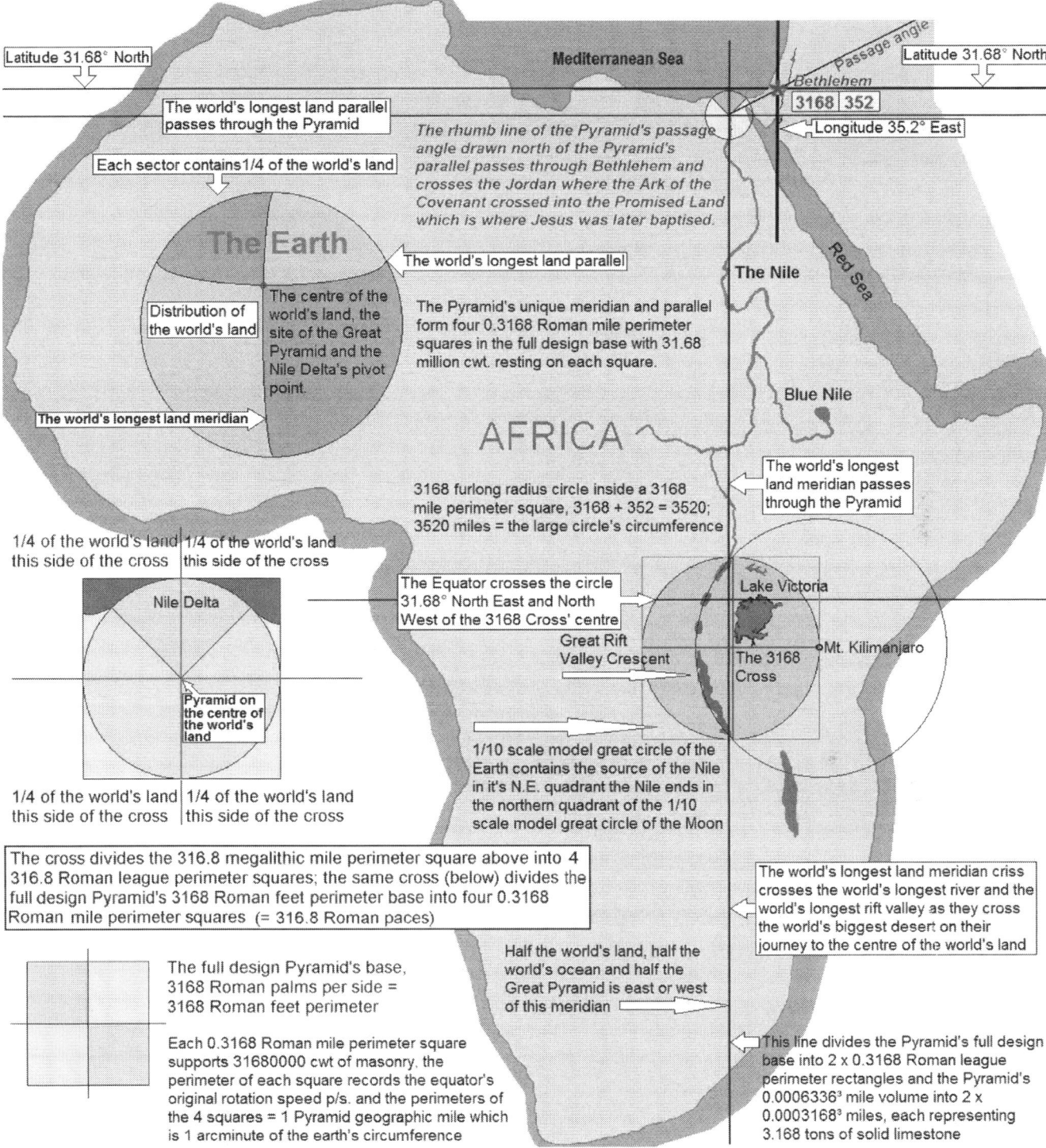

The source and terminus of the Nile River are contained inside quadrants of the 1/10th scale model great circles of the earth and the moon, the earth model like the moon model is based on regional geographic features and is aligned to several interacting geometric designs that are the subject of this and following pages. The Great Pyramid occupies the centre of the model great circle of the moon and is built on the pivot point of the Nile Delta quadrant which is the delta's geographic and geometric centre. This is one of the few places on the earth's surface that can support the Pyramid's great weight; this is also the centre of the world's land where the world's longest land meridian crosses the world's longest land parallel. This is where the world's longest river leaves the world's biggest desert and enters the world's most fertile land the Nile Delta which fills and defines the northern quadrant of the 1/10th scale model great circle of the moon, this has been featured in many books and magazines. Now the perfect partner for the moon model is revealed at the other extremity of the Nile which is the 1/10th scale model great circle of the earth that encloses Lake Victoria the source of the Nile in its north east quadrant.

The 3168 Cross is the governor and interpreter of the region's geometric geographic designs, the 3168 Cross supplies the exact orientation and measurements needed to interpret the geographic design, potential areas of confusion are eliminated or bypassed by calculating how the region's geometric features relate to the 3168 Cross. The 3168 Cross' lines reach 3168 furlongs North, South, East and West from where 3.168° south crosses 31.68° east, the Great Rift Valley Crescent starts and ends on 31.68° east and 3.168° south is the crescent's only line of symmetry. The 1/10 scale model great circle of the earth placed around the 3168 Cross perfectly encloses the Great Rift Valley Crescent and also perfectly encloses the western quadrant of the larger circle which occupies exactly half of the smaller circle's area. Lake Victoria the source of the Nile is contained in the north east quadrant of the smaller circle therefore the Nile's extremities are enclosed in 1/10 scale model great circles of the Earth and the Moon and each circle conforms to regional geographic features.

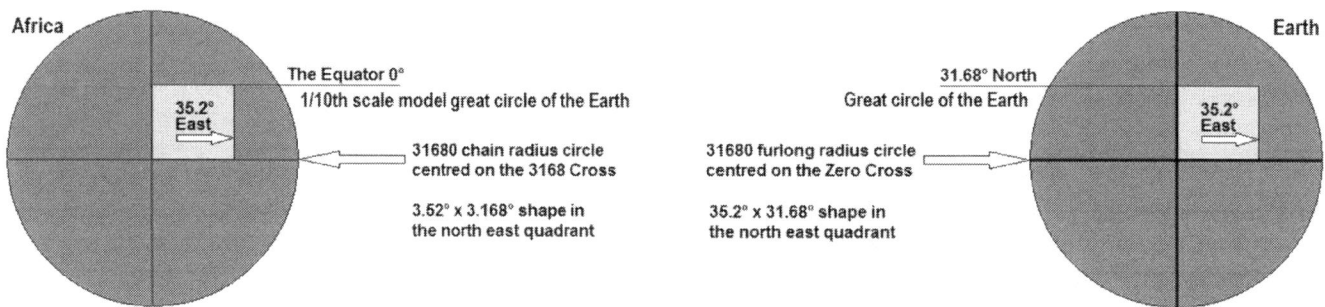

The 1/10th scale model of the large latitude//longitude design is formed inside the 1/10th scale model great circle of the earth by the 3168 Cross, the equator and longitude 35.2°.

To transform the 1/10th scale model into full scale we first change the coordinates of the 3168 Cross into the coordinates of the Zero Cross, then change the model's 31680 chain radius to 31680 furlongs radius and change the model's 0° parallel to latitude 31.68° North this completes the transformation. Bethlehem's longitude 35.2° East remains unchanged because it is a feature of both designs, this confirms that the model's prime meridian and prime parallel i.e. the lines of the 3168 Cross really do symbolize the earth's Prime Parallel and Equator. This is more powerful evidence that the Creator ordained where the Prime Meridian of the latitude longitude system would be located which is also supported by the numbers of the Musical Scale Chart.

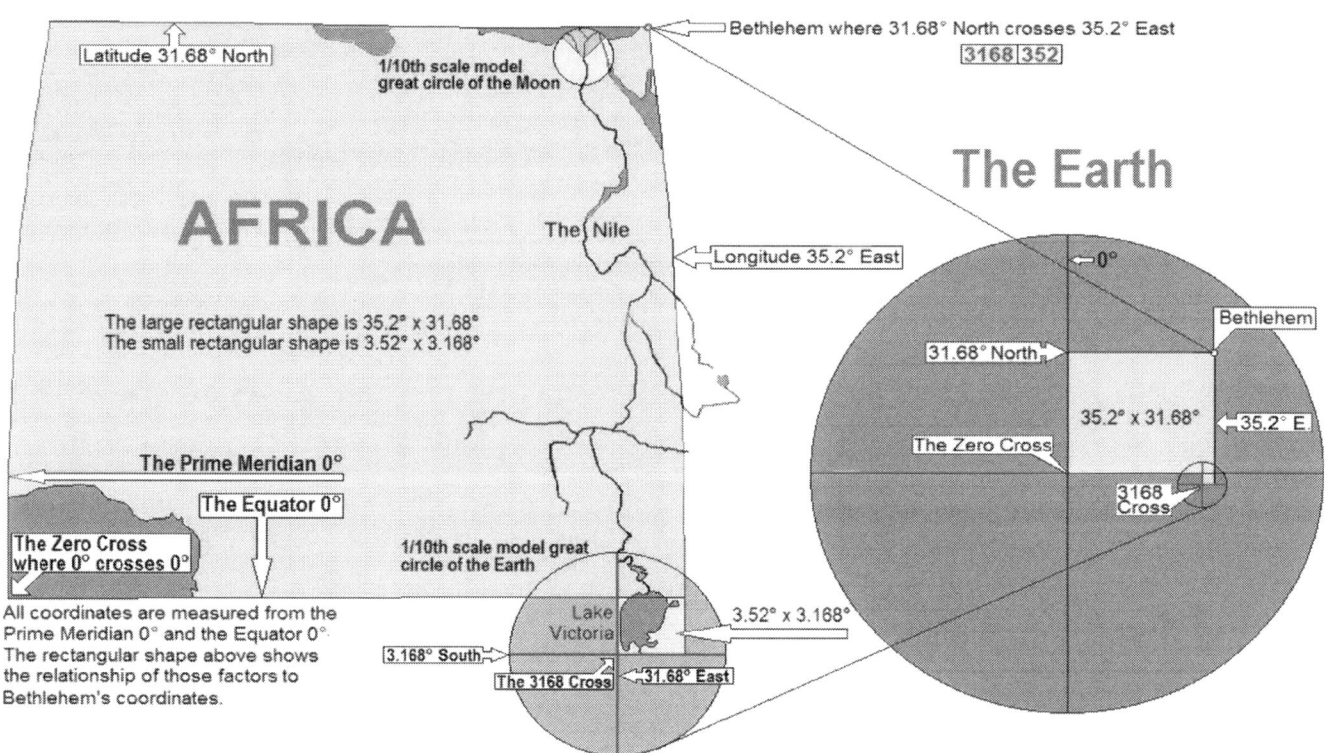

The yellow square below represents the 3168 Roman feet perimeter full design Pyramid's base.

The red lines of the cross inside the base = 0.3168 Roman miles and form four 0.3168 Roman mile perimeter squares in the base and remember the vertical section is an analogue of a 0.3168 Roman mile circumference circle which is a 1 inch to 1 Roman league scale model of a great circle of the earth.

The great circle distance from the Pyramid to where Jesus was baptised is appropriately 316.8 Roman miles (the product of 0.3168 Roman miles x 1000).

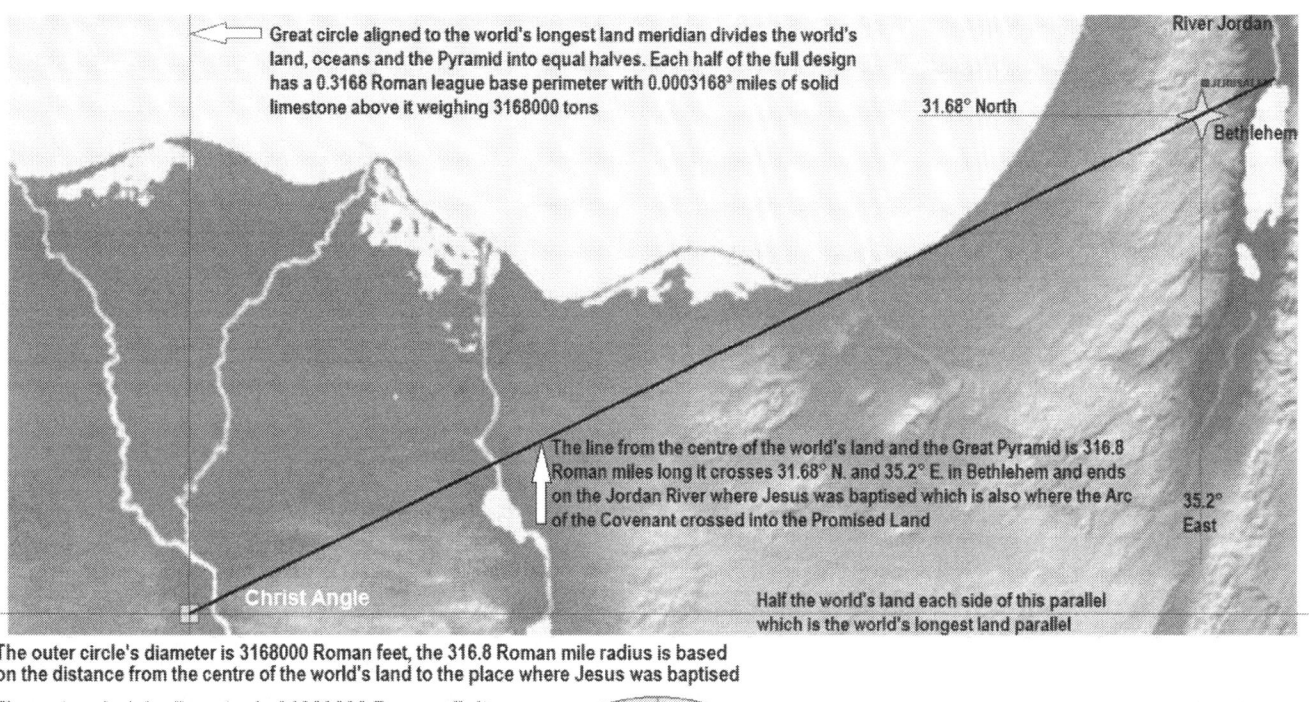

The outer circle's diameter is 3168000 Roman feet, the 316.8 Roman mile radius is based on the distance from the centre of the world's land to the place where Jesus was baptised

The outer circle's diameter is 31680000 Roman digits or 316800 fathoms more than the inner circle's diameter.

The ring between the circles is 316800 yards wide.

The 316.8 megalithic mile perimeter square placed around the Delta Circle is divided into four 316.8 Roman league perimeter squares by the unique factors of its global position.

The Nile Delta represents the northern quadrant of the inner circle which is a 1/10th scale model great circle of the moon.

Red lines on the world's longest land parallel and the world's longest land meridian mark the centre of the world's land.

Jesus baptised here

The Christ Angle rhumb line is the slope angle of the Pyramid's entrance passage, the line passes through Bethlehem and ends where Jesus was baptised.

Jabal al Lawz the real Mount Sinai

The distance from the Pyramid to Bethlehem is the same as the distance from the Pyramid to the real Mount Sinai.

The lines of the 3168 Cross extend 3168 furlongs north, south, east and west from where latitude 3.168°south crosses longitude 31.68°east

Africa's two largest lakes in the headwaters of Africa's two longest rivers are in opposite quadrants of the circle around the 3168 Cross

Portion of the Music Chart

3168	352	3520

3168 mile + 352 miles = 3520 miles, 3168 miles = the perimeter of the square below and 3520 miles = the larger circle's circumference.

The 2 lines of the 3168 Cross = 1584 miles or 12672 furlongs 1 line = 792 miles or 6336 furlongs. The lines divide each other into 3168 furlong lengths or 396 miles = 198000 reeds, we have just used all the numbers in the Musical Scale Chart's column G.

G	Column G from the Musical Scale Chart
198	
396	396 miles = radius of smaller circle
792	792 miles = small circle's diameter = 1 side of square
1584	1584 miles = the lines of the 3168 Cross
3168	3168 miles = perimeter of the square (blue lines)
6336	6336 miles = diameter of circle that encloses the large
12672	latitude longitude design (not shown on this page)

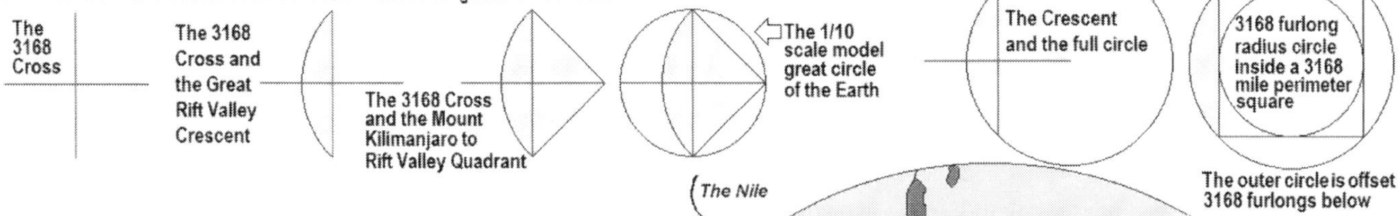

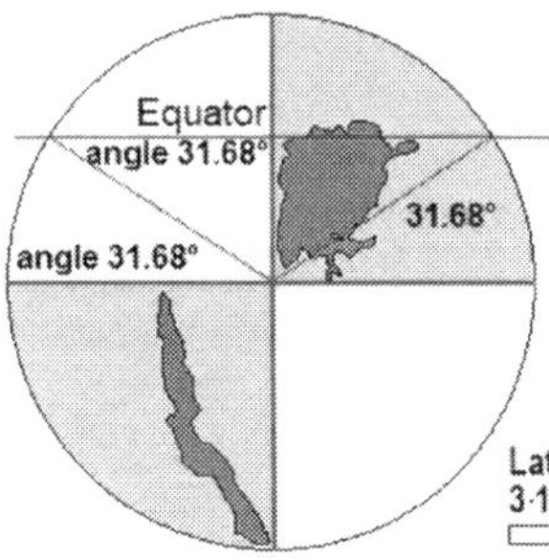

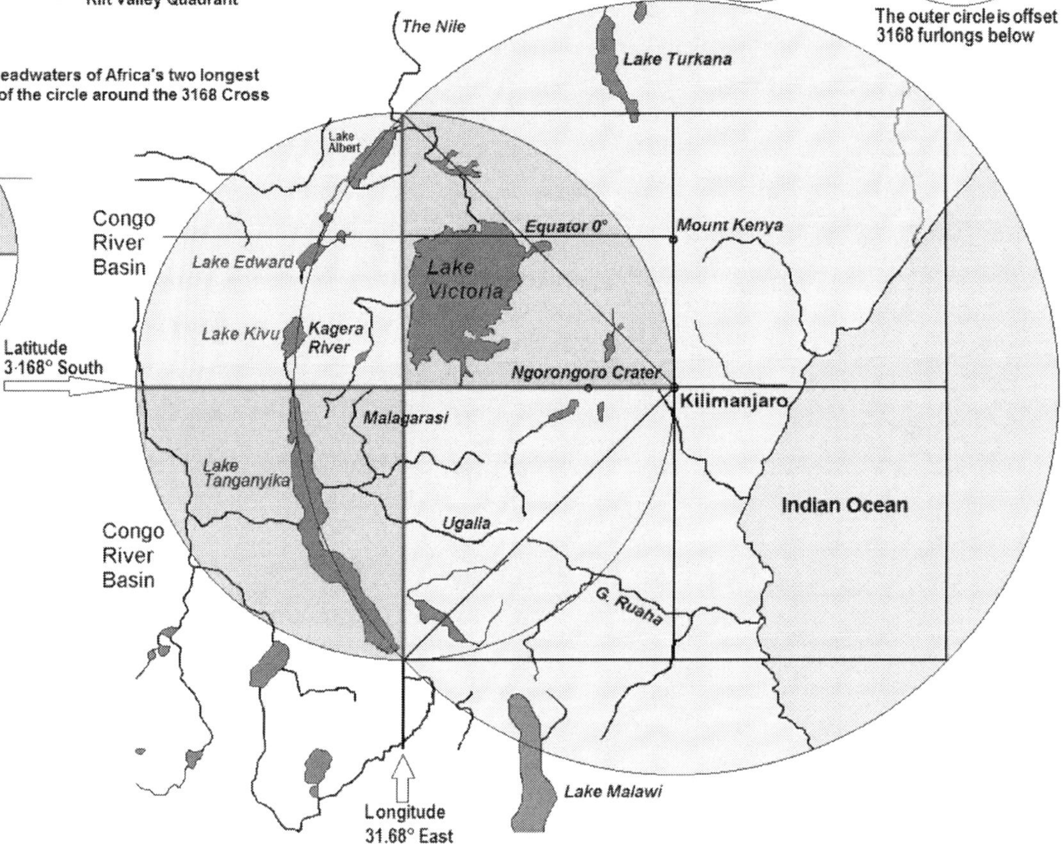

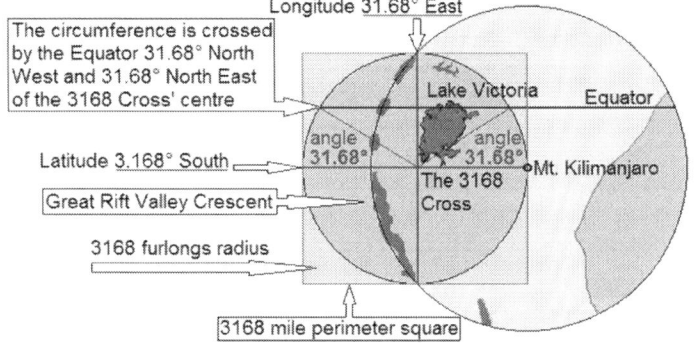

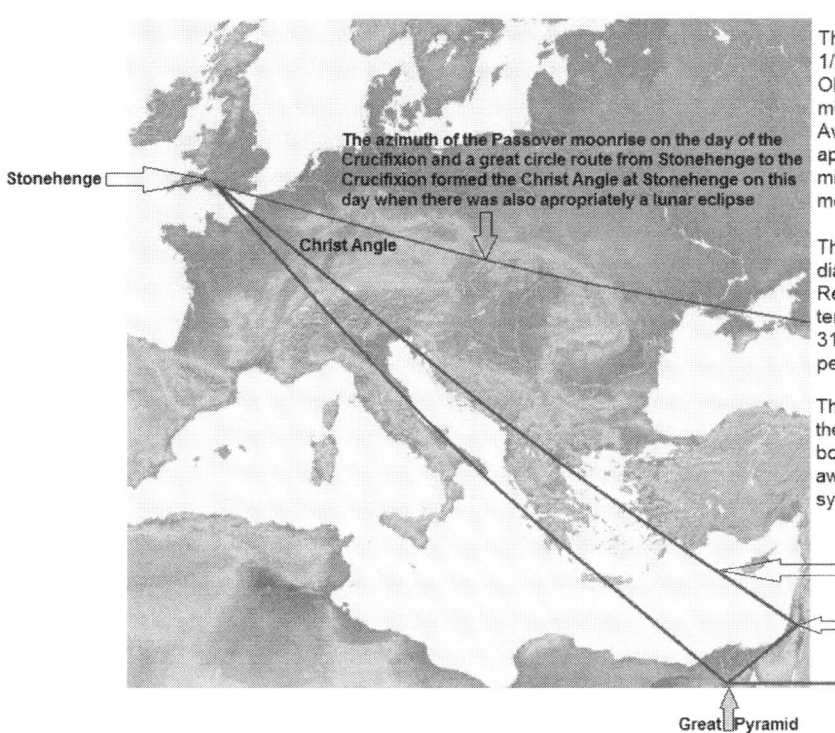

The great circle route on azimuth 316.8° from north starts 1/2 way between Bethlehem and Jerusalem and is 3.168 Old English miles from either point. The terminus is the midpoint between the centres of Stonehenge and the Avebury monument, the centres are 6.336 megalithic miles apart which records the equatorial rotation speed per minute. Therefore the midpoint terminus is 3.168 megalithic miles from the centre of each monument.

The azimuth of other great circle route is aligned to one diagonal of Stonehenge's 3168" long Station Stone Rectangle which has a 3168 Roman palm perimeter. The terminus is the Great Pyramid and the full design base is a 3168 Roman palm square with a 3168 Roman feet perimeter.

The Christ Angle is produced by drawing a rhumb line from the Pyramid to (or through) Bethlehem where Jesus was born on latitude 31.68° North. The Crucifixion was 31680' away outside the old city of Jerusalem. Therefore 31680' symbolizes the life of Jesus from birth to the Crucifixion.

Great circle route on azimuth 316.8° from North

Christ Angle rhumb line from the Pyramid to Bethlehem on 31.68° N. The Crucifixion was 31680' away outside Jerusalem; 1/2 way was 3.168 Old English miles from both places and from there to the English monuments Centre Point is on azimuth 316.8° from North.

Latitude 31.68° North crosses longitude 35.2° East in Bethlehem, the latitude longitude design below is based on 31.68° and 35.2° of latitude and longitude and it is anchored to the Bethlehem coordinates, 31.68° + 35.2° = 66.88° the design is 66.88° x 66.88° and is placed between the latitudes 31.68° North and 35.2° South the longitudes 31.68° E and 35.2° E. are features of the design.

The Western border is fixed to the Prime Meridian 0°, the Equator 0° is also a design feature.

The design is the product of four overlapping 35.2° squares and no other lines were needed to complete the design, it was placed inside a 3168 mile radius circle (on the 31680 furlong radius Earth) simply because it fits.

Notice that overlapping the four 35.2° squares created a 31.68° squares in each corner of the design, the borders of the North West 31.68° square are on latitude 31.68° North and longitude 31.68° East, the 2 other borders are on the Prime Meridian 0° and the prime parallel 0° called the Equator.

Because the design is on the Earth's surface it contains no true squares or rectangles but we call them that for the sake of convenience.

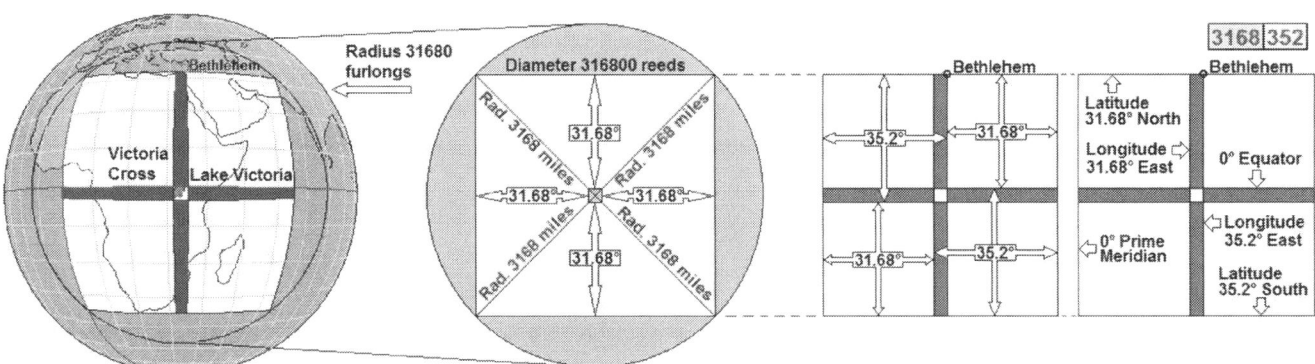

The latitude longitude design focuses attention on Lake Victoria which appears to be displaced, and in a way it is, but if it was placed inside the 3.52° centre square of the Victoria Cross we would not have the amazing Displacement Design shown below.

We can't move Lake Victoria further South without moving it out of the North East quadrant of the 1/10 scale model great circle of the Earth around the 3168 Cross.

If we moved the pink 3.52° square containing Lake Victoria into alignment with the other 3.52° square resting on 3.52° South then the lake would be where it seems to belong, at the centre of the Victoria Cross but it's displacement creates the Displacement Design which is a 1/10 scale model of the North West portion of the large latitude longitude design and even occupies the same relative position in the 1/10 scale model great circle of the Earth which is illustrated below.

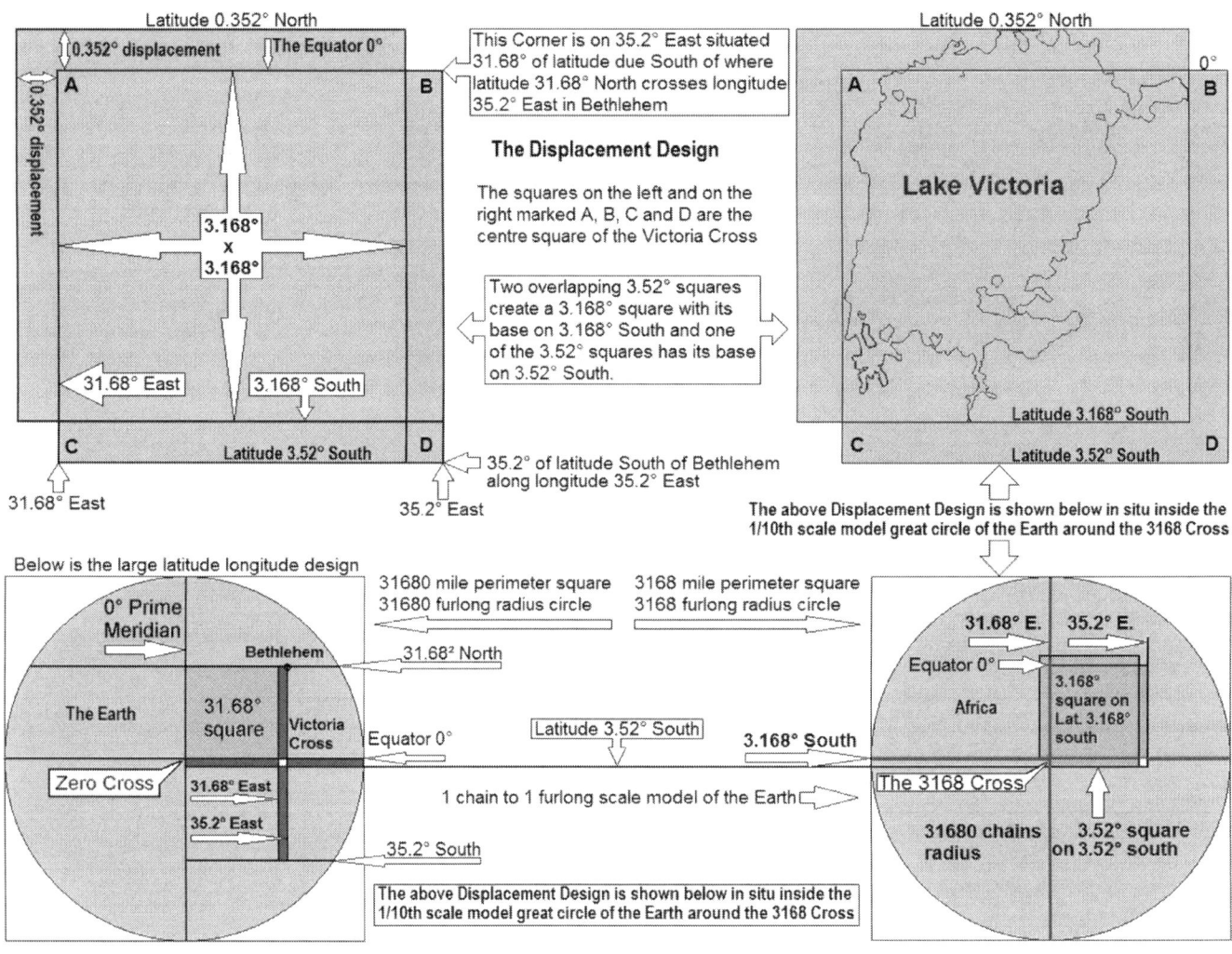

At the centre of the Victoria Cross is a 3.52° square resting on latitude 3.52° south which is 35.2° of latitude south of Bethlehem, the east side is on longitude 35.2° east, the centre of the square is 35.2 Pyramid geographic leagues south of the equator 0° which is the square's northern border which is 35.2% of a great circle's circumference south of Bethlehem. The square's North West corner is 35.2% of the equator's circumference east of the Prime Meridian 0°.

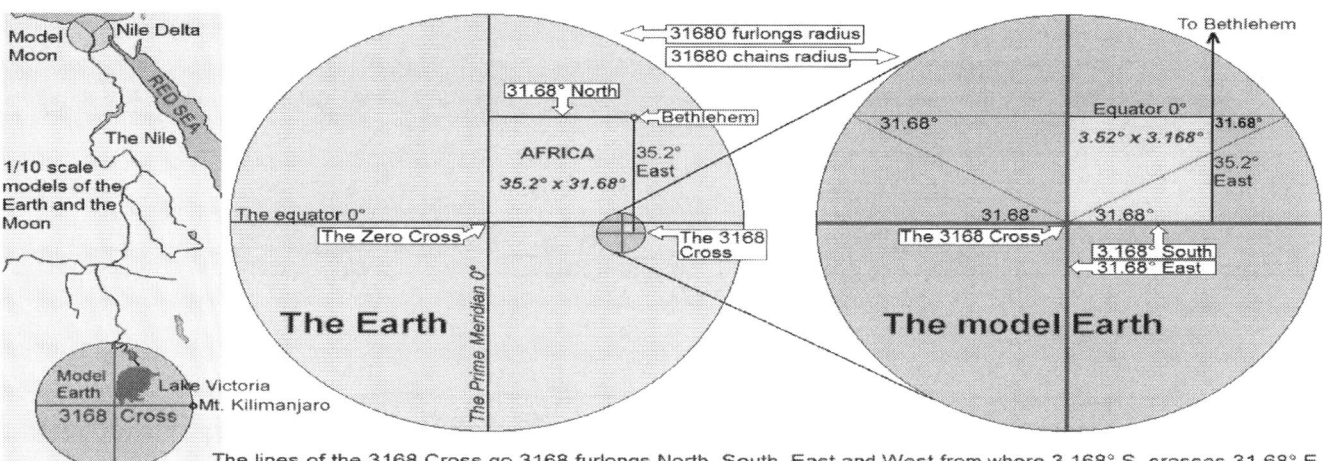

The message embedded in the units of measure is the same as the message embedded in the map of Africa and cosmic design therefore it is appropriate to use these preselected units of measure to reveal the message embedded in the map of Africa. I think you can read the message without my help but note that the 3168 signature is like the signature on a letter or a painting claiming this is my work or I wrote this. Therefore the claim is made via verifiable supernatural factors that the earth etc. is designed and has a message embedded in the design that is addressed to anyone and everyone who is interested in the highest level of truth, we are not the result of unintelligent blind undirected forces creating intelligent life out of dead materials, we are here for a reason so we should not be too busy being here to spare a thought about why we are here. Would you like to know if the message is addressed to you and you're not just reading someone else's mail? If the message is not addressed to a person it will repel them and they will reject it, unaware that the repelling force of the message actually rejected them first, because they have not received the love of truth they flee from it and take refuge in lies (a strong delusion). Obviously the message is not new for many have already received the same message delivered by other means from the same Source.

The message of the earth, the moon and the map of Africa cannot be read via the metric system, one has to use the appropriate units of measure. The squares and circles based on units of linear measure reveal independently the same message that is embedded in the cosmic design and the map of Africa but more power is added by showing they are all interacting factors of one great design.

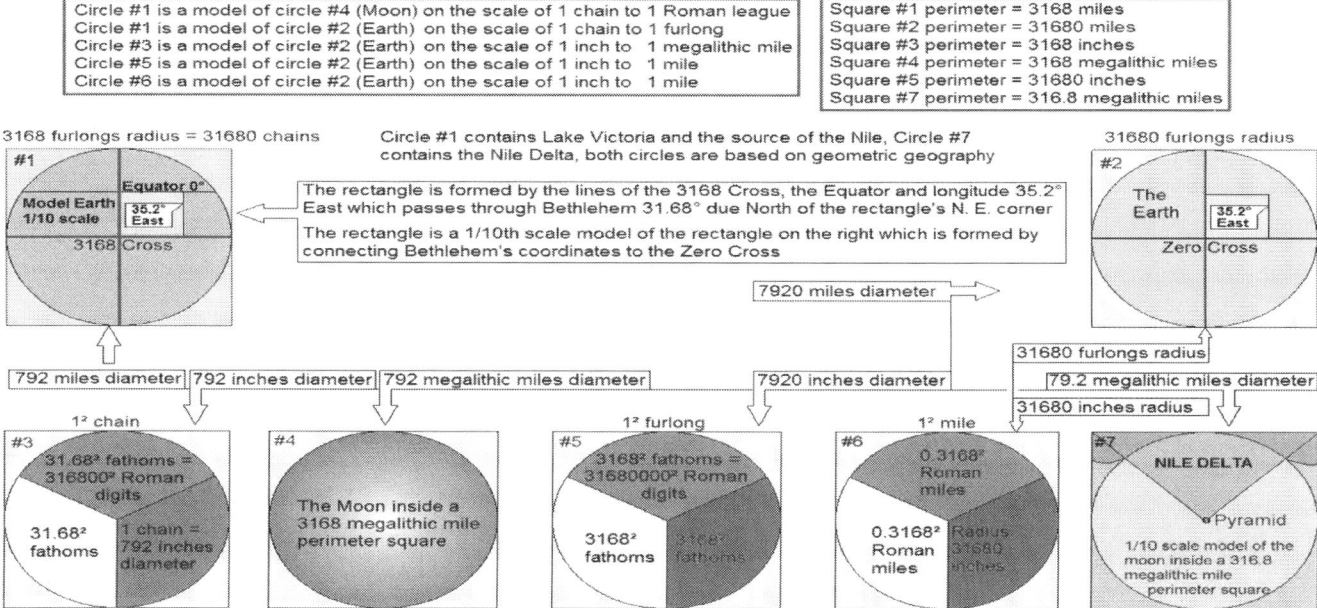

The calculations below measure the great circle diagonal distances between the corners of the latitude longitude design and the answers were supplied by the American Government's FCC site, they say the diagonals are 6335.057 miles, we want 6336 miles so this is very close, the reason it is not exactly 6336 miles is probably because they're using a variable measure for the earth's radius or some other method to compensate for the earth not being a perfect sphere.

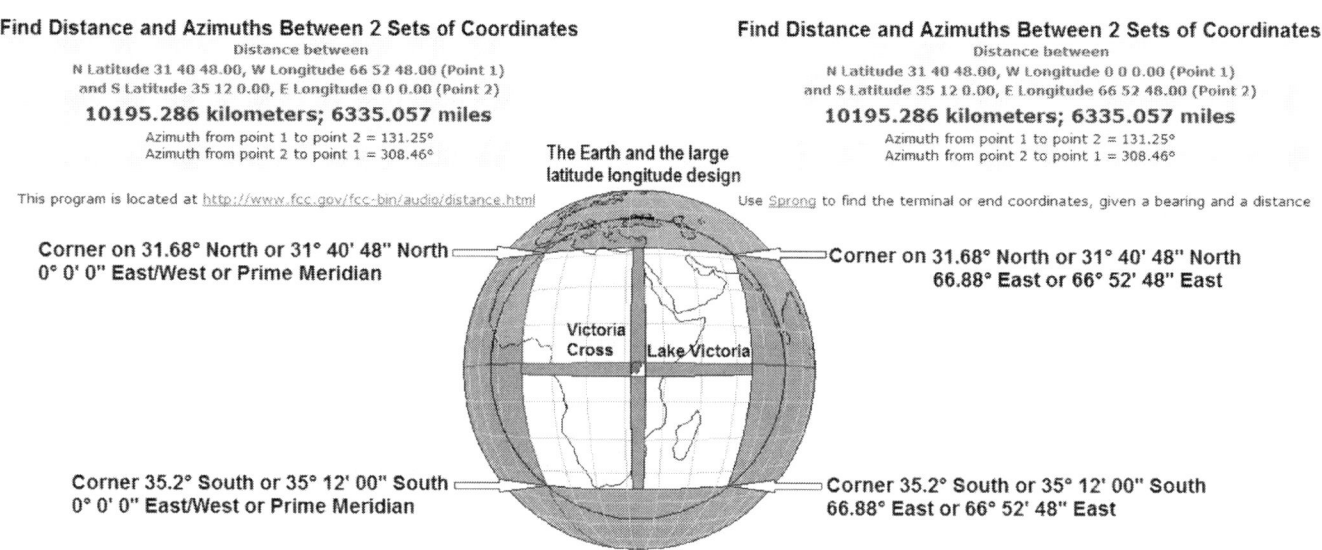

The 6336 mile North South diameter of the circle reaches from 44.07333335° North to 47.59333335° South and if you use the FCC site it says this is 6332.705 miles but it is 91.6666667° the same as the great circle distance of the diagonals and would have the same mileage if they weren't using a variable radius. Note that the above information informs us that although the FCC calculations are good they are not exact concerning the Great Design which has factors based on the earth's original perfect condition.

Bethlehem's latitude 31.68° north and longitude 35.2° east coordinates where Jesus was born are a strategic global position and the key to revealing the geographic designs shown in following pages. The message encoded in the numbers of the Musical Scale Chart reveals predetermined cosmic design, the cosmos did not create itself spontaneously out of nothing, the Earth is not a piece of debris flung out from the scene of an accident caused by nothing exploding into everything. The earth was created to designated predetermined measurements that are recorded in the numbers of music, we will see that the Earth's dimensions were designed and the numbers of music bear witness of this. The Bethlehem coordinates on the Musical Scale Chart show that the Earth was not only designed to predetermined dimensions but it's surface was also charted and the Greenwich Prime Meridian was predestined.

The Musical Scale Chart

The natural tuning called Just Musical Scale Chart, middle A was determined to be 440 Hz. (Hertz) by international agreement in 1955

	C		D		E		F		G		A		B	
1 below	132	16.5	148.5	16.5	165	11	176	22	198	22	220	27.5	247.5	16.5
Middle	264	33	297	33	330	22	352	44	396	44	440	55	495	33
1 above	528	66	594	66	660	44	704	88	792	88	880	110	990	66
2 above	1056	132	1188	132	1320	88	1408	176	1584	176	1760	220	1980	132
3 above	2112	264	2376	264	2640	176	2816	352	3168	352	3520	440	3960	264
4 above	4224	528	4752	528	5280	352	5632	704	6336	704	7040	880	7920	528
5 above	8448	1056	9504	1056	10560	704	11264	1408	12672	1408	14080	1760	15840	1056

The conversion charts are incomplete because all the numbers except the 2 blue numbers are taken from the Musical Scale Chart, rows 7 and 11 were left out because they are blank which is a common phenomena when using the numbers of music with units of measure yet in other ways the numbers 7 and 11 dominate the chart, their are 7 rows of 7 notes and 7 columns of 7 intervals and every number is a multiple of 11.

Conversion Charts

	Reeds				Great Cubits		
	FEET	YARDS	INCHES		INCHES	FEET	YARDS
1	10.56	3.52	126.72		21.12	1.76	
2	21.12	7.04			42.24	3.52	
3	31.68	10.56			63.36	52.8	1.76
4	42.24	14.08	633.6		84.48	7.04	
5	52.8	17.6			105.6	8.8	
6	63.36	21.12			126.72	10.56	3.52
8	84.48	28.16				14.08	
9	95.04	31.68	1140.48		190.08	15.84	5.28
10	105.6	35.2	1267.2		211.2	17.6	
12	126.72	42.24				21.12	7.04

Bethlehem's 35.2° longitude = 2112' or 126720" and suggested making the conversion charts because the reed is 3.52 yards or 126.72 inches and the great cubit is 21.12 inches. Note that six numbers in rows 9 and 10 relate to the Bethlehem coordinates stated in degrees, minutes or seconds.

31.68° = 1900.8 minutes = 114048 seconds
35.2° = 2112 minutes = 126720 seconds

352	3168	Gematria
		His Name = 352 (Hebrew)
		Lord Jesus Christ = 3168 (Greek)

Two factors dictate the numbers on the Musical Scale Chart which are the Laws of Musical Harmony and the length of the second.

If for any reason the second was changed then the numbers on the chart would also have to be changed pro rata to conform to the Laws of Musical Harmony.

They say the second was invented some five thousand years ago by the Sumerians but the Musical Scale Chart informs us who the Source is.

The second is also a design factor of the English monuments and the Pyramid.

In the Bible's Hebrew text His Name has a Gematria of 352 and in the Greek text The Way has a Gematria of 352 so the red numbers on the Musical Scale Chart translated via Gematria say His Name, Lord Jesus Christ, The Way.

Christianity was called The Way in the time of the apostles and was first called Christianity in Antioch.

The international agreement in 1955 defining the note middle A as 440 Hertz would not have happened if the second was not used universally.

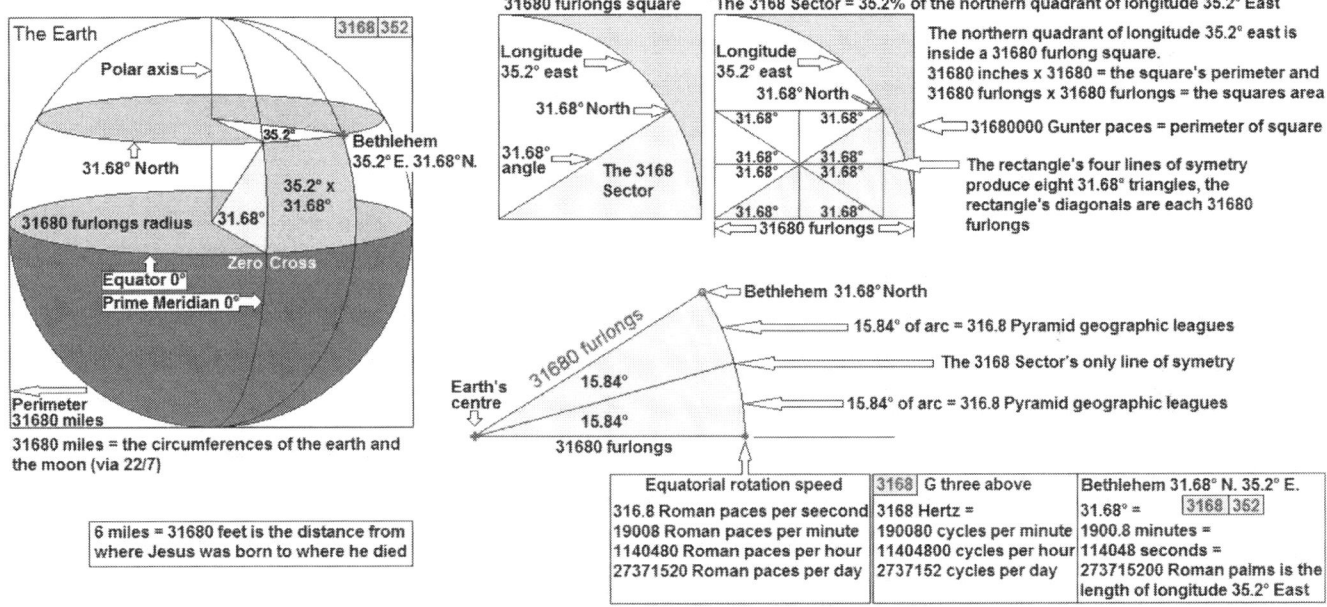

The secular world has tried and failed to explain the reason why music exists but the explanation is simple, music is a gift from God, the numbers on the Musical Scale Chart bear witness of this.

This shows that when looking for the answers no portion of the big picture should be designated taboo, but secular experts can wreck their careers by finding the right answers is the wrong places unless they keep silent.

The Bethlehem coordinates can be translated via Gematria to say, His Name, Lord Jesus Christ or alternatively His Name = 3168.

Anthony Morris

This line is on azimuth 316.8° from North, it begins 1/2 way between Bethlehem and Jerusalem which are 31680 feet apart so the starting point is 3.168 Old English miles from both places, the line terminates 3.168 megalithic miles from the centres of Stonehenge and the Avebury monument. The line starts and ends in the centre of two geometric designs which are shown on other pages.

Stonehenge
Bethlehem
Pyramid
Arc of the 3168 Sector
31.68° Rectangle (centre)

Great circle of the earth aligned to longitude 35.2° East

Polar Axis
Bethlehem
31680 furlongs
31.68°
31.68°
The 3168 Sector
Latitude 31.68° North
Arc = 316.8 Pyramid geographic leagues
Arc = 316.8 Pyramid geographic leagues
Equator 0°
Centre of the Earth
31680 furlongs
31680 furlongs

31680000 Gunter paces = the 2 lines of the cross which form four 31680 furlong squares with 31680000 Gunter pace perimeters inside the 31680 mile perimeter square around the great circle of the earth

The line from Stonehenge to the Pyramid is on a great circle route and is aligned to one diagonal of Stonehenge's Station Stone Rectangle

Rhumb line = the Christ Angle
Bethlehem and 31.68° North
The Pyramid in the centre of the 1/10th scale model great circle of the Moon

Diameter 31.68°
Radius 316.8 Pyramid geographic leagues

Red line = 31.68° of latitude and is the arc of the 3168 Sector

31.68° long line from the Pyramid to the centre of the 31.68° Rectangle

Below is the 1/10th scale model of the earth which contains the source of the Nile, the 31.68° Rectangle is shown in situ notice that there is a 1/10th scale model of the 3168 Sector. The equator crosses the circle's circumference at an angle of 31.68° East and West of the 3168 Cross' centre.

Longitude 31.68° East
The Equator 0°
31.68° angle
31.68°
31.68°
31.68°
31.68°
31.68°
31.68°
31.68°
31.68°
31.68 Pyramid geographic leagues
31.68 Pyramid geographic leagues
31.68° angle
Latitude 3.168° South
3168 furlong radius circle around the 3168 Cross

Red line = 31.68° and is the arc of the 3168 Sector

The Equator 0°

Lake Victoria and the Great Rift Valley Crescent inside the 3168 furlong radius circle around the 3168 Cross

The Displacement Design and the 31.68° Rectangle share lines with each other and the 3168 Cross

It is called the 31.68° Rectangle because it contains eight 31.68° triangles and one side of the rectangle is on longitude 31.68° East

The 31.68° Rectangle
Longitude 31.68° East
Equator 0°
31.68 Pyramid geographic leagues
31.68°
31.68°
31.68°
31.68°
3.168° of latitude
31.68 Pyramid geographic leagues
31.68°
31.68°
31.68°
31.68°
Latitude 3.168° South
Two 3168 furlong diagonals

From the Great Pyramid to the centre of the rectangle is 31.68° on a great circle, from the Northern border of the rectangle to Bethlehem is 31.68° of latitude, the eastern border is on longitude 31.68° East. The rectangle contains eight 31.68° triangles which is why it is called the 31.68° Rectangle.

31.68° South of Bethlehem on the same longitude

Below shows how the 31.68° Rectangle shares lines with the Displacement Design, note that the 3.168° lavender square resting on latitude 3.168° South shares 3 lines with the 31.68° Rectangle and 2 of these lines are also shared with the 3168 Cross

The blue 3.52° square rests on latitude 3.52° South

The blue square is the centre of the Victoria Cross

The perimeters of the 2 blue squares of the Displacement Design overlap and create the 3.168° lavender square on 3.168° South

35.2° South of Bethlehem on longitude 35.2° East

Blue 3.52° square on latitudes 3.168° South

Lavender 3.168° square on latitude 3.168° South

It is interesting that the Ngorongoro Crater is on the same line of longitude as the Sea of Galillee. They are 36° of latitude from centre to centre which is 1/10 of the earth's circumference or 2488.32 miles (= 9123.84 MMi. = 2737.152 RMi. etc.).

Both places are in the Great Rift Valley and the Sea of Galillee almost fits inside the crater's floor and would fit inside the rim.

I don't know what this means but it has something to the do with the 1/10 scale model great circle of the earth around the 3168 Cross.

Because the 1/10 scale model earth around the 3168 Cross is on the surface of a globe the diameter is correct but the circumference is not 2488.32 miles so the 2488.32 miles from the centre of the Ngorongoro Crater to the Sea of Galilee on the east west line of the 3168 Cross might be a compensation factor.

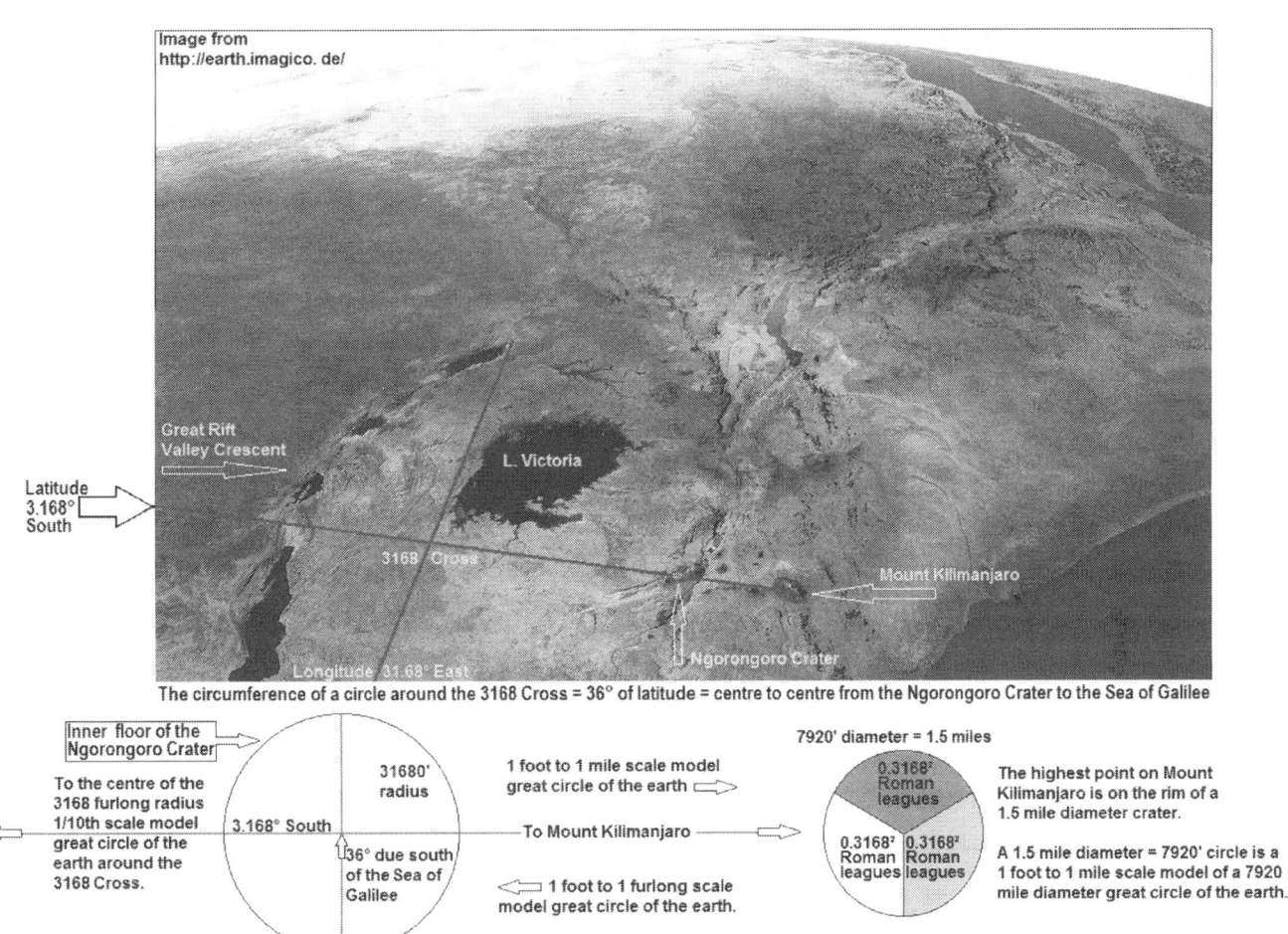

The amazing symbolism and proportions of a 1 mile diameter sphere

The 1 mile diameter sphere is an inch to the furlong scale model of the earth

The Sphere
31680 inch radius 63360 inch diameter
The Earth
31680 furlongs radius 63360 furlongs diameter

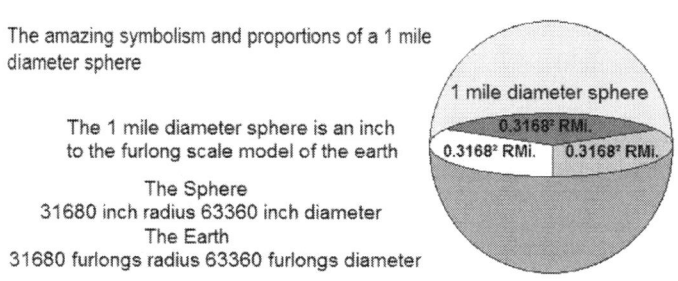

1 Pyramid geographic mile x 1 megalithic mile = the surface area of the 1 mile diameter sphere

The sphere's equator is a 1 mile diameter circle that divides it into north and south hemispheres the area of each hemisphere is the product of 1 Pyramid geographic mile x 1 Roman league

1 megalithic mile x 0.3168 Roman miles (the earth's equatorial rotation speed per second) = the 1 mile diameter circle's area and each 1/3rd sector's area is the product of 1 Roman mile x 0.3168 Roman miles

Therefore the earth and the 1 mile diameter sphere feature the Prime Measure of this investigation which is 0.3168 Roman miles

The ancient 864/275 method for pi is a major factor in the Great Pyramid's full design and is also a factor in the formatting of the Pyramid geographic mile.

The earth's 7920 mile diameter x 864/275 = 24883.2 miles or 21600 Pyramid geographic miles, there are 21600 minutes in a circle so 1 Pyramid geographic mile = 1 minute of the earth's circumference.

Therefore it is significant that 1 Pyramid geographic mile x 1 megalithic mile = $864/275^2$ miles which is the surface area of the 864/275 mile circumference sphere or in other words the 1 mile diameter sphere above.

The rectangles below demonstrate the remarkable commensurate factors of the 1 mile diameter sphere.

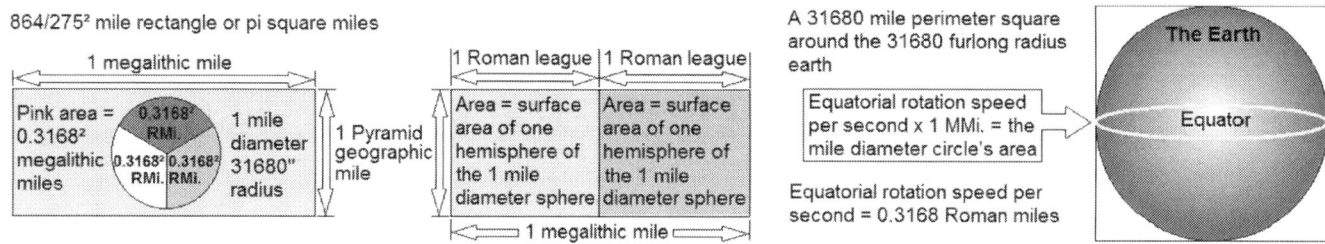

The 2 mile diameter circle below has an area of $864/275^2$ miles therefore we could say that it represents the surface area of the 1 mile diameter sphere in circular 2 dimensional format and the outer rectangles above and below represent the same in rectangular format.

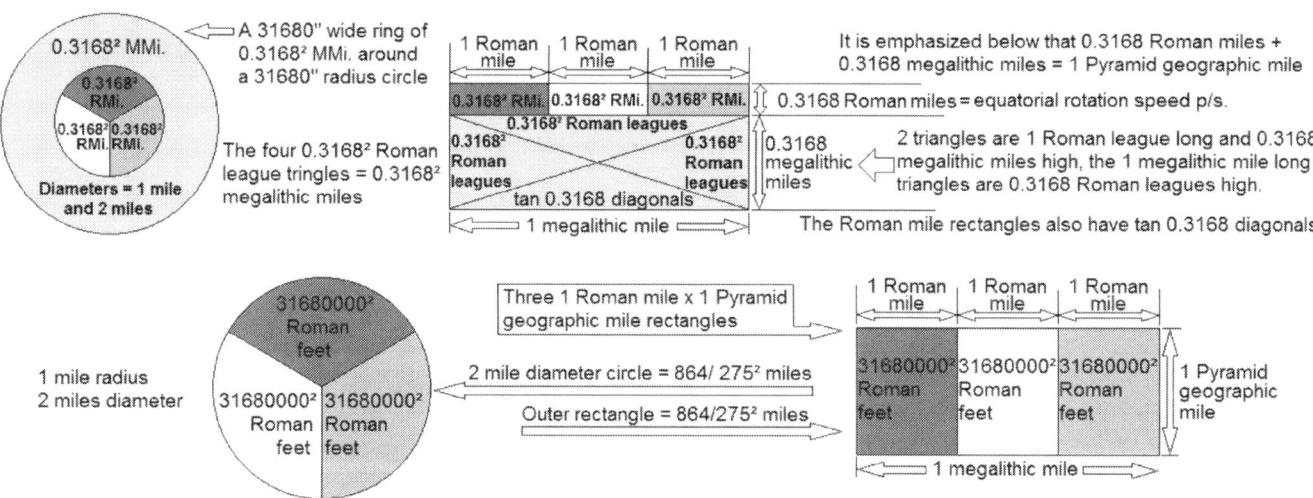

The tan 0.3168 diagonals inside the column of 0.3168² Roman mile rectangles cross each other 9123.84" or 3168 Roman palms from each rectangle's top and bottom which gives us the interesting set of measurements to the right of the column.

For example the Roman measurements = 1, 2, 3 and 4 sides of the full design Great Pyramid's base.

The inch measurements i.e. 9123.84", 18247.68" and 27371.52" are used by the designs of the monuments to represent the earth's 9123.84 megalithic mile circumference = 18247.68 Roman leagues or 27371.52 Roman miles.

This was the size of the earth's equator before the bulge appeared and turning once a day = 18247.68" or 0.3168 Roman miles per second.

The measurements below also emphasize that the Pyramid geographic mile is the product of 3168 Roman feet + 3168 Roman feet = 6336 Roman feet = 1 Pyramid geographic mile. Interesting because they are features of the 31680" radius sphere = 63360" diameter = 1 mile.

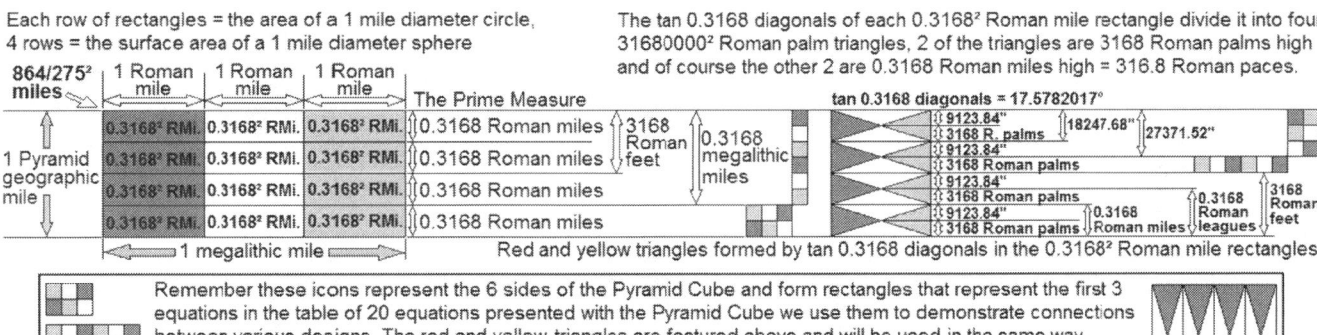

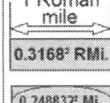
One 0.3168² Roman mile rectangle fits tangent around a 0.248832² Roman mile ellipse and symbolizes a 31680 mile perimeter square fitted tangent around the earth's 24883.2 mile circumference, but really refers to the equator before the bulge appeared. The 6336 Roman palm width of the ellipse refers to the earth's 63360 furlong diameter, 6336 Roman palms = 18247.68" which simultaneously refers to the 18247.68" per second rotation speed when the earth had a 18247.68 Roman league circumference.

864/275² statute miles + 0.03168² Roman miles = 3.168² statute miles

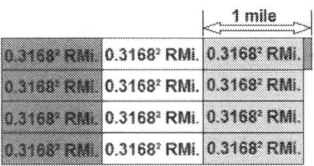
As each row of rectangles represents the area of a 1 mile diameter circle it seems appropriate to make one of the rectangles 1 mile long which produces interesting results, it adds 0.03168² Roman miles to the 864/275² mile rectangle's area changing it into a 3.168² mile polygon. This has created an interesting 1 mile long rectangle which reconnects back to the 1 mile diameter sphere and features symbolic factors concerning the earth's diameter, circumference and rotation speed.

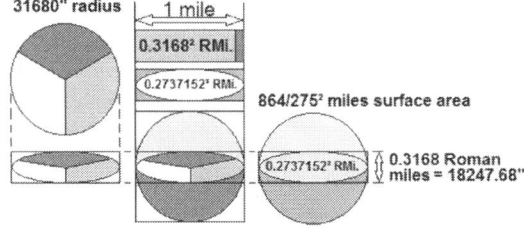
The 63360" length of the ellipse symbolizes the earth's 63360 furlong diameter = 7920 miles, multiplied by 864/275 = 24883.2 miles circumference = 27371.52 Roman miles which is symbolized by the 0.2737152² Roman mile area of the ellipse. The 0.3168 Roman mile width of the ellipse = 18247.68" which is the rotation speed per second of the pre bulge equator when it had a 18247.68 Roman league circumference equator and the 18247.68 Roman league circumference is symbolized by the 18247.68" width of the ellipse. Finally the 864/275 mile circumference sphere's surface area is 864/275² miles which is where we began on the previous page.

In this investigation we have encountered a few interesting scale model great circles of the earth based on 1 unit to another 1 unit scale and the 1 mile diameter sphere is the star performer.

The hill top design of Silbury Hill and Stonehenge's Sarsen Circle are each analogues of a 633.6" diameter sphere inside a 633.6" cube they are scale models of a 1 mile diameter sphere inside a cubic mile.

The dimensions of one model cube are 633.6" x 633.6" x 633.6" and multiplying the cube's volume x 633.6 = 0.0006336^3 miles which is the volume of the Great Pyramid's full design.

The Pyramid's full design represents 6.336 million tons of solid limestone and has a 0.6336 Roman mile or 633.6 Roman pace base perimeter, he vertical section represents 1 quadrant of a 0.6336 Roman mile circumference circle.

There are many more connections of this calibre in the designs of the monuments that demonstrate they had the same Designer.

They say it is 1 mile from the centre of Silbury Hill to the centre of the Avebury monument so the monuments have been placed on opposite sides of a 1 mile diameter circle below.

The star design inside the circle is based on the mile and the tan 0.576 slope angle of Silbury Hill notice that the tan 0.576 lines go from the centre of each monument to the corners of the 1 mile x 1 Pyramid geographic mile rectangle.

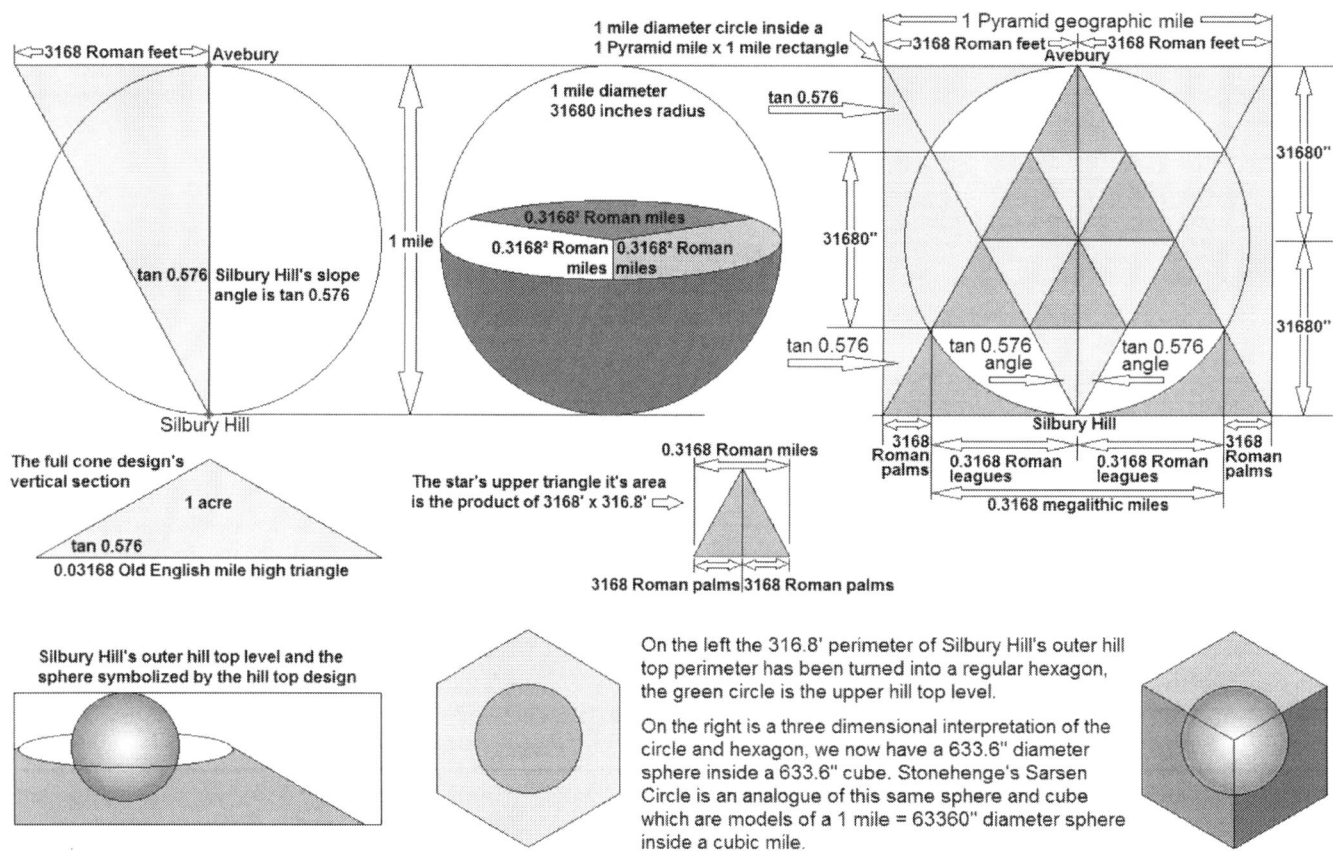

The diameters, radii and circumferences of Stonehenge's Outer Bank, Ditch, the Inner Bank and the Sarsen Circle conform to the factors of the 3 hexagons which symbolize the silhouettes of the 3 cubes. Cube B and Cube C have a remarkable relationship to Silbury Hill's design. The Great Pyramid's full design has a remarkable relationship to Cube A and Cube C.

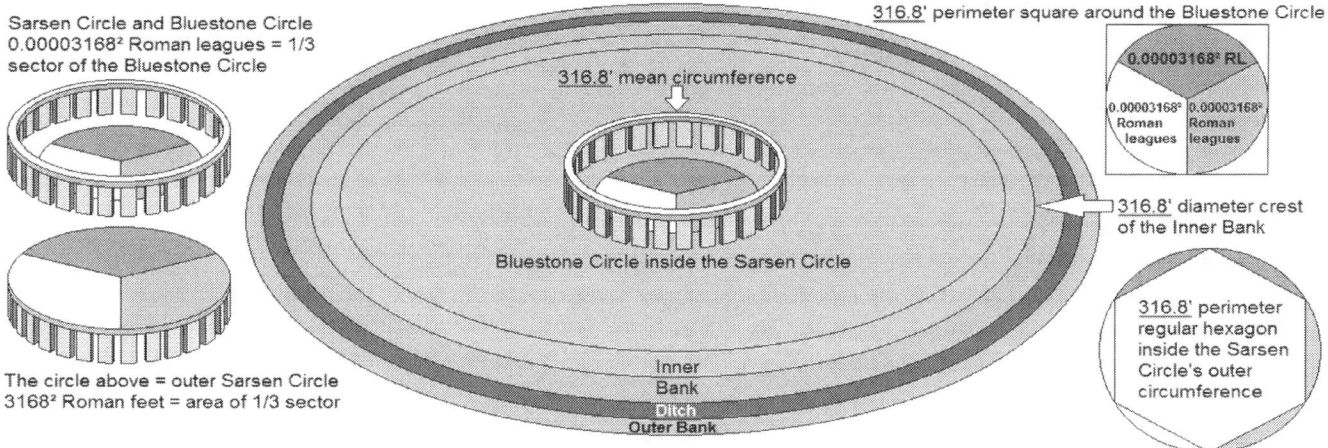

Hexagon A is the only regular hexagon that fits tangent inside the 3168 Roman digit radius Outer Bank, each face of the hexagon is also 3168 Roman digits, the hexagon's 1140.48' perimeter = the Outer Bank's inner circumference which is also the Ditch's outer circumference, the hexagon's area = the area of the hexagon = the area of a 345.6'+ diameter circle and 345.6' is the inner diameter of the Ditch and also the outer diameter of the Inner Bank.

Hexagon B is the only regular hexagon that fits tangent inside the 316.8' diameter crest of the Inner Bank, each side of the hexagon is 0.03168 Old English miles same as the crest's radius, the hexagon's perimeter = 316.8 yards which is the mean circumference of the Inner Bank's inner face (important) the hexagon's area = the area of a 3456"+ diameter circle and the Bank's inner diameter is 3456".

Hexagon C is the only regular hexagon that fits tangent inside the Sarsen Circle's outer circumference, it's 316.8' perimeter = the Sarsen Circle's mean circumference, the hexagon's area = the area of a 96.0'+ diameter circle, the Sarsen Circle's inner diameter = 96.0'+.

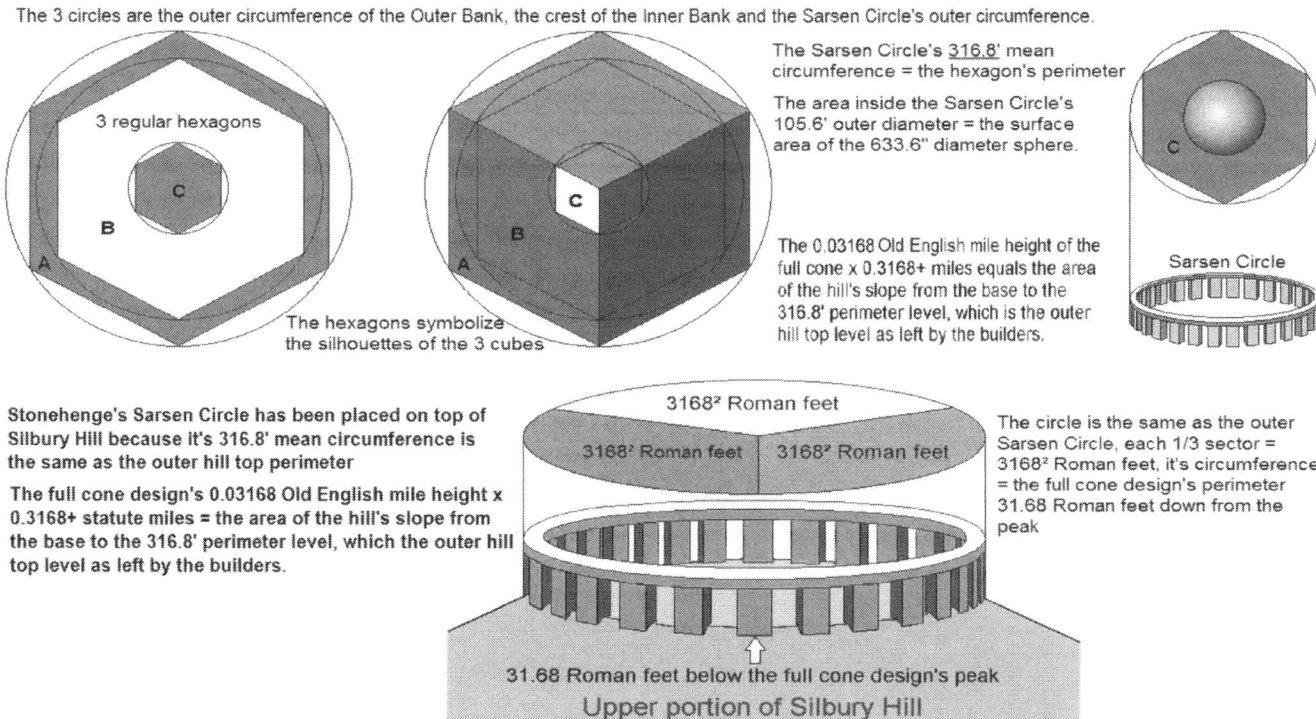

The outer perimeter of Silbury Hill's hill top as left by the builders was a 316.8' circumference circle and the centre of the hill top was 31.68" or 1 Gunter pace higher, the outer level's circumference was the same as the mean circumference of Stonehenge's Sarsen Circle and symbolizes the same thing which is the 316.8' perimeter of a regular hexagon that symbolizes the silhouette of a 633.6" cube (Cube C).

A plan of Stonehenge's Inner Bank's crest, the Station Stone Rectangle and the outer circumference of the Sarsen Circle are centred on the centre base of Silbury Hill's vertical section, the cube and sphere symbolized by the Sarsen Circle's design are also shown.

The two spheres below right demonstrate the height of the full cone design, the size, the shape and elevation of the upper hill top level which is the same as the top 31.68" of the 3168" diameter sphere which protrudes 31.68" through the 316.8' perimeter outer hill top level, the circumferences of the two spheres and the base of the upper level are all exactly the same on the elevation of the 316.8' perimeter level.

Note that the centre of the 316.8" radius sphere is on the centre of the hill top as left by the builders which is 316.8" below the full cone design's peak and the sphere's circumference on the outer level = the base perimeter of the upper level, these are the factors of the hill top design that symbolize the 316.8" radius sphere.

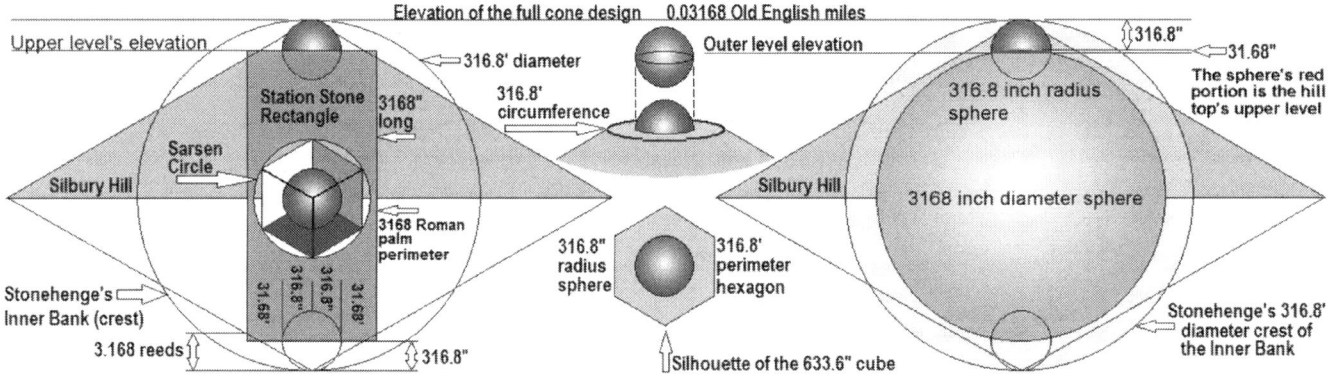

The surface area of the 316.8" radius sphere = the area of a 105.6' diameter circle, the outer 105.6' diameter Sarsen Circle's area symbolizes the sphere's surface area.

The 633.6" cube around the 633.6" diameter sphere below is the same as Stonehenge's Cube C which is symbolized by the Sarsen Circle and remember it is a model of a 1 mile = 63360" diameter sphere inside a cubic mile which is an inch to the furlong scale model of the earth.

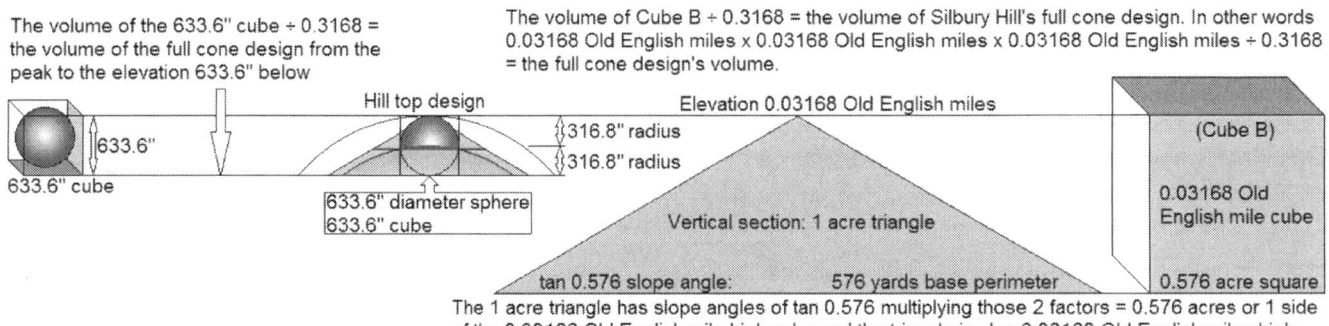

The 1 acre triangle has slope angles of tan 0.576 multiplying those 2 factors = 0.576 acres or 1 side of the 0.03168 Old English mile high cube and the triangle is also 0.03168 Old English miles high.

The standard way to calculate the volume of a cone is pi x radius squared x height then dividing by 3 = volume.

Therefore if we multiply the area of a 1/3 base sector by a cone's height we produce the cone's volume, this is interesting concerning Silbury Hill's full cone design, because it is 0.03168 Old English miles high and 0.003168² Old English miles is the area of a 1/3rd base sector so 0.03168 Old English miles x 0.003168² Old English miles produces the full cone design's volume.

Silbury Hill has a special slope angle and to calculate the volume of any cone with a tan 0.576 slope angle you just cube the height and divide by 0.3168, so to produce the volume of the full cone design is 0.03168 Old English miles x 0.03168 Old English miles x 0.03168 Old English miles ÷ 0.3168 = volume.

Do you get the impression the Designer of Silbury Hill is trying to tell us something?

Note that Silbury Hill's design focused attention on 1/3 sectors of circles which are a mathematical factor of any cone this gives us an excuse (if we need one) to divide circles into 1/3 sectors which is why we divided all those circles in previous pages into 1/3 sectors and the Great Pyramid's full design told us to look at quadrants of circles.

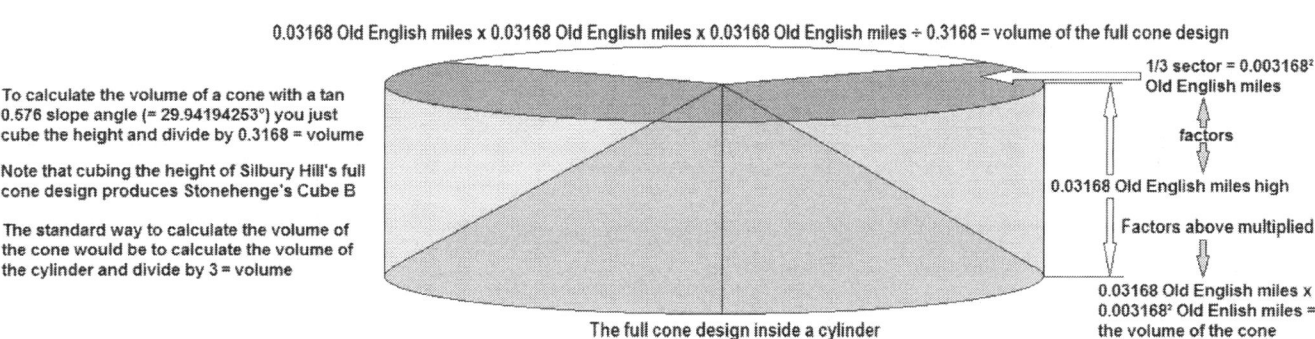

Stonehenge's design symbolizes the two cubes below left which were named Cube B and Cube C, Silbury Hill's hilltop design also symbolizes Cube C and Cube B has an interesting geometric and mathematic relationship to Silbury Hill's full cone design.

It is obvious that cube C is a scale model of a 1 mile cube containing a 1 mile diameter sphere and on the same scale Cube B is a model of a cubic league or significantly a 3.168 Old English mile cube.

On the right below are full size versions of the two cubes as symbolized by Stonehenge's design they are placed with Bethlehem and Jerusalem on opposite sides of a 31680' diameter circle which is Stonehenge's crest of the Inner Bank enlarged on the same scale as the cubes.

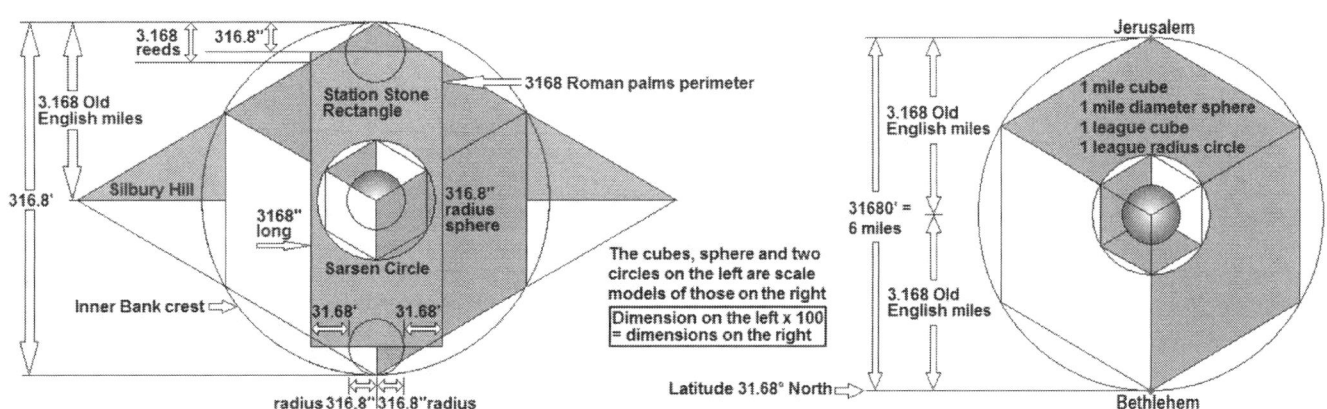

From the centre of the 31680" radius sphere (1 mile diameter) to the centre point of the Stonehenge group of monuments is on azimuth 316.8° from north, the centre point is ½ way between the centres of Stonehenge and Avebury which is 3.168 megalithic miles from both centres, therefore the centres are 6.336 megalithic miles = 91238.4 inches apart which records the (original) equatorial rotation speed per minute of the 9123.84 megalithic mile circumference equator.

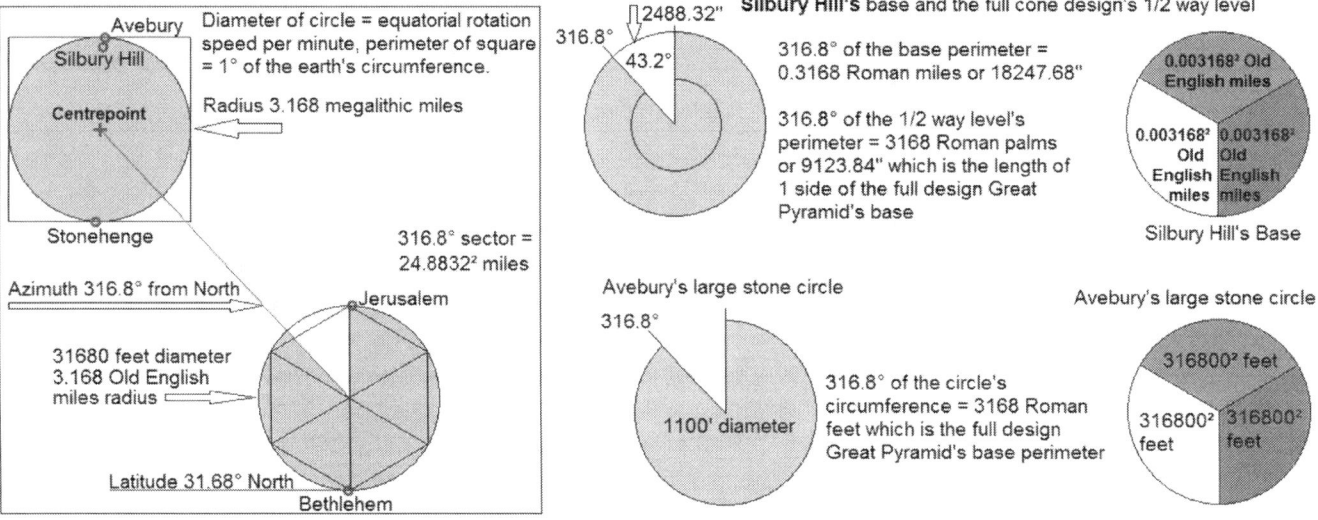

The star design is based on the 6 megalithic miles = 31680 MY. distance between the centres of Stonehenge and Avebury, and the tan 0.44 diagonals of Stonehenge's Station Stone Rectangle.

On the 1 foot per second scale the 3168 fathom wide triangles = 316.8 minutes and the outer rectangle represents 1 day x 21.12 hours, interesting because the earth rotates 316.8° every 21.12 hours and 316.8° is an interesting feature of Silbury Hill (see above) more importantly azimuth 316.8° from North connects Bethlehem and Jerusalem to the English monuments.

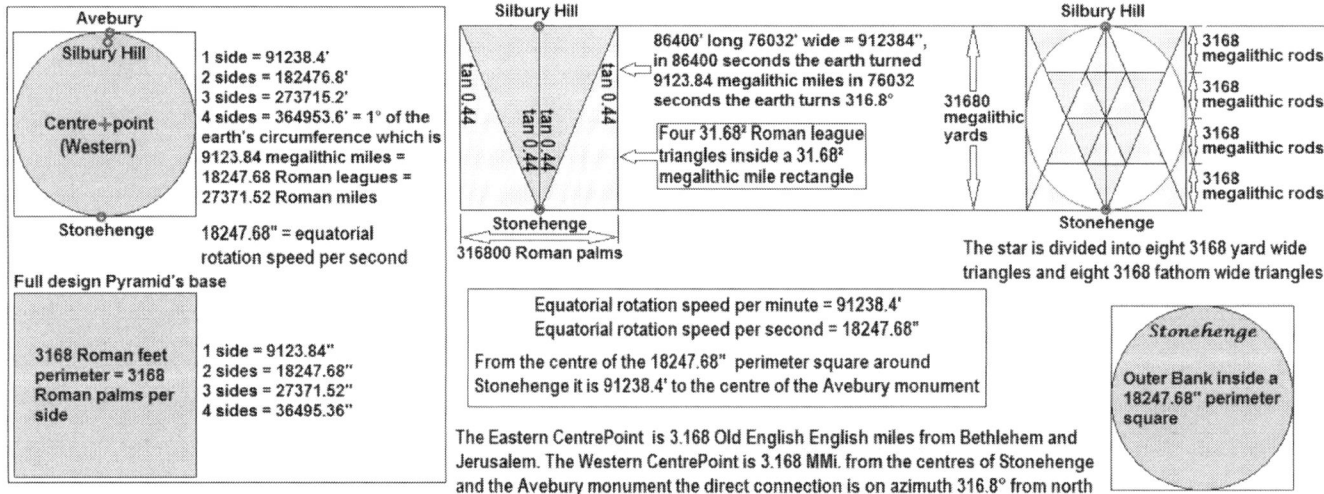

The interaction between the geometry and symbolism of Stonehenge and Silbury Hill has been demonstrated and how the designs relate to where Jesus was born and where he was crucified has been shown but we will take another look to emphasize these design features.

The people trying for centuries to answer the big question of why Stonehenge was built have been hampered by their own assumptions that being so old it has to be some form of pagan religion and that the builders only had primitive intelligence, tools and knowledge.

People generally believe that Christianity is a comparatively new religion, but the Saviour (Jesus) was prophesized in great detail before any other religions were invented even long before Moses started to write the Bible some 3.5 thousand years ago.

True Christianity recognises that the long awaited Saviour turned up at the appointed time and place and fulfilled the many ancient and detailed prophecies about why he came and what he would do, so why do people think that belief in the prophesized Saviour only began after he came, the roots of Christianity predate all other religions (including Judaism) which is demonstrated in this investigation.

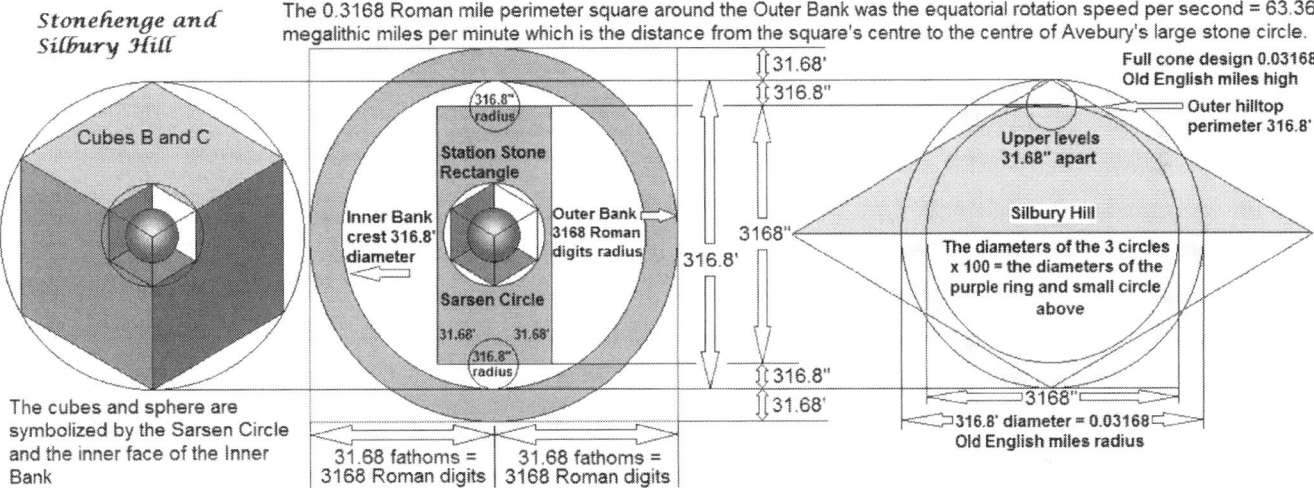

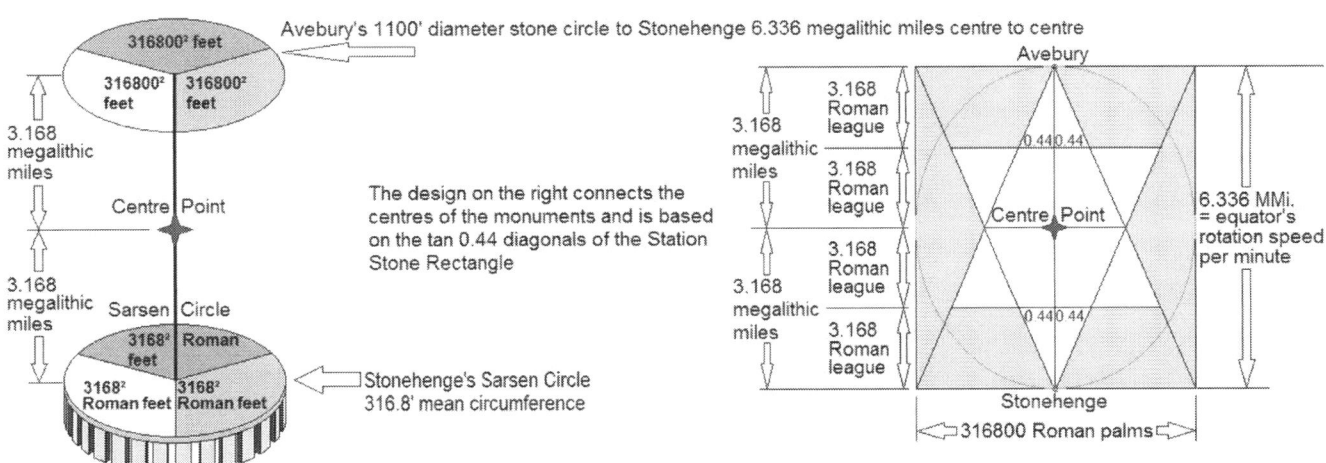

I thought that the tallest mountain in the world might have the 3168 signature and found it was 5.5 miles high which reminded me of Silbury Hill's 550' diameter base and a circle, 5.5 miles x 864/275 = 17.28 miles or 6.336 MMi. circumference which is the distance between the centres of Avebury and Stonehenge. Five miles = 316800" add 31680" = 5.5 miles add 31680" = 63360' or 6 miles which is the distance from Bethlehem to Jerusalem, below we show how the factors of Silbury Hill's design shown above relate to Mount Everest and how they also relate to the 6 mile diameter circle used to connect Bethlehem and Jerusalem symbolizing the life of Jesus.

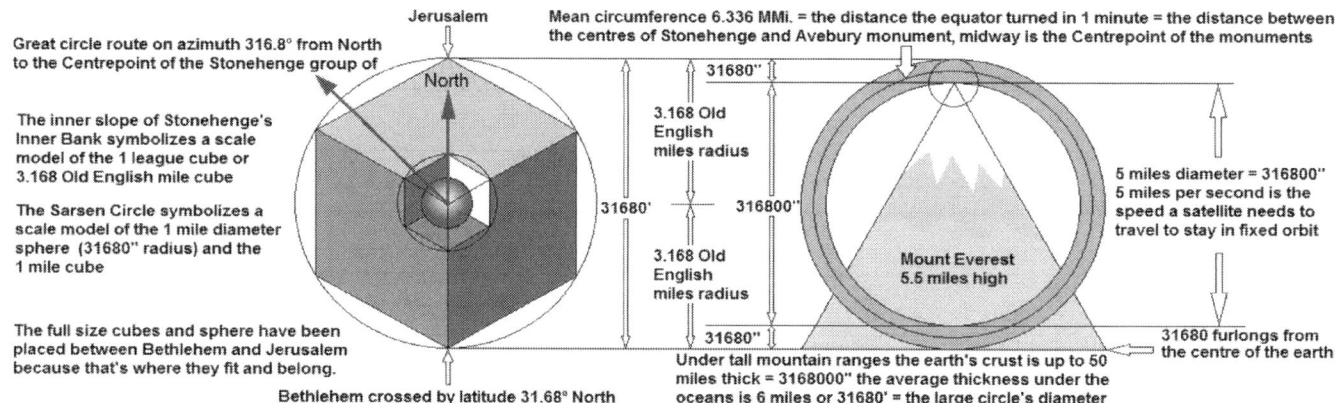

A 0.3168 Roman mile perimeter square fits tangent around Stonehenge's 0.248832 Roman mile circumference Outer Bank and symbolizes a 31680 mile perimeter square fitted tangent around the earth's 24883.2 mile circumference, also the Outer Bank's 3168 Roman digit radius and 6336 Roman digit diameter refer to the earth's 31680 furlong radius and 63360 furlong diameter.

The most significant day of the year at Stonehenge is the summer solstice or midsummer's day when crowds gather to watch the sunrise, the azimuths of the summer solstice sunrise and moonrise cross each other at right angles on Stonehenge's parallel and this is a feature of the monument's design that fixes the monument to this parallel. The exact position on this parallel is fixed to one tan 0.44 diagonal of the Station Stone Rectangle being on the azimuth of a great circle route to the Great Pyramid and each side of the rectangle is aligned to the above summer solstice azimuths. Stonehenge was designed to be built exactly where it is, another cosmic and global factor that dictates the monument's position on the earth's surface is that on the day of the Crucifixion the azimuth of the Passover moonrise and a great circle route to the Crucifixion formed the Christ Angle at Stonehenge, (the Christ Angle is the Great Pyramid's passage angle).

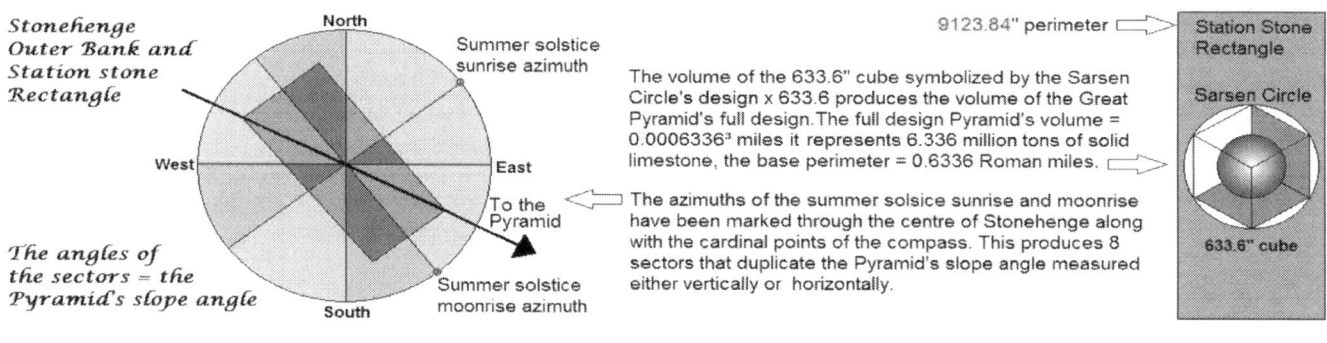

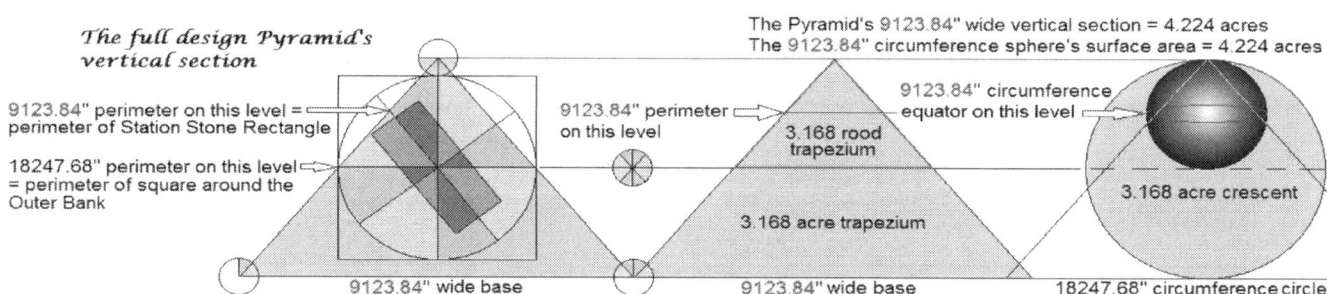

The 9123.84" circumference sphere is a 1 inch to 1 megalithic mile scale model of the 9123.84 megalithic mile circumference earth, this was the equator's circumference before the bulge appeared and it turned 18247.68" per second which is very appropriately the perimeter of ½ way level at the sphere's base.

The base of the sphere above is on the centre of a 18247.68" circumference circle and resting on the centre of the 18247.68" perimeter square ½ way level.

It is interesting that the 18247.68" circumference of the circle above records the equatorial rotation speed per second of the original perfect earth's equator when it had an 18247.68 Roman league circumference, this circle is also a 1 inch to 1 Roman league scale model of a great circle of the earth.

The full design Pyramid's vertical section is an analogue of this circle but the design is very multifunctional and the vertical section is also an analogue of a quadrant of a 0.6336 Roman mile circumference circle as well as the sphere above.

Below: The Great Pyramid's design demonstrates the knowledge of its unique global position by incorporating it into the design, global factors divide the full design base into four 18247.68" perimeter squares and each square fits tangent around Stonehenge's Outer Bank, the ½ way level is also an 18247.68" perimeter square.

One diagonal of Stonehenge's Station Stone Rectangle pointed the way to the Great Pyramid so we compared the relationship of the Pyramid's full design to Stonehenge's Outer Bank and Station Stone Rectangle which demonstrated another way in which the monuments interact and that they have the same Designer.

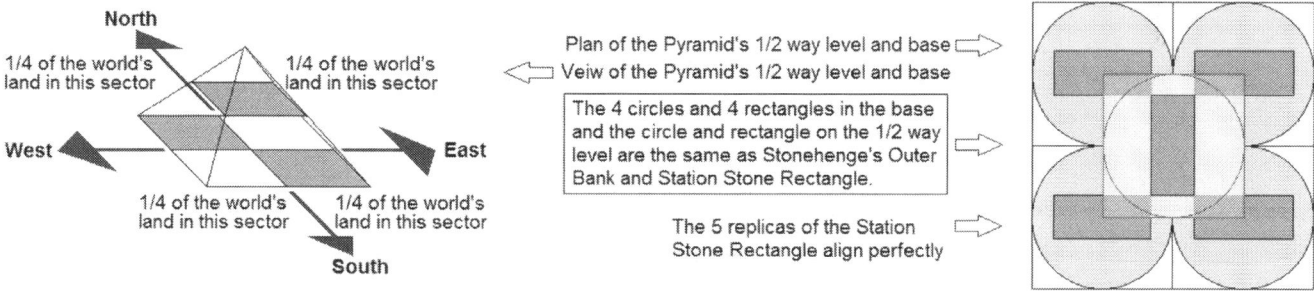

Next we will look at a plan of the 5 squares shown above on the left with Stonehenge's Outer Bank placed in the middle square and see how this relates to the African geographic design centred on Mount Kilimanjaro.

The outer squares in the illustrations below have perimeters featuring the number 3168 and each square is divided into 4 squares that also feature the number 3168; illustration #3 is produced by combining the first 2 illustrations and illustration #3 belongs in the centre of illustration #4.

Illustration #5 has a profound partnership with illustration #4 these 2 designs enclose the extremities of the Nile River inside quadrants of 1/10 scale model great circles of the earth and the moon, both designs are based on geographical features.

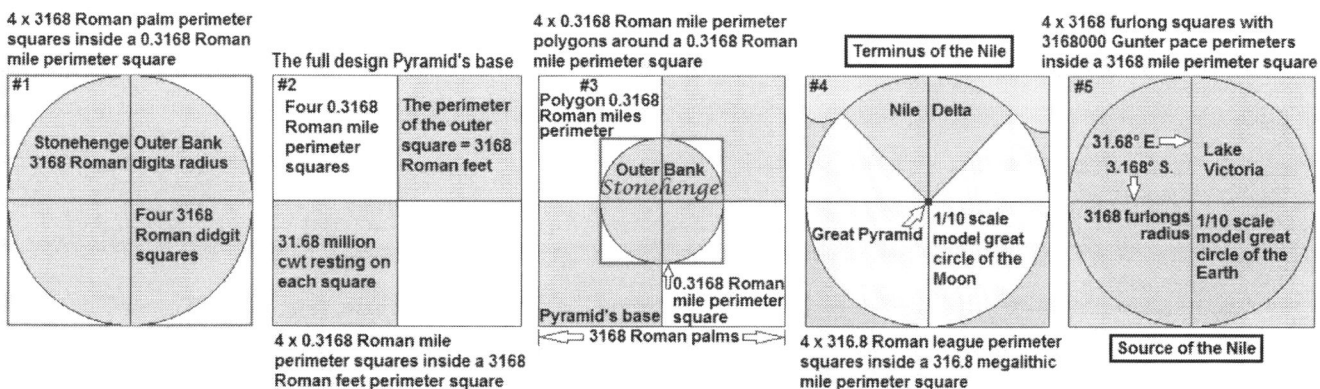

Below we are comparing the #3 design above with the geographic design centred on Mount Kilimanjaro.

The design composed of basic design factors of Stonehenge and the Great Pyramid's full design is an analogue of the Mount Kilimanjaro design.

This confirms that the designs of the two monuments and the geographic designs have the same Source.

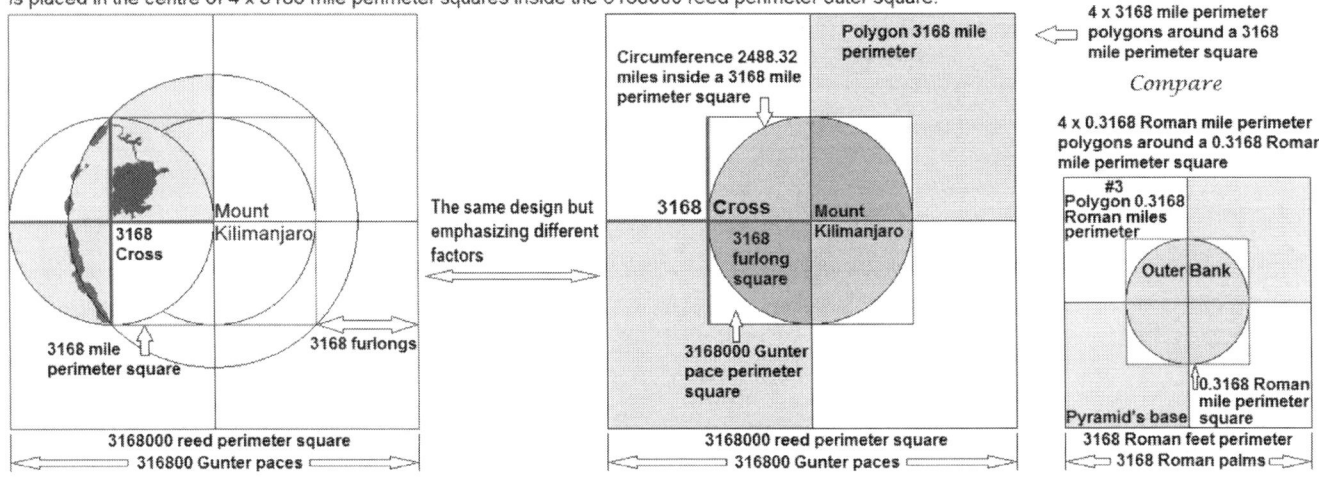

The circle enclosing the large latitude longitude design below has a diameter of 6336 miles = 3168000 reeds and a radius of 3168 miles, those 2 measurements are used as the perimeters of the 2 squares above that are centred on Mount Kilimanjaro and 3168000 reeds is also the length of one diagonal of the latitude longitude design.

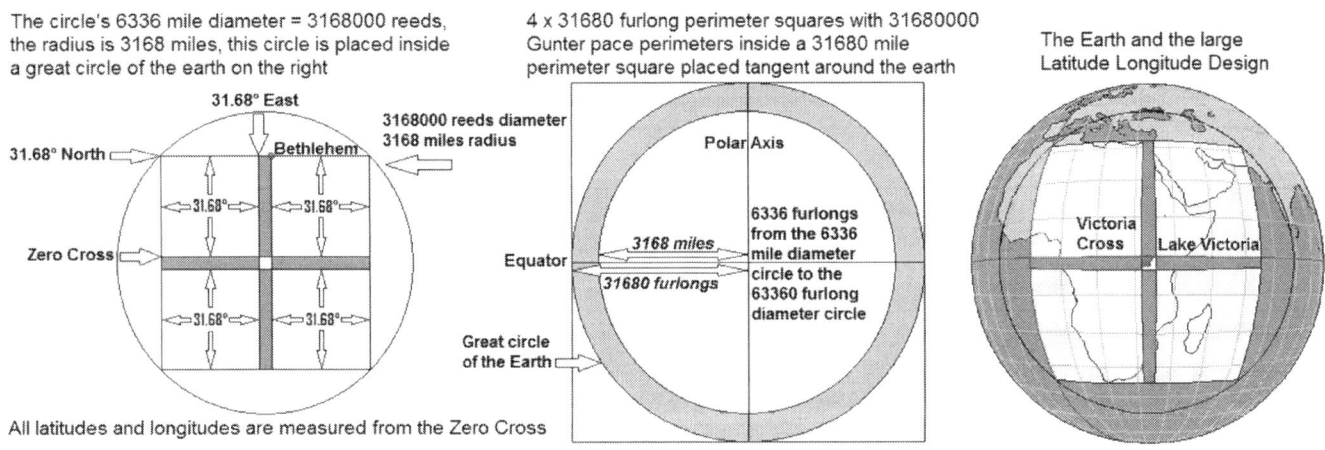

The Numerical Universe

The numbers of music meaning the numbers on the Musical Scale Chart are a feature of every page of this investigation but it is mostly not commented upon the most important numbers are generally found in Column G and their importance relates to their proximity to the number 3168. The number 198 is the furthest away and is not used as much as the other numbers, the number 396 is used a bit more but not as much as the next number 792 which has increased in importance and the rest of Column G's numbers are generally even more important. For example the full design Pyramid's volume of 79200^3 reeds is worthy of note but is much more significant when defined as 0.0006336^3 miles and it is worthy of note concerning column G numbers that the reed is 126.72" and the mile is 63360".

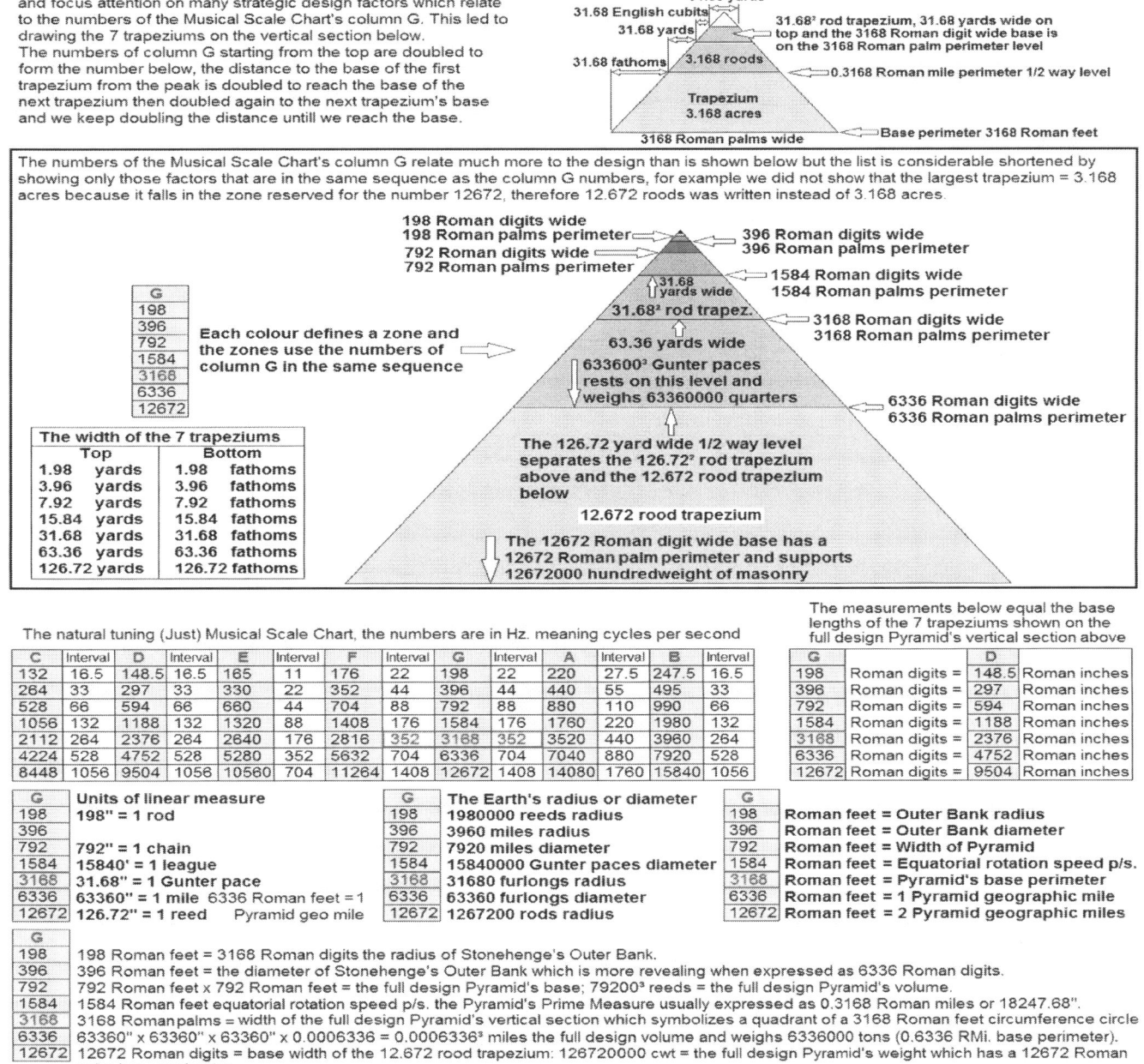

The rod = 198" and the reed = 126.72" column G begins with 198 and ends with 12672 the earth's radius of 1980000 x 126.72" long reeds = 1267200 x 198" long rods or in other words the earth's radius of 1980000 reeds = 126720000 rods. Other details not shown in the column of linear units of measure are the link = 7.92" and the furlong = 7920"; the league of 15840' = 3.168 Old English miles; the Pyramid geographic mile = 63366 Roman feet, the Greek geographic stade = 633.6 Roman feet; the Greek geographic cubit = 1.584 Roman feet or 0.3168 Roman paces and the Greek geographic great cubit = 0.0003168 Roman leagues. Note, 10 Greek geographic stades = 1 Pyramid geographic mile.

The numbers of the Musical Scale Chart are prophetic of the chain mapping scales used in the Enclosure Maps of England and Wales 1595-1918, the same scales were used all over the world in English speaking countries for cadastral maps etc. Edmund Gunter born 1581 was professor of astronomy at Gresham College London, he invented and made navigation instruments, printed the first table of logarithms and laid the groundwork for the slide rule. He introduced the 22 yard chain of 100 links, each 7.92" link was a solid bar, he also introduced the 31.68" Gunter pace. In Subsequent centuries the chain was used to produce maps and for obvious reasons we are interested in maps drawn using the 1:3168 scale (1" to 3168" or 1" to 4 chains). When we investigate how the 1:3168 scale relates to Stonehenge's design and the Great Pyramid's full design it reveals that the monuments are prophetic of the 1:3168 scale and of course the inch and the chain along with the Roman digit, the Roman palm and the Roman foot but we already knew all this except for the 1:3168 connection.

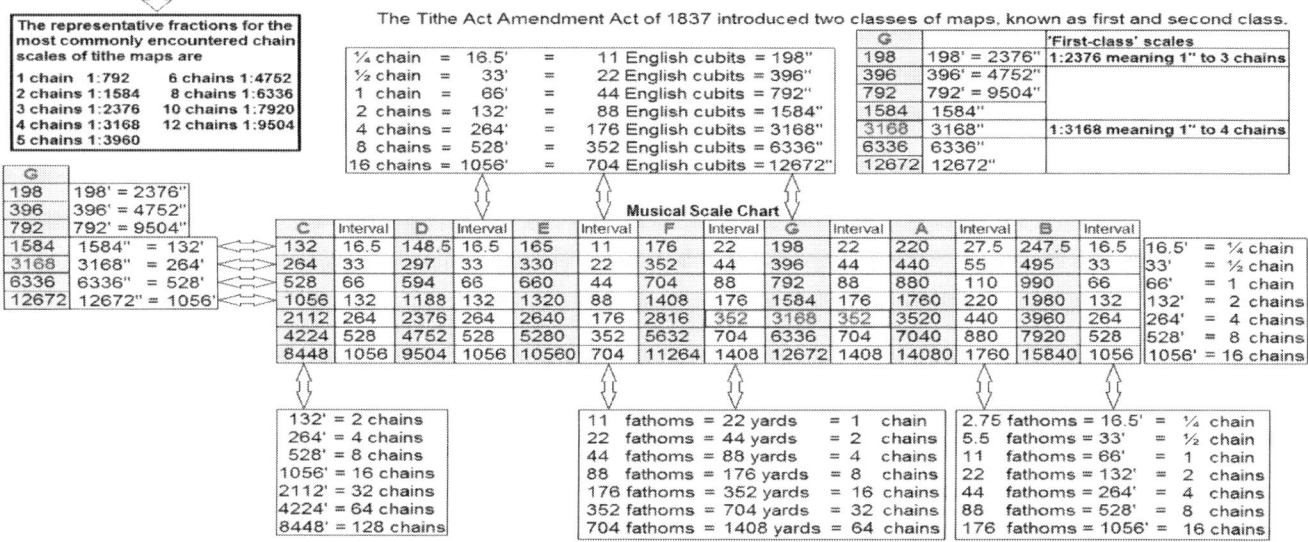

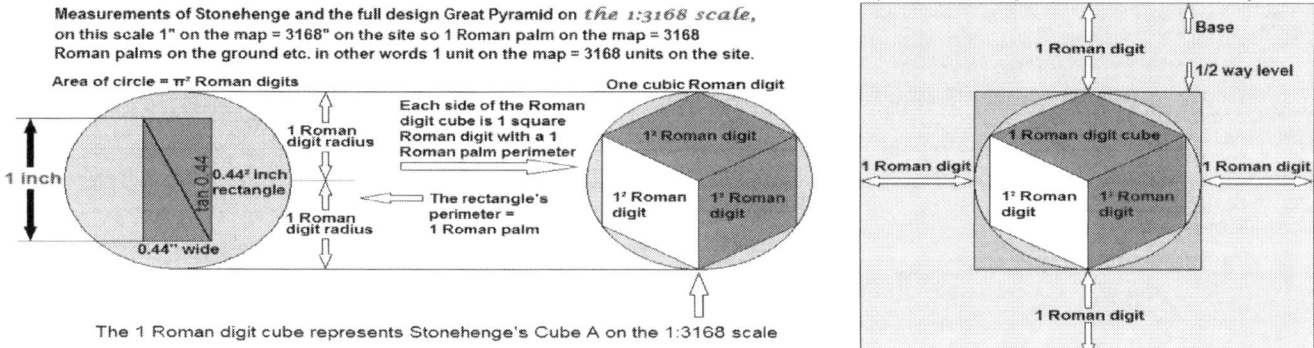

In 1884 the north-south line at Greenwich was by international agreement designated the prime meridian 0° it is the reference meridian for all the other meridians of longitude which are numbered east or west of it. The Greenwich meridian also serves as the basis for the world's standard time zone system. The mean solar time at Greenwich is now called Universal Time and was formerly called Greenwich Mean Time. Because the earth turns 15° per hour standard time theoretically becomes successively one hour earlier at each 15° longitude west of the Greenwich meridian and one hour later at each 15° longitude east. We have previously demonstrated how the reference to the Bethlehem coordinates on the Musical Scale Chart is prophetic of the above longitude system and reveals why it came into universal use. The latitude longitude system is a factor of comprehensive regional maps, national maps and especially world maps etc. and the Musical Scale Chart's numbers are also prophetic of the mapping done via the various chain scales featured on this page. Theoretically the Prime Meridian (0°) is arbitrary but the Prime Parallel (0°) is not because it is the equator, when the international representatives met to decide which meridian to use as the Prime Meridian they probably had no idea the final decision had been made a very long time before they were born.

The Prime Measure of this investigation is 0.3168 Roman miles and the second most important measure is 3168 Roman palms, so it is especially interesting that 0.3168 Roman miles = 316.8° of Silbury Hill's base perimeter and 316.8° of the full cone design's 316.8 megalithic yard perimeter 1/2 way level = 3168 Roman palms.

Below is demonstrated some of the monuments harmonious interaction which use several methods and one method is numbers as in Gematria and the common use of the Musical Scale Chart's numbers and also the constant use of the same strategic numbers in different forms i.e. with a floating decimal point.

Another method of interaction is geographic such as global positioning and the geometric interaction is awesome, geometry is a form of mathematics and other methods of mathematic interaction are used.

There is a common modus operandi in the designs such as for example the way many different units of measure are used.

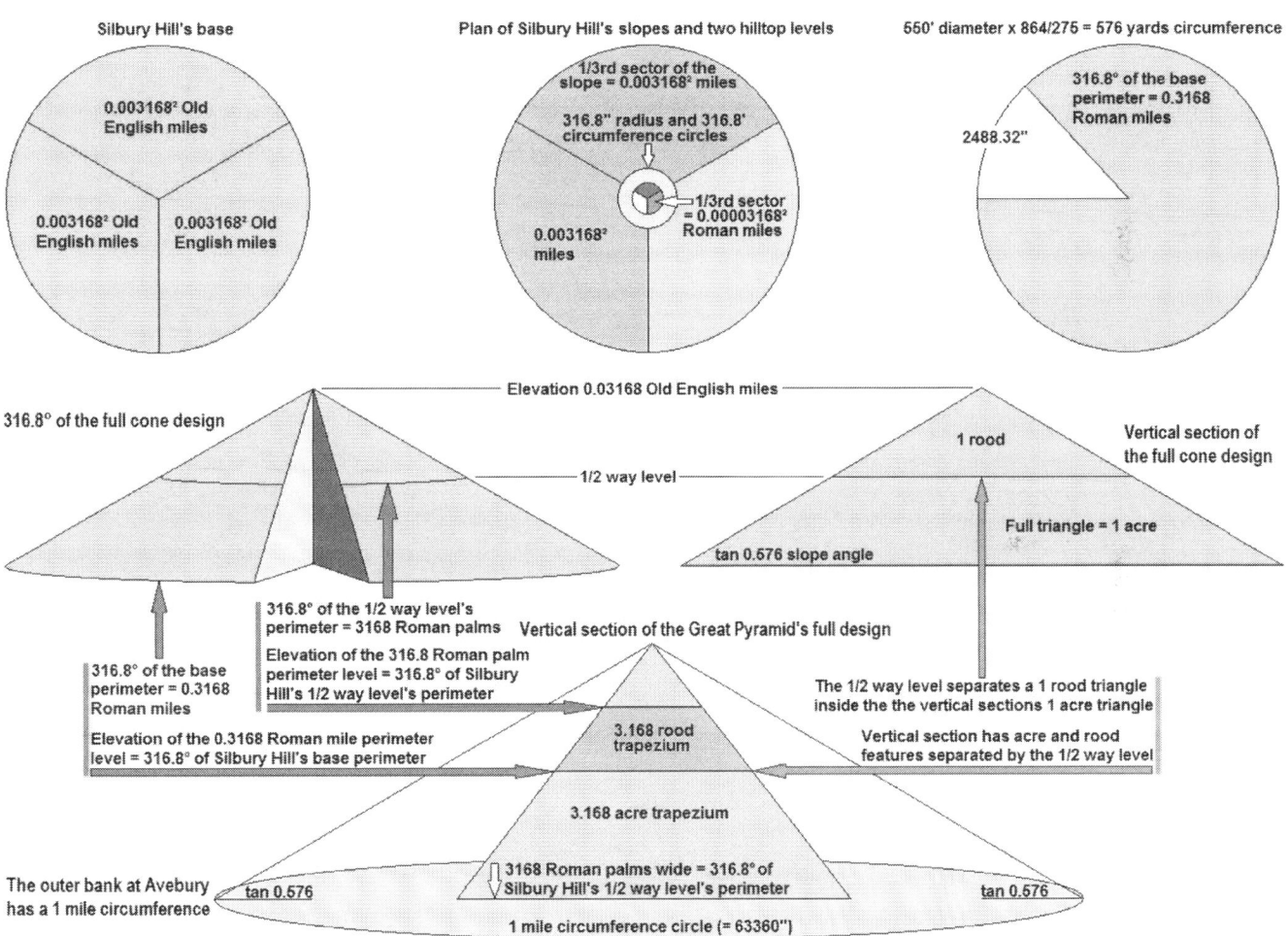

A Pyramid and a cone are two examples of simple solid geometric shapes and there are only a few, by simple I mean they have simplest formulas and methods for calculating their volumes and surface areas other examples are a sphere, a cube, a prism and a cylinder.

The Great Pyramid and Silbury Hill as left by the builders were both simple solid geometric shapes with their peaks missing, a true pyramid represents a series of squares decreasing in size from the base to the peak and a cone represents a series of ever decreasing circles from the base to the peak. The Pyramid is geometrically and mathematically linked to the cube or cuboid and the cone has the same relationship to a cylinder, the vertical sections of a cone and a pyramid are triangles, the slope angles of Silbury Hill and the Great Pyramid are both very special.

Below demonstrates how the designs of Silbury Hill and the Avebury monument interact with each other and produce basic measurements relating to the full design Pyramid's base perimeter and its height. The 1 mile or 63360" circumference circle shows the relationship of Silbury Hill's slope angle to the full design Pyramid it is especially interesting because of the relationship of 63360" to the full design's volume i.e. 63360" cubed x 0.0006336 = the volume.

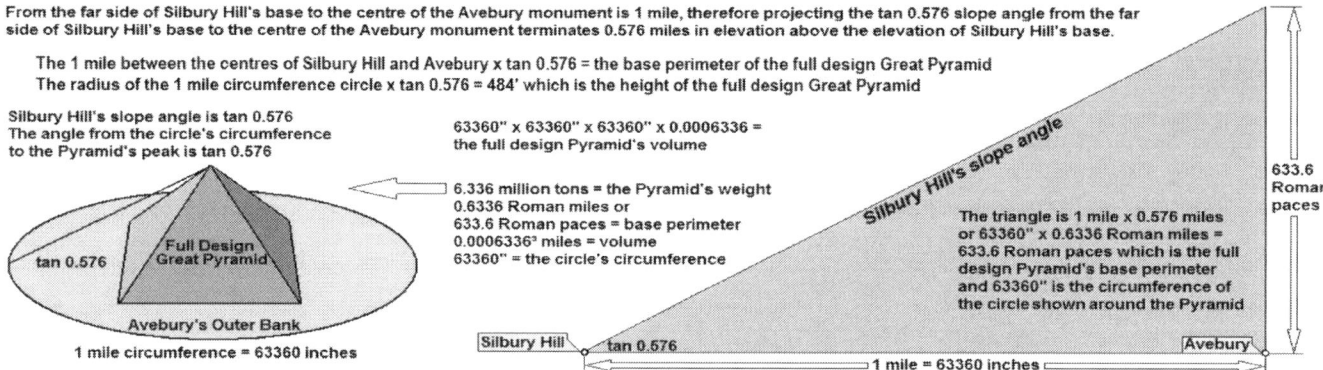

Note that the distances between Stonehenge, Silbury Hill and the Avebury monument produce six different measurements and each one is a profound design factor.

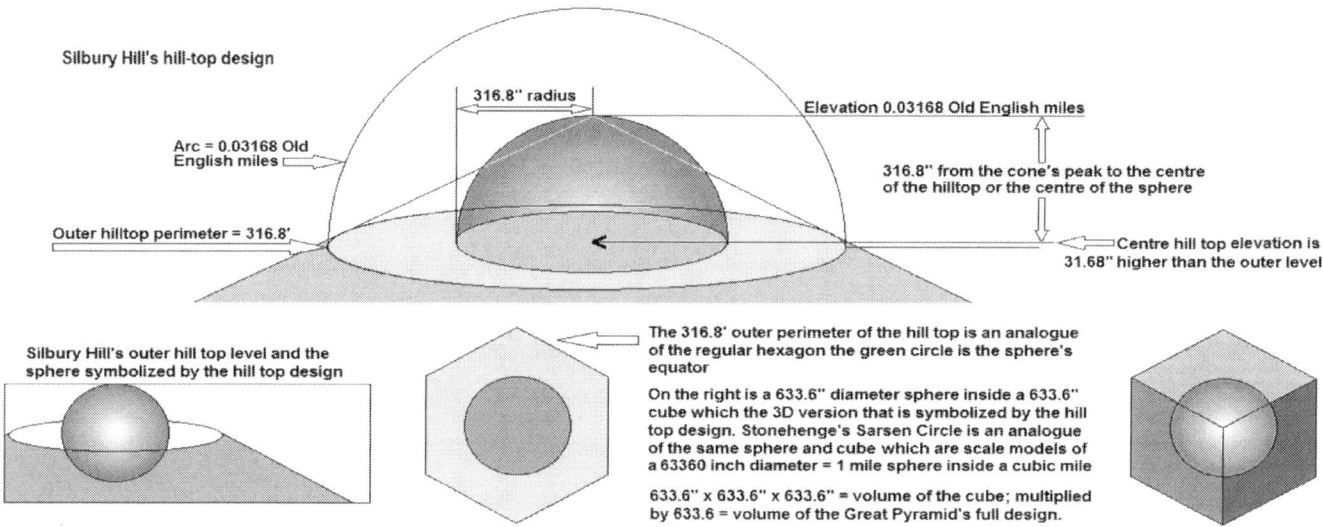

Summer solstice or midsummer's day is the big day of the year at Stonehenge when thousands gather to see the sunrise and Stonehenge's Station Stone Rectangle is aligned to the summer solstice sunrise and moonrise azimuths so it must mean something. If we mark the solstice azimuth and the cardinal points of the compass through the centre of the monument it produces 8 sectors and the angle of each sector is the same as the Great Pyramid's slope angle measured either horizontally or vertically. The shortest distance between any two points on the earth's surface is on the circumference of a great circle and one diagonal of the 9123.84" perimeter Station Stone Rectangle is on the azimuth of a 9123.84 megalithic mile circumference great circle connecting Stonehenge to the 9123.84" wide full design Great Pyramid. Note that 9123.84" = 3168 Roman palms so we connected the 3168 Roman palm perimeter Station Stone Rectangle to the Pyramid's 3168 Roman palm square base by the 31680 furlong radius great circle of the earth.

These factors are connected, Silbury Hill's base perimeter = 633.6 megalithic yards and the full cone design's 1/2 way level has a 316.8 megalithic yard perimeter. The Pyramid's full design has a 0.6336 Roman mile perimeter = 633.6 Roman paces and the 0.3168 Roman mile perimeter of the ½ way level = 316.8 Roman paces. Stonehenge's Outer Bank has a radius of 3168 Roman digits or 31.68 fathoms and the 6336 Roman digit diameter = 63.36 fathoms. The mile is the product of 31680" + 31680", the 31680 furlong radius earth has a 63360 furlong diameter, the Pyramid geographic mile = 6336 Roman feet or 2 x 3168 Roman feet which is 2 x the full design Pyramid's base perimeter. The designs of the monuments thoroughly exploit the fact that 4 x 0.3168 Roman miles = 1 Pyramid geographic mile, 0.3168 Roman miles + 0.3168 megalithic miles = 1 Pyramid geographic mile and 3168 Roman feet + 3168 Roman feet = 1 Pyramid geographic mile.

Author Bonnie Gaunt states the crest of the Inner Bank had a 316.8 feet diameter which is significant and very important concerning the geometric harmony of the whole design, the crest marks the mean diameter which is the same as the mean circumference of the Sarsen Circle and a 316.8 feet perimeter square fits tangent around the Bluestone Circle. Author professor Raymond Capt said the Inner Bank was once at least 6 feet high and that the Outer Bank now almost worn away had a diameter of some 380 feet and was once about 2 feet high, geometric design analysis shows the Inner Bank was 6.336 feet high, the Outer Bank had a 380.16 feet diameter and was 1.9008 feet high (1.9008 feet x 100 = the radius). If the Inner Bank was at least 6 feet high the nearest height to this with any meaning is 6.336 feet and to calculate the width of the base all we need is the slope angle, fortunately the only real candidate is tan 0.44 which is the diagonal angle of the Station Stone Rectangle and using this angle also led to revealing the original width, height and depth of the Ditch and Outer Bank as you can see for yourself below.

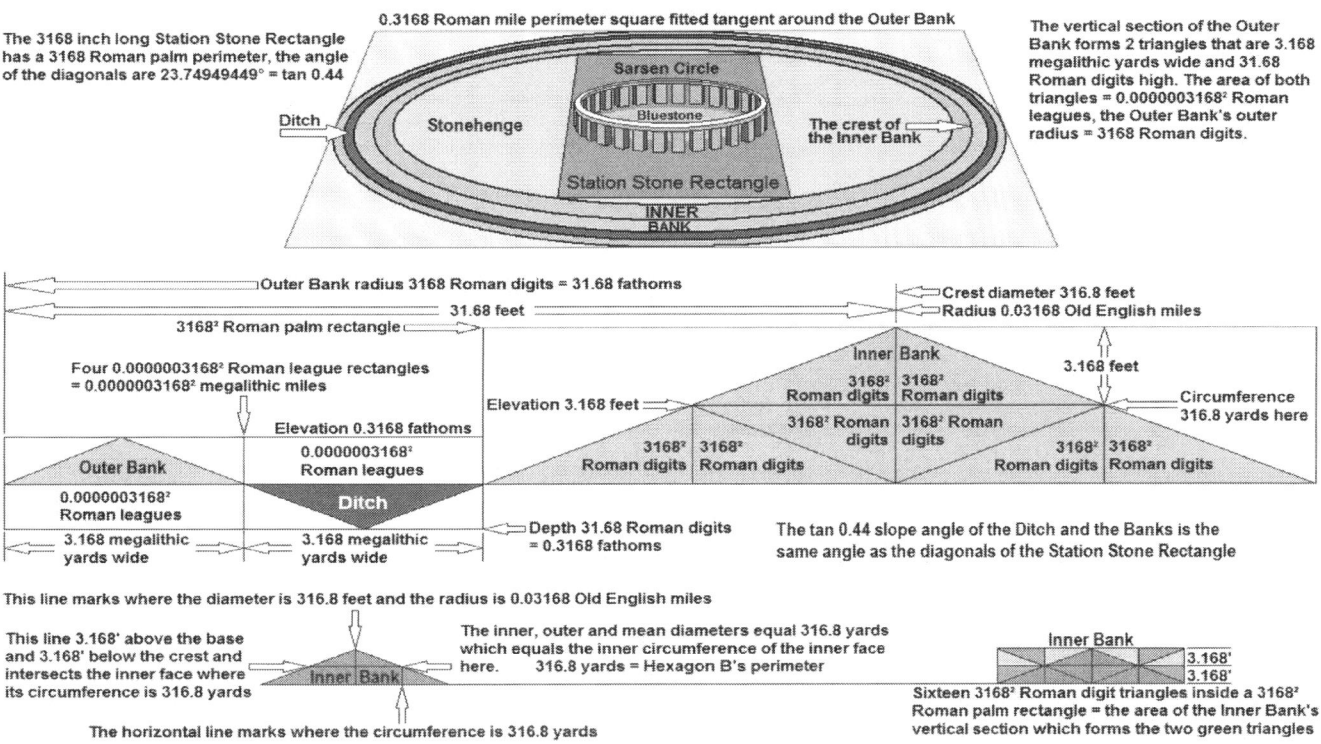

Note that below and on the following page the Outer Bank, the Ditch and the Inner Bank are shown on a larger scale than their diameters so that details can be seen better or we could say the diameters have been shortened to fit on the page, this has no effect on the geometric calculations.

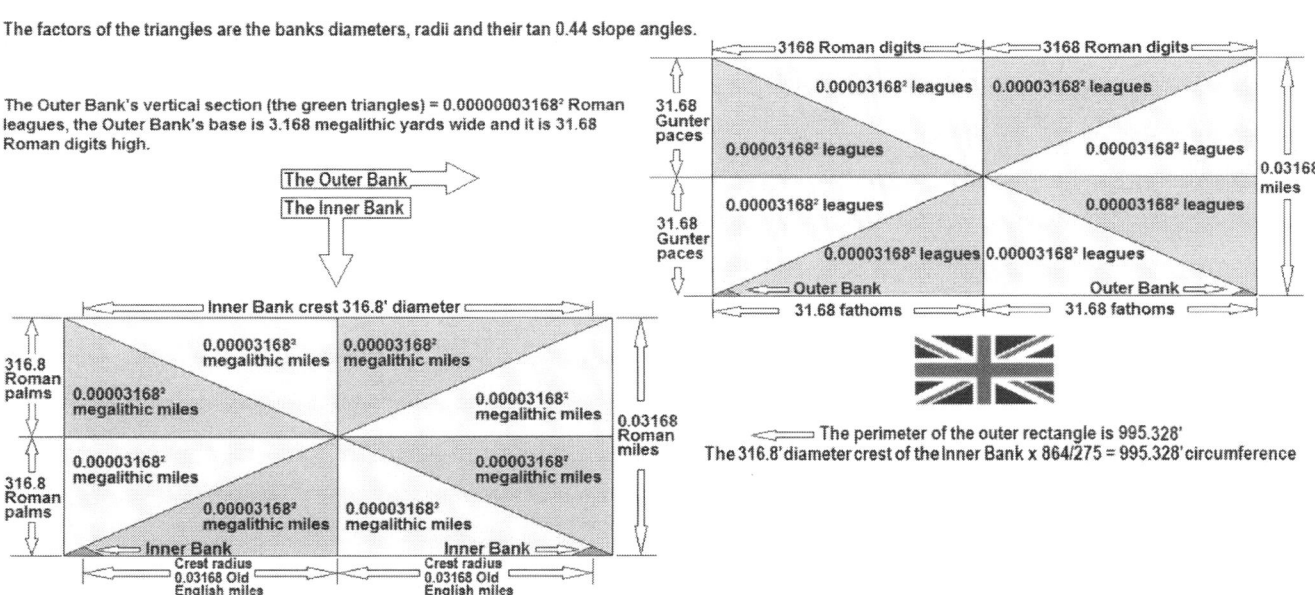

Using the number 3456 to demonstrate another layer of design harmony and confirm we have the exact dimensions of the original design, the Sarsen Circle's outer circumference is 345.6 Roman feet, the Inner Bank is 345.6 inches wide, it's inner diameter = 3456 inches, it's outer diameter is 345.6 feet, 34.56 x 345.6 inches = the mean circumference and the area covered by its base is the product of 345.6 inches x 345.6 inches x 345.6. The inner face is an analogue of a cube with a 3.456 acre surface area (Cube B). Adding 34.56 feet to the Inner Bank's 345.6 feet outer diameter produces the Outer Bank's outer diameter and it's circumference is the product of 34.56 x 34.56 feet. All circles encountered in this investigation that have circumferences of 3456 units or 34.56 units etc. are significant i.e. concerning their 1/3 sectors.

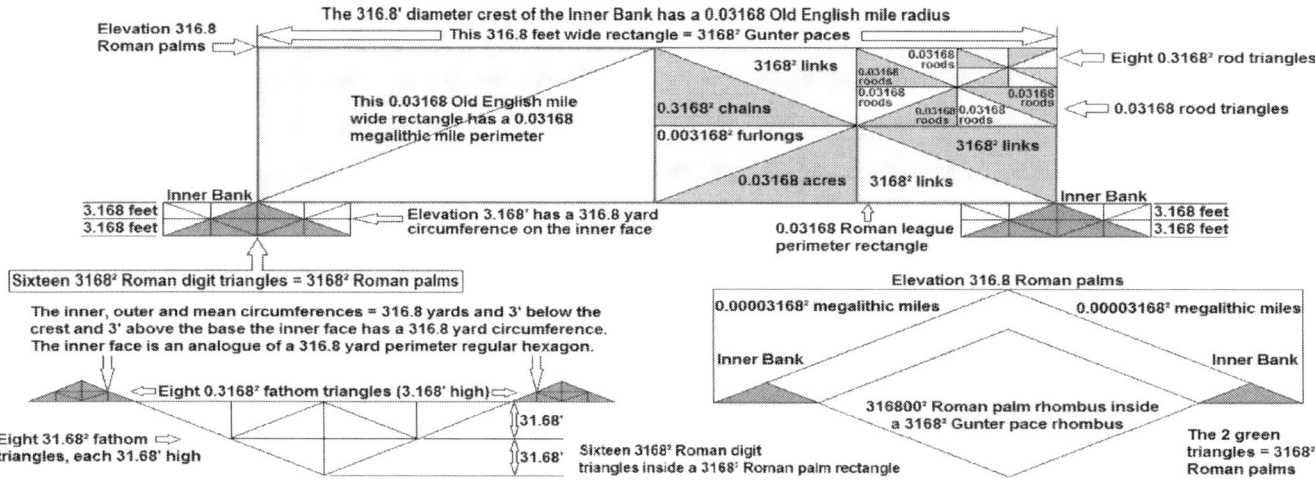

The 3168 signature is placed on many units of linear measure and for the area measures called the acre and the rood this is a common theme.

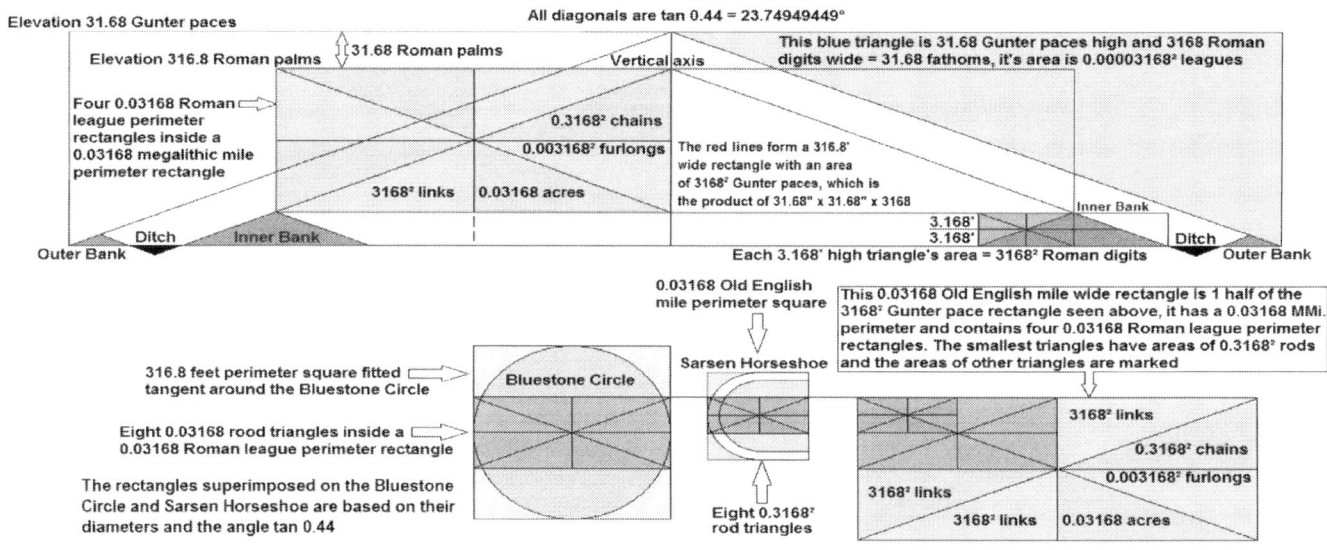

The 3168 signature is placed on many units of linear measure and the area measures called the acre and the rood this is a common theme of the Great Pyramid's full design, the design of Silbury Hill, cosmic design and simple geometric factors of the units themselves.

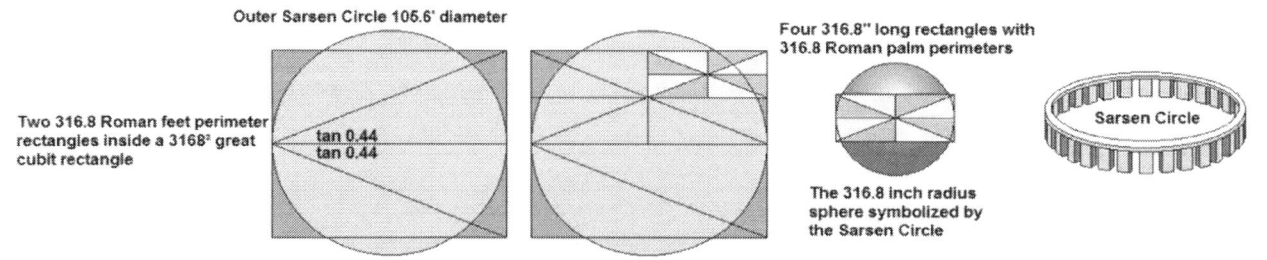

The Numerical Universe

The illustrations below demonstrate more examples of how Stonehenge's design relates to the number 3168 and various units of measure.

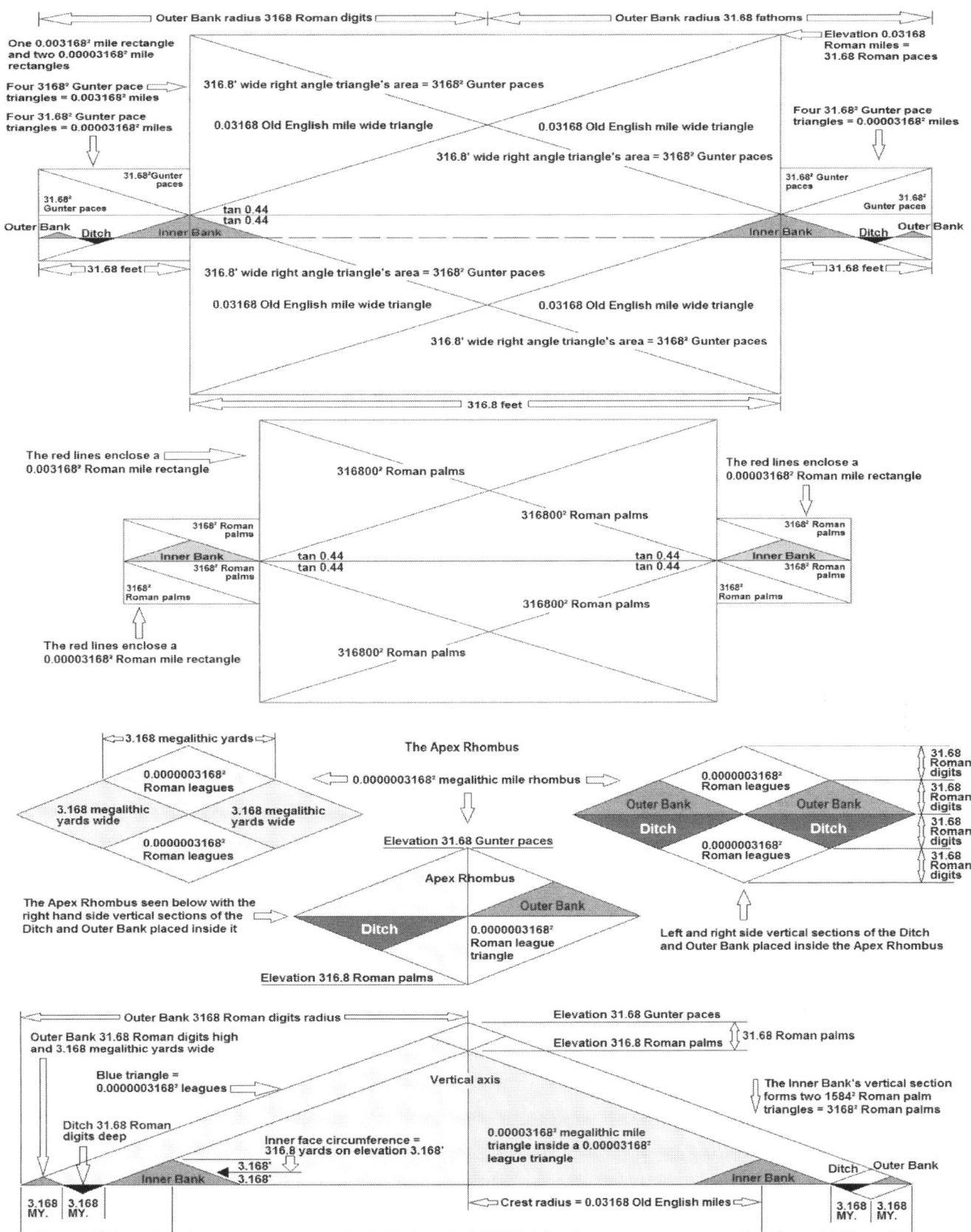

The Outer Bank at Stonehenge is almost worn away and author Raymond Capt said it has a diameter of some 380 feet, but geometric design analysis in several ways refines it to exactly 380.16 feet diameter which is very significant whereas 380 feet diameter is meaningless as are any other estimates in the near vicinity. The Station Stone Rectangle indicates in different ways the exact diameter and circumference of the Outer Bank for example the rectangle's length + breadth produces the 380.16 feet diameter of the Outer Bank.

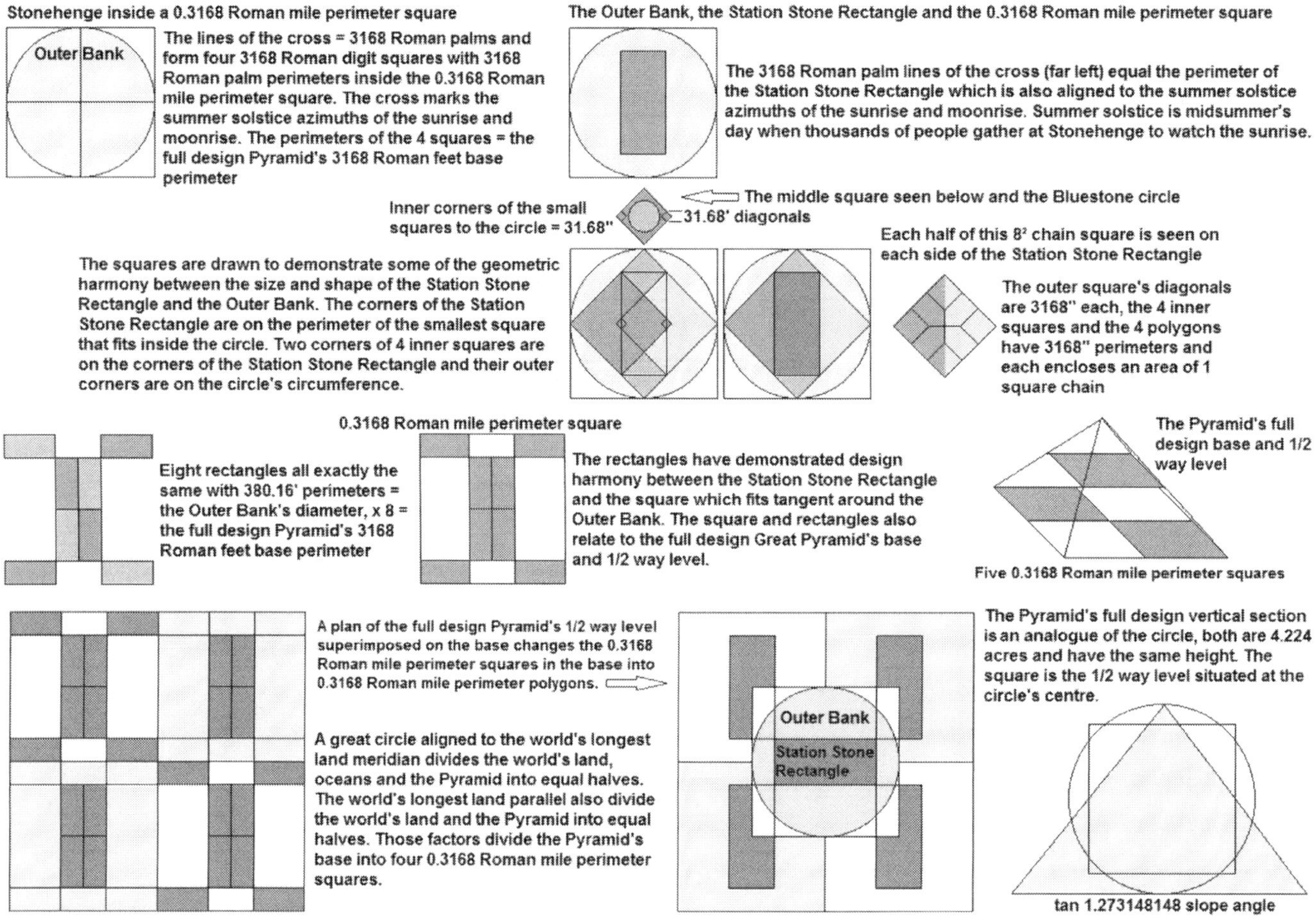

Stonehenge inside a 0.3168 Roman mile perimeter square

The lines of the cross = 3168 Roman palms and form four 3168 Roman digit squares with 3168 Roman palm perimeters inside the 0.3168 Roman mile perimeter square. The cross marks the summer solstice azimuths of the sunrise and moonrise. The perimeters of the 4 squares = the full design Pyramid's 3168 Roman feet base perimeter

The Outer Bank, the Station Stone Rectangle and the 0.3168 Roman mile perimeter square

The 3168 Roman palm lines of the cross (far left) equal the perimeter of the Station Stone Rectangle which is also aligned to the summer solstice azimuths of the sunrise and moonrise. Summer solstice is midsummer's day when thousands of people gather at Stonehenge to watch the sunrise.

Inner corners of the small squares to the circle = 31.68"
The middle square seen below and the Bluestone circle
31.68' diagonals

The squares are drawn to demonstrate some of the geometric harmony between the size and shape of the Station Stone Rectangle and the Outer Bank. The corners of the Station Stone Rectangle are on the perimeter of the smallest square that fits inside the circle. Two corners of 4 inner squares are on the corners of the Station Stone Rectangle and their outer corners are on the circle's circumference.

Each half of this 8² chain square is seen on each side of the Station Stone Rectangle

The outer square's diagonals are 3168" each, the 4 inner squares and the 4 polygons have 3168" perimeters and each encloses an area of 1 square chain

0.3168 Roman mile perimeter square

Eight rectangles all exactly the same with 380.16' perimeters = the Outer Bank's diameter, x 8 = the full design Pyramid's 3168 Roman feet base perimeter

The rectangles have demonstrated design harmony between the Station Stone Rectangle and the square which fits tangent around the Outer Bank. The square and rectangles also relate to the full design Great Pyramid's base and 1/2 way level.

The Pyramid's full design base and 1/2 way level

Five 0.3168 Roman mile perimeter squares

A plan of the full design Pyramid's 1/2 way level superimposed on the base changes the 0.3168 Roman mile perimeter squares in the base into 0.3168 Roman mile perimeter polygons.

A great circle aligned to the world's longest land meridian divides the world's land, oceans and the Pyramid into equal halves. The world's longest land parallel also divide the world's land and the Pyramid into equal halves. Those factors divide the Pyramid's base into four 0.3168 Roman mile perimeter squares.

The Pyramid's full design vertical section is an analogue of the circle, both are 4.224 acres and have the same height. The square is the 1/2 way level situated at the circle's centre.

tan 1.273148148 slope angle

We have introduced a 0.3168 Roman mile circumference circle to the 0.3168 Roman mile perimeter square via the Great Pyramid and now we will see how the circle and square relate to Stonehenge's design. In the first illustration below the triangle is formed by drawing tan 0.44 lines (the angle of the Station Stone Rectangle's diagonals) from the left side corners of the 0.3168 Roman mile perimeter square, the lines meet each other on the circumference of the 0.3168 Roman mile circumference circle, this is repeated on each side of the square to produce the designs below. An eight pointed star is produced and the 4 points that pierce the circle are contained by the square and the 4 points that pierce the square are contained by the circle.

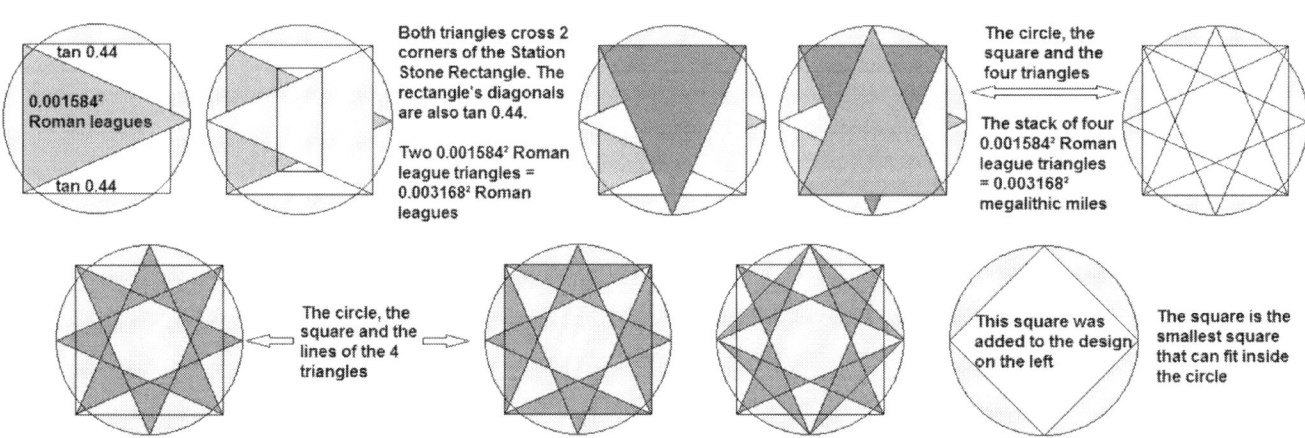

tan 0.44
0.001584² Roman leagues
tan 0.44

Both triangles cross 2 corners of the Station Stone Rectangle. The rectangle's diagonals are also tan 0.44.

Two 0.001584² Roman league triangles = 0.003168² Roman leagues

The circle, the square and the four triangles

The stack of four 0.001584² Roman league triangles = 0.003168² megalithic miles

The circle, the square and the lines of the 4 triangles

This square was added to the design on the left

The square is the smallest square that can fit inside the circle

The 8 pointed star is exactly 483.84 feet wide and 483.84 feet high and 483.84 feet x 22/7 = 1520.64 feet or 0.3168 Roman miles so the 4 points of the star that protrude 864/275 rods beyond the 0.3168 Roman mile perimeter square fit tangent inside a 0.3168 Roman mile circumference circle. The 0.3168 Roman mile circumference circle symbolized above by the Pyramid's vertical section has a 484 feet diameter and 484 feet x 864/275 = 0.3168 Roman miles circumference. If we assign the line of the Stonehenge's circle a thickness of 1 Roman inch and the inner circumference via 22/7 = 0.3168 Roman miles then the outer circumference via 864/275 would also be 0.3168 Roman miles which brings those circle and square features of the two monuments into perfect harmony. This is an interesting rectification factor using the Roman inch, the 0.3168 Roman mile Prime Measure of this investigation and the 22/7 with the 864/275 methods for pi.

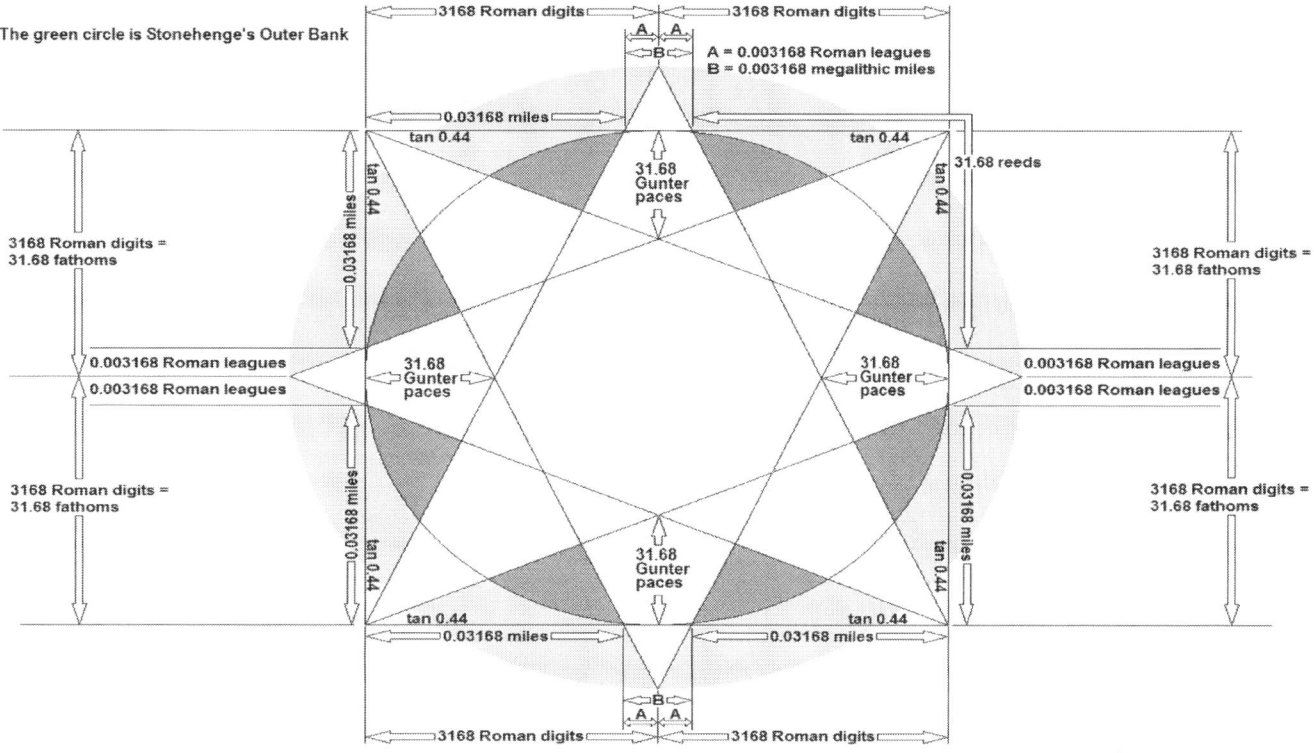

Notice how the 1/2 circles on the right instantly reveal the design harmony between the size and shape of the Station Stone Rectangle and the Outer Bank's outer circumference demonstrating that the outer diameter has to be exactly 380.16 feet or the harmony would not exist.

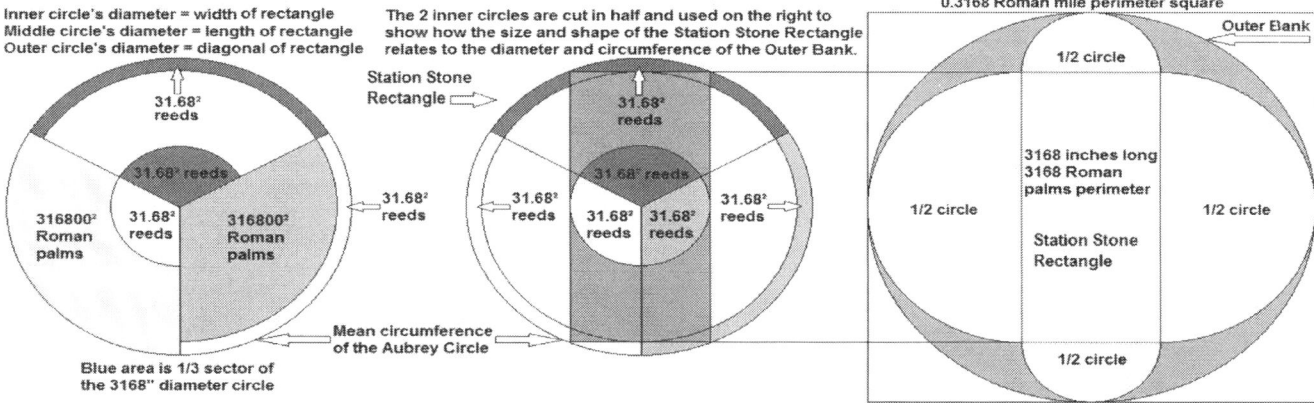

The sides of the Station Stone Rectangle are aligned to the summer solstice azimuths of the sunrise and moonrise which cross each other at right angles on Stonehenge's parallel, the design would not function properly if placed North or South of this parallel. One diagonal is aligned to a great circle route to the Pyramid which fixes the monuments East West position, therefore the exact global position of Stonehenge is fixed by the above factors. On the day of the Crucifixion the Passover moonrise azimuth at Stonehenge formed the Christ Angle with a great circle route to the crucifixion site, this also fixes the global position of Stonehenge. There was a lunar eclipse on this day which is very appropriate and can only happen on a full moon and the feast day of Passover was always held on a full moon, this feast day commemorated the Angel of Death passing over the Israelites in Egypt and was a prophecy of the sacrifice of the spotless Lamb of God being sacrificed at the Crucifixion. If Jesus had premarital sex with Mary Magdaline as some claim then he would not have been the spotless lamb demanded by the law and would certainly not have bothered going through with an invalid sacrifice.

The 86400 feet or 31680 megalithic yards from the centre of Stonehenge to the centre of Silbury Hill and the tan 0.44 diagonals of the Station Stone Rectangle are the prime factors of the designs below.

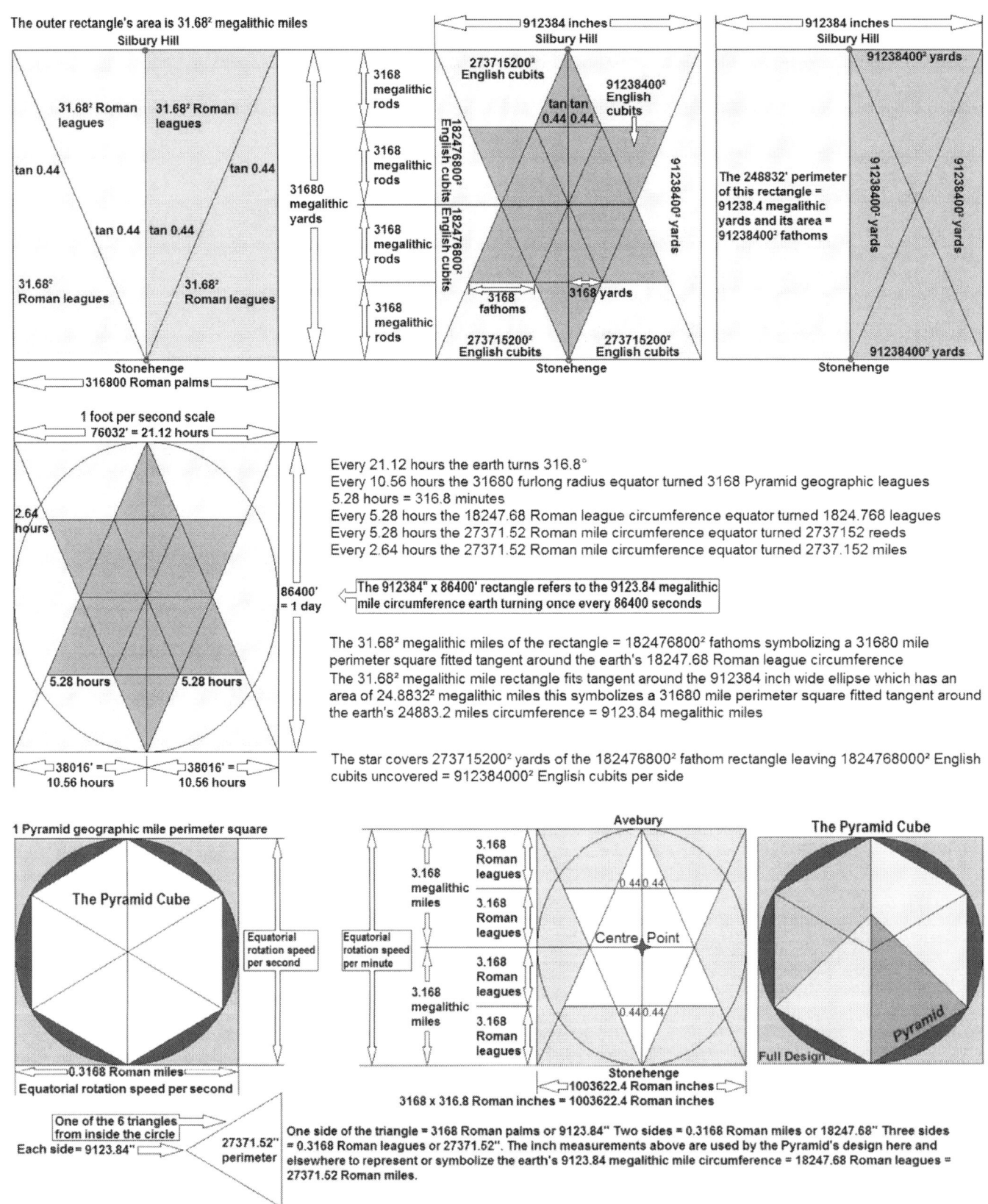

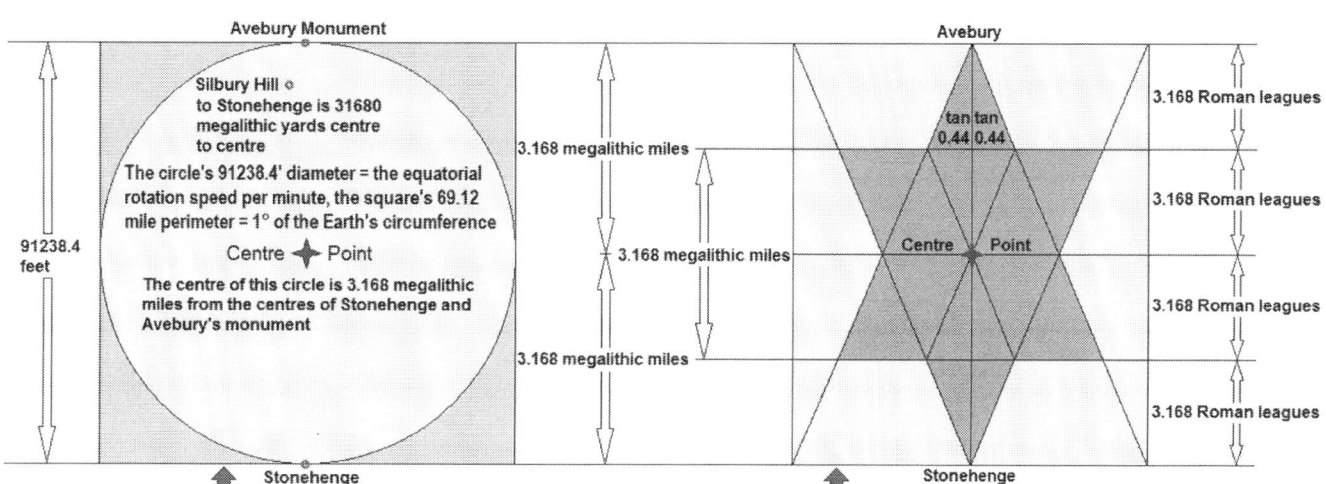

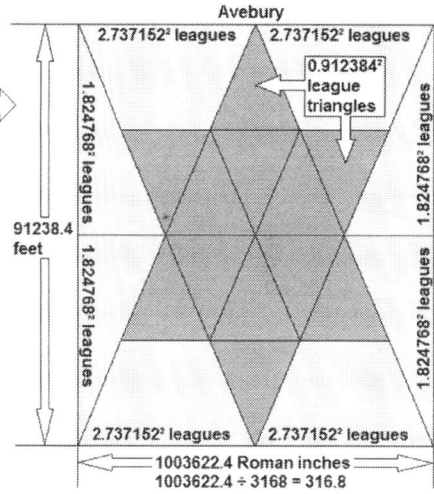

My source said it is 1 mile from the centre of Avebury's monument to the centre of Silbury Hill but when I checked I found it is 1 mile to the far side of Silbury Hill which is no problem because the mile is still featured and the offset works perfectly with other design factors. For example if we project the slope angle from the far side of Silbury Hill over Avebury it increases 3168 Roman feet in elevation above the centre of Avebury's monument which equals the full design Pyramid's base perimeter.

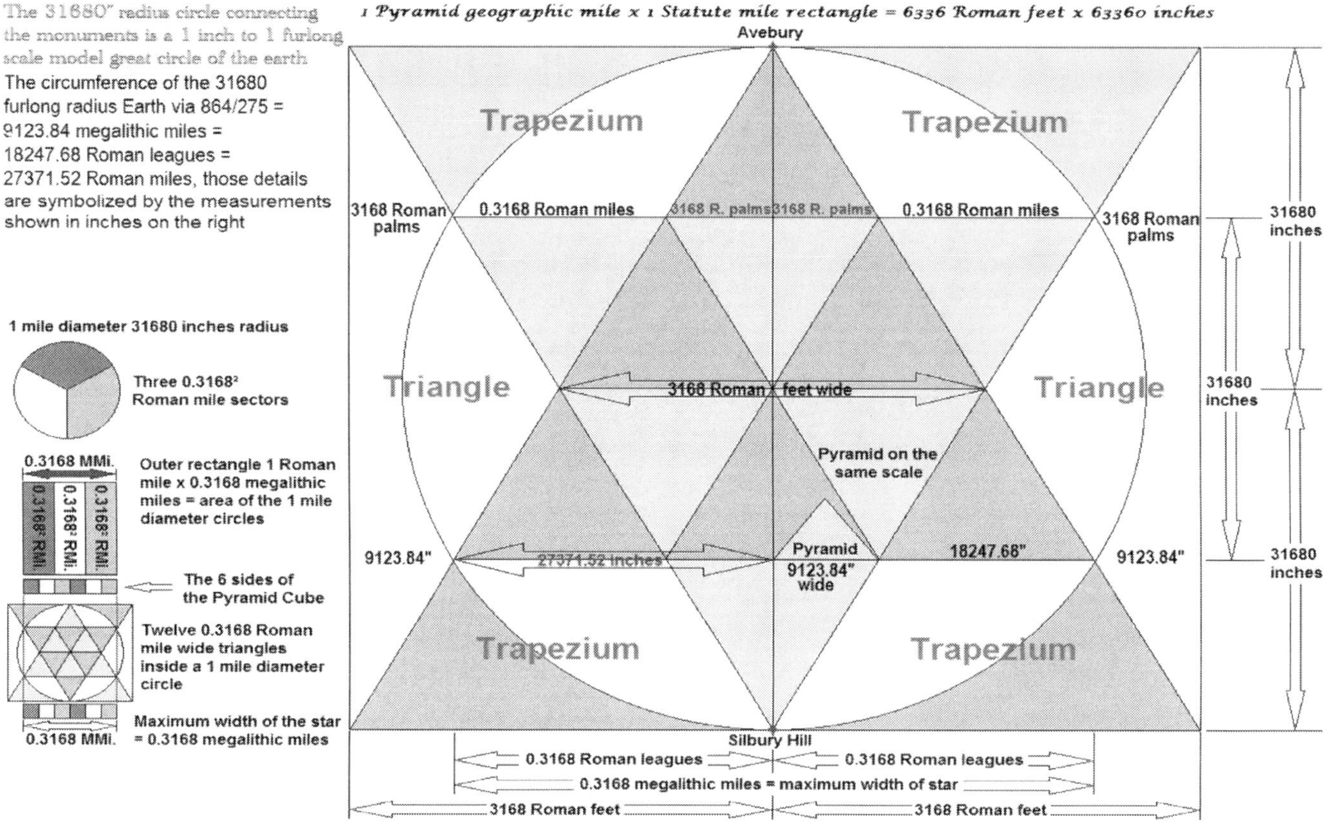

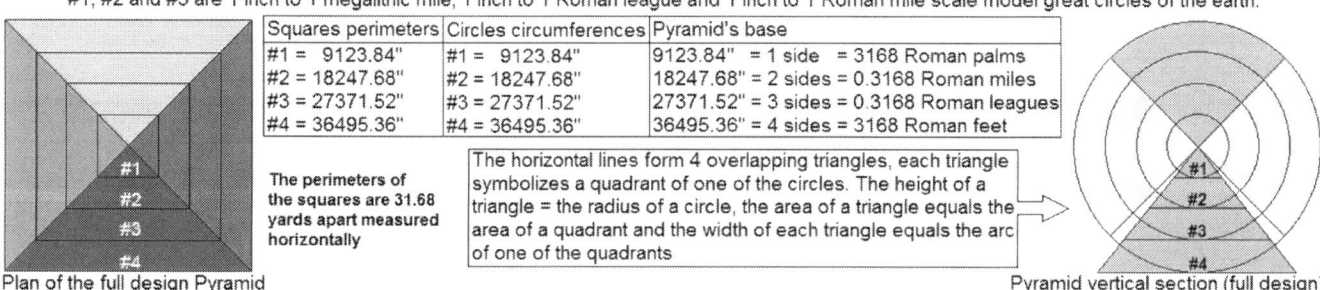

0.3168 Roman miles x 31680 inches = area of 1 **Triangle**	3168 yards x 316.8 feet = area of 1 **Trapezium**	0.3168 Roman mile wide triangle's area is the product of 3168 feet x 316.8 feet	0.3168 Roman miles = width of eight of the triangles inside the star
0.3168 megalithic miles x 1 Roman mile = area of the circle	0.3168 megalithic miles x 31680 inches = area of the star	3.168 Old English miles = (1 league) x 3168 Roman palms = area of the star	0.3168 Roman miles x 1 megalithic mile = area of the circle

The four measurements featured below are also featured inside the 1 inch to 1 furlong scale model great circle of the earth above and three of the circles below are also scale model great circles of the earth on the scale of 1 inch to 1 megalithic mile, 1 inch to 1 Roman league and 1 inch to 1 Roman mile.

The Pyramid's full design base actually focuses attention on the circles and squares which selects them as features of interest. The circles #1, #2 and #3 are 1 inch to 1 megalithic mile, 1 inch to 1 Roman league and 1 inch to 1 Roman mile scale model great circles of the earth.

Squares perimeters	Circles circumferences	Pyramid's base
#1 = 9123.84"	#1 = 9123.84"	9123.84" = 1 side = 3168 Roman palms
#2 = 18247.68"	#2 = 18247.68"	18247.68" = 2 sides = 0.3168 Roman miles
#3 = 27371.52"	#3 = 27371.52"	27371.52" = 3 sides = 0.3168 Roman leagues
#4 = 36495.36"	#4 = 36495.36"	36495.36" = 4 sides = 3168 Roman feet

The perimeters of the squares are 31.68 yards apart measured horizontally

The horizontal lines form 4 overlapping triangles, each triangle symbolizes a quadrant of one of the circles. The height of a triangle = the radius of a circle, the area of a triangle equals the area of a quadrant and the width of each triangle equals the arc of one of the quadrants

Plan of the full design Pyramid   Pyramid vertical section (full design)

The measurements #1, #2 and #3 refer to the earth's circumference and #4 is 36495.36 inches or 3168 Roman feet which refers to a 31680 mile perimeter square fitting tangent around the earth's circumference, also the English monuments have shown that the 36495.36 inch perimeter of the full design Pyramid's base (square #4) refers to 1° of the earth's circumference = 36495.36 feet, it is also interesting that 36495.36 miles = 31680 Pyramid geographic miles.

The Numerical Universe

The design below connecting Silbury Hill and Stonehenge is based on the distance from the centre of Silbury Hill to the centre of Stonehenge and the tan 0.576 slope angle of Silbury Hill.

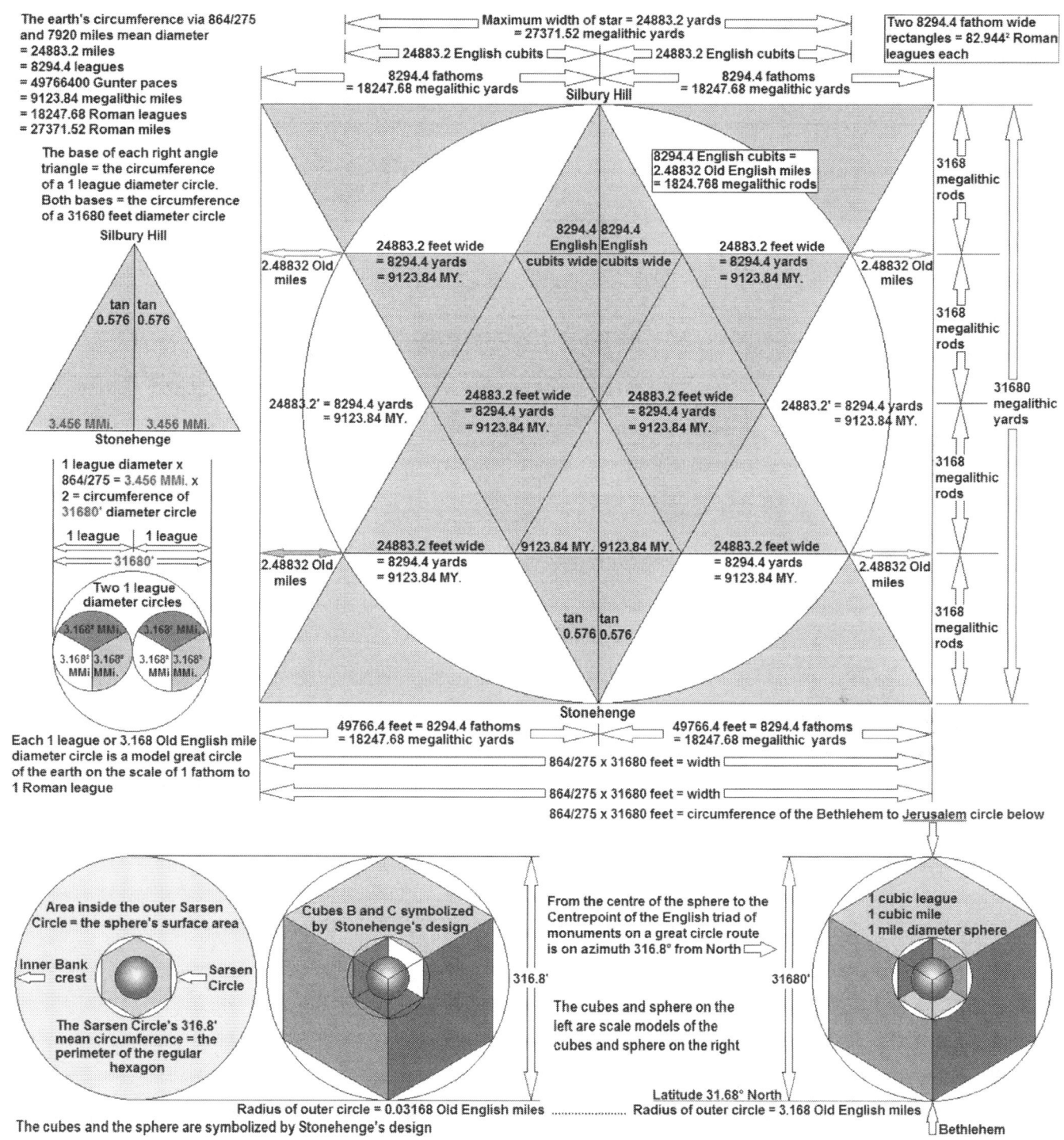

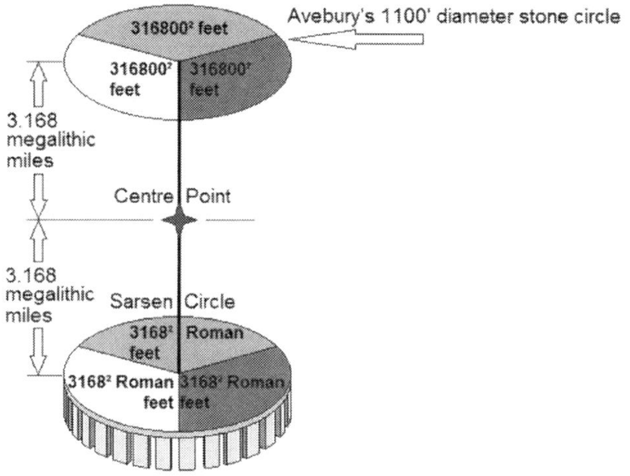

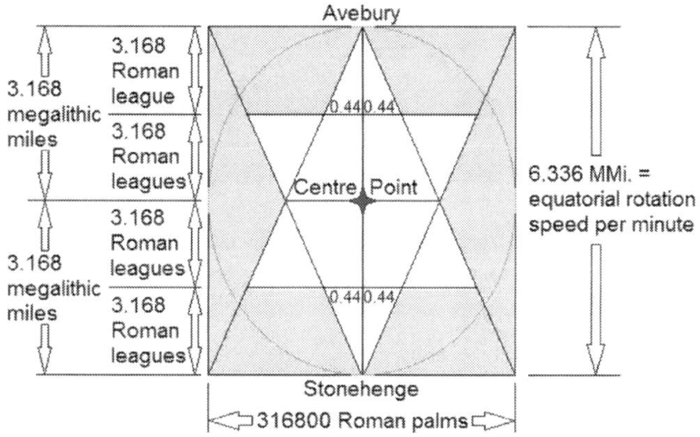

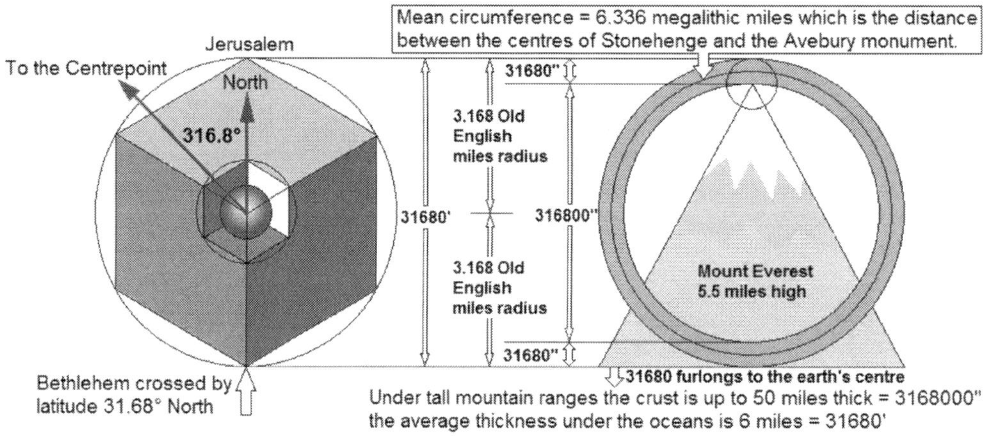

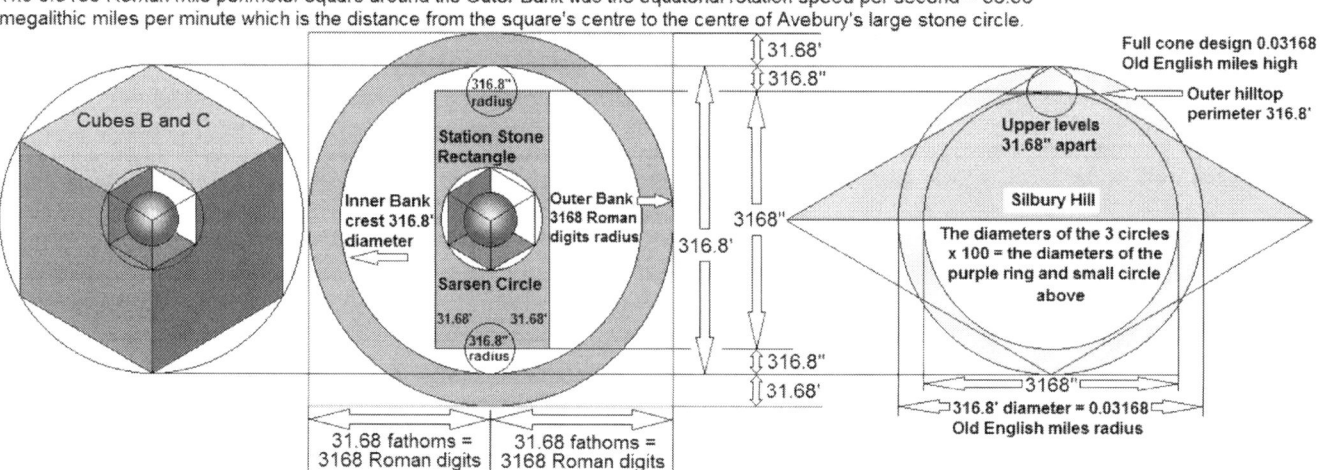

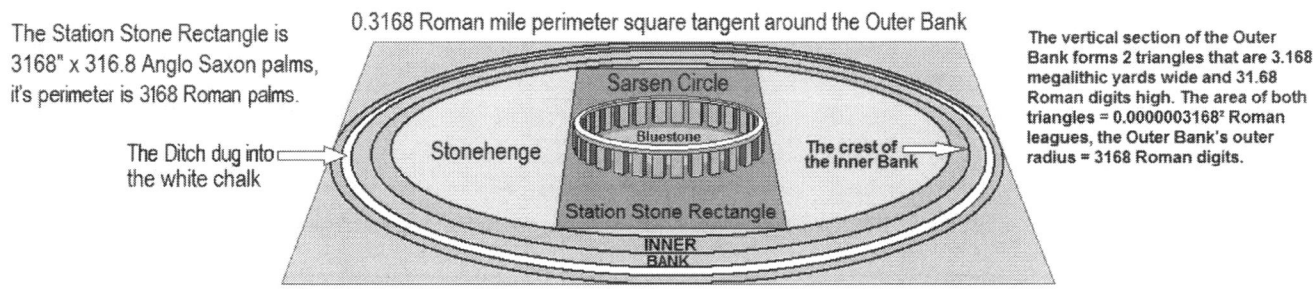

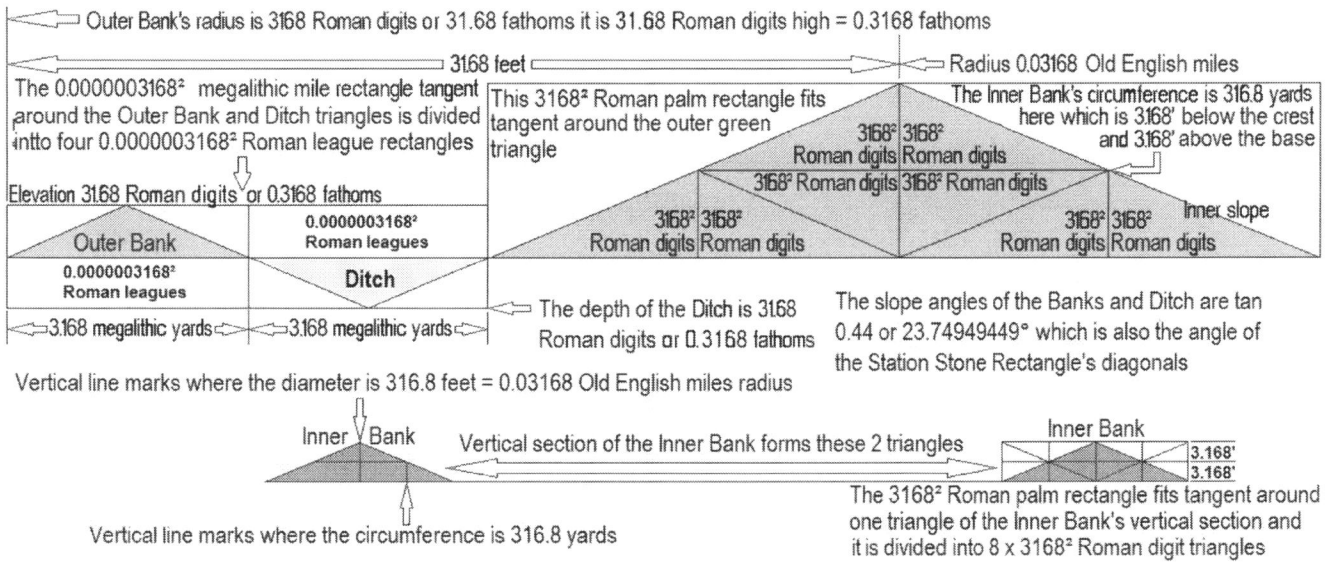

Drawing the circle above with a 1 Roman digit thick line added 1 Roman digit to the radius, if instead we added 1 Roman digit to the diameter and calculated the circle's area using 22/7 we get 9.504 roods = 3 x 3.168 roods which is the area inside the mean circumference of the 1 Roman digit thick line used to draw the circle. The rectification factors are not merely fix it solutions but remarkable added features that enrich the monument's design. The red lines of each 120° angle = 6336 Roman digits, each contains a 3.168 rood sector with an arc of 6336 Roman digits the outer diameter of the Outer Bank is 6336 Roman digits, these are features of design harmony signifying we are on the right path, the message of the monuments are revealed by following paths marked by signs.

1 side of Cube A squared x pi = the area inside the Outer Bank's outer circumference, 1 side of Cube B squared x pi = the area inside the Inner Bank's crest, 1 side of Cube C squared x pi = the area inside the Sarsen Circle's outer circumference.

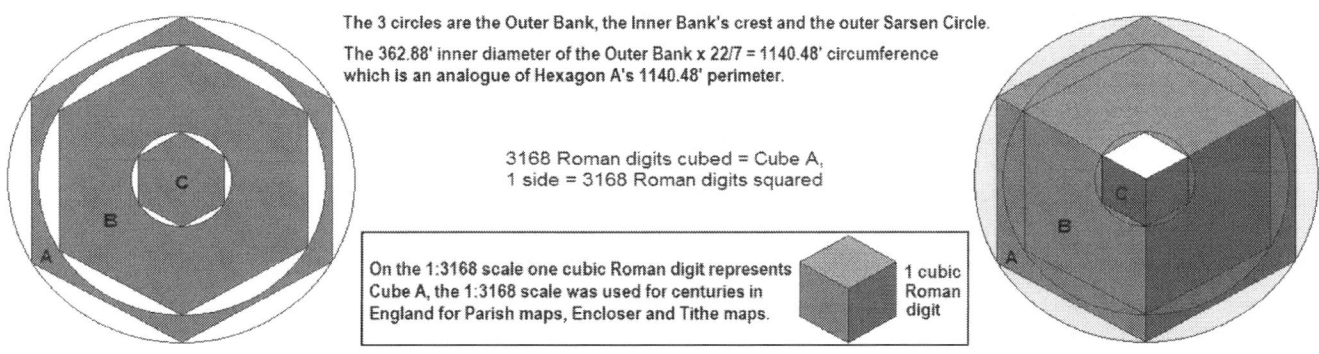

The 1:3168 scale means 1" on the map represents 3168" or 4 chains on the ground therefore using this scale the 3168" long Station Stone Rectangle would be 1" on the map its 3168 Roman palm perimeter would be 1 Roman palm on the map and the 3168 Roman digit radius would be 1 Roman digit on the map.

Below: The 0.3168 Roman mile perimeter square fitted tangent around Stonehenge's 0.248832 Roman mile circumference Outer Bank refers to a 31680 mile perimeter square fitted tangent around the earth's 24883.2 mile circumference, the Outer Bank's 3168 Roman digit radius relates to the earth's 31680 furlong radius.

Before the equator bulged it had a 24883.2 mile circumference = 18247.68 Roman leagues and 18247.68" is how far it turned in 1 second which is recorded by the square's 18247.68" perimeter (= 0.3168 Roman miles).

If the outer circle on the left was drawn with a 1 Roman inch thick line and an inner diameter of 483.84' the outer diameter would be 484' = outer circle on the right.
If the inner circle on the left was drawn with a 1 Roman digit thick line and an inner diameter of 362.88' the outer diameter would be 363' = inner circle on the right.

Stonehenge

The outer circles on the left and on the right	The inner circles on the left and on the right
483.84' diameter x 22/7 = 1520.64' circumference	362.88' diameter x 22/7 = 1140.48' circumference
484' diameter x 864/275 = 1520.64' circumference	363' diameter x 864/275 = 1140.48' circumference

The Great Pyramid

484' high
363'

tan 0.44   tan 0.44   864/275 rods

3.168 roods

3.168 roods   3.168 roods

The 18247.68" circumference circle and 18247.68" perimeter square have their centres on the 18247.68" perimeter 1/2 way level

The area of the 363' diameter circle and the 363' high triangle are the same and the centre of this 1140.48' circumference circle is on the 1140.48' perimeter level

The Station Stone Rectangle's diagonals are tan 0.44. The tan 0.44 lines drawn from each corner of the square pass through a corner of the rectangle and end on the outer circle's circumference forming two points 483.84 feet apart

tan 0.44   tan 0.44   864/275 rods

3.168 rood sector

3.168 rood sector   3.168 rood sector

The area of the Pyramid's 484' high vertical section and the 484' diameter circle are the same

The 18247.68 inch circumference circles above are models of great circles of the earth on the scale of 1 inch to 1 Roman league, the 27371.52 inch circumference circle below is a model great circle of the earth on the scale of 1 inch to 1 Roman mile.

The earth's 7920 mile diameter x 864/275 = 24883.2 miles circumference = 18247.68 Roman leagues or 27371.52 Roman miles.

The outer circumference of Stonehenge's Outer Bank encloses 2.605949673 acres; multiply by the Pyramid's slope angle tan 1.273148148 = 3.31776 acres which is the area of one of the 0.3168 Roman mile perimeter squares above, one square fits tangent around Stonehenge and the other square is the Pyramid's ½ way level.

One square's 3.31776 acres x tan 1.273148148 = 4.224 acres which is the area of the Pyramid's full design vertical section or the area of one of the 0.3168 Roman mile circumference circles.

The earth's 24883.2 mile circumference x tan 1.273148148 = 31680 miles which equals the circumferences of the earth and the moon (via 22/7) and 31680 miles is also the perimeter of a square fitted tangent around the earth.

Stonehenge's Outer Bank has a 0.248832 Roman mile circumference; multiply by tan 1.273148148 = 0.3168 Roman miles which is the perimeter of the square fitted tangent around the Outer Bank.

The original equatorial rotation speed per second calculated via 864/275 was 0.3168 Roman miles = 18247.68" this was when the equator had no bulge and had a 18247.68 Roman league circumference.

Each 18247.68" circumference circle represents a scale model 18247.68 Roman league circumference great circle of the earth, an 18247.68" circumference circle = 4.224 acres via 864/275 and equals the area of the full design vertical section, 4.224 acres x 864/275 = 1327104 acres which is the area of the 3168 Roman palm square base which has a 3168 Roman feet perimeter.

The 13.27104 acres of the base x tan 1.273148148 = 16.896 acres and this is the area of a 3168 Roman feet circumference circle, the full design vertical section is an analogue of one quadrant of the 3168 Roman feet circumference circle.

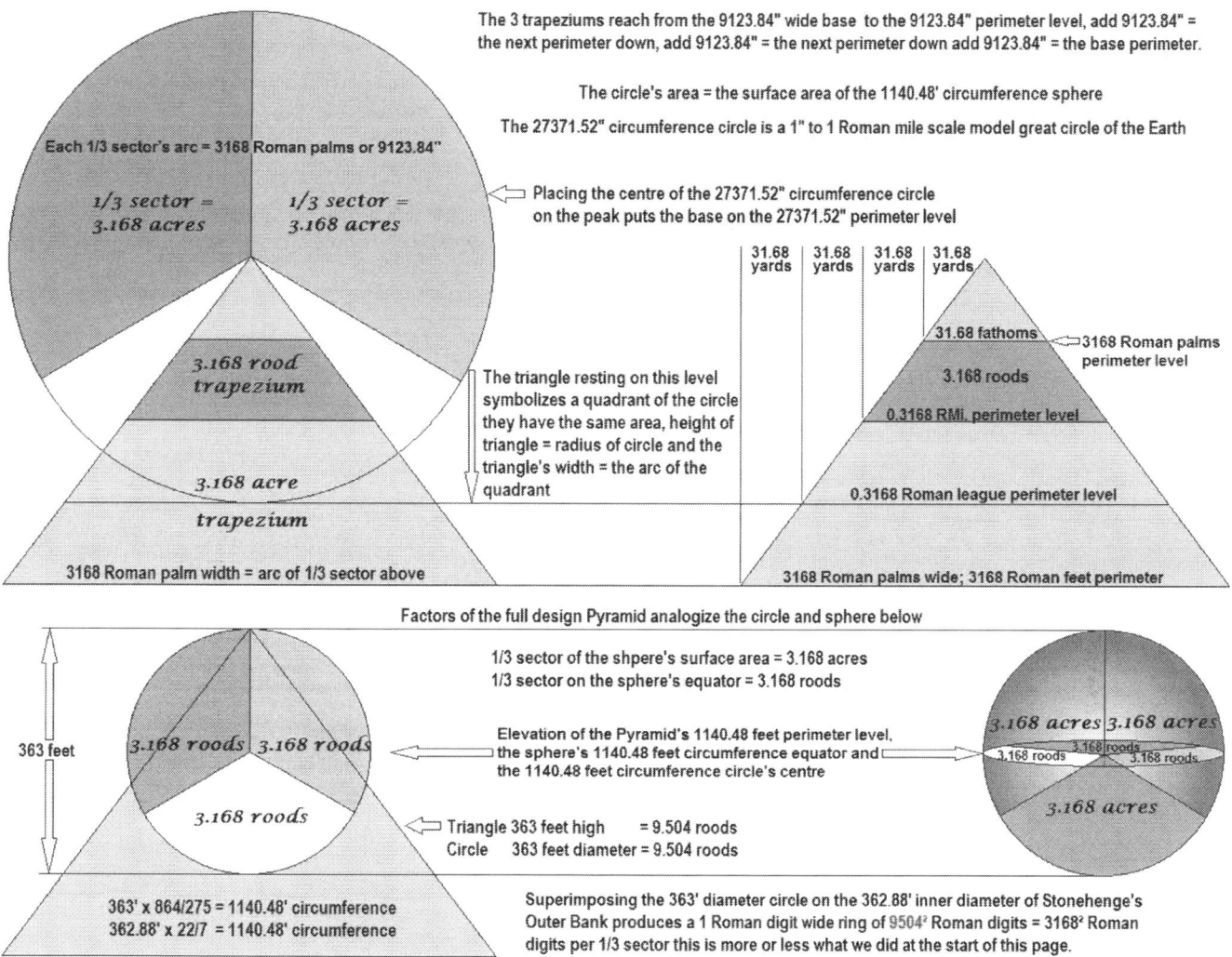

2 Chronicles 3:3 Now these *are the things wherein* Solomon was instructed for the building of the house of God. The length by cubits after the first measure *was* threescore cubits, and the breadth twenty cubits.

Ezekiel 43:13 And these *are* the measures of the altar after the cubits: The cubit *is* a cubit and an hand breadth; even the bottom *shall be* a cubit, and the breadth a cubit,

1 Kings 6:2 And the house which king Solomon built for the LORD, the length thereof was threescore cubits, and the breadth thereof twenty cubits, and the height thereof was thirty cubits. The dimensions above along with other Biblical information inform us that the House of the Lord is comprised of the Holy of Holies (Oracle), the Upper Chamber and the Holy Place, so the Porch and the 3 tiers of chambers around the building are not included when the House of the Lord is referred to.

The Bible verse above also provides the necessary information (with the help of other Biblical details) to determine the size and whereabouts of the Upper Chamber and that it is on top of the Holy of Holies (Oracle) and its length and breadth are the same as the Oracle and its height + the Oracle's height = the height of the Holy Place.

The description of the House of the Lord's dimensions allows zero thickness for the wall separating the Oracle from the Holy place and zero separation between the Oracle and the Upper Chamber this must be to focus on the spiritual symbolism of the measurements and for some reason the walls do not count.

It is important to note that I have drawn the Biblical description and not the actual Temple that was built which would look more pleasing to the eye when the thick walls and pitched roof etc. were included.

I believe the cubit referred to in the Bible concerning Solomon's Temple is the great cubit of 1.76 feet, also I believe the great cubit is referred to in the description of Noah's Ark, the Tabernacle in the Wilderness, the Ark of the Covenant and Ezekiel's visionary temple etc. Using the great cubit produces many profound results instead of the mostly meaningless results produced using any of the other versions of the cubit.

Each long side of the House of the Lord has a 316.8' perimeter, the base perimeter of the House of the Lord with it's Porch = 316.8',
The the Sarsen Circle's 316.8' mean circumference is an analogue of the 316.8' perimeter regular hexagon.

The Sarsen Circle's 633.6" outer radius and 1267.2" outer diameter equal the height of the House of the Lord and it's length, the Temple's 35.2' width probably equals the width of the avenue and the gaps in the banks at Stonehenge.

The Bible defines the House of the Lord as the Holy Place and the Oracle with the Upper Chamber above it, the Porch and chambers round about are not included.

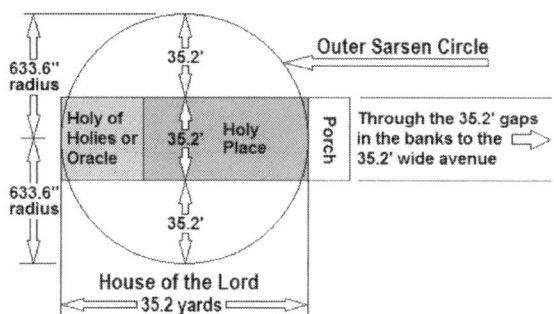

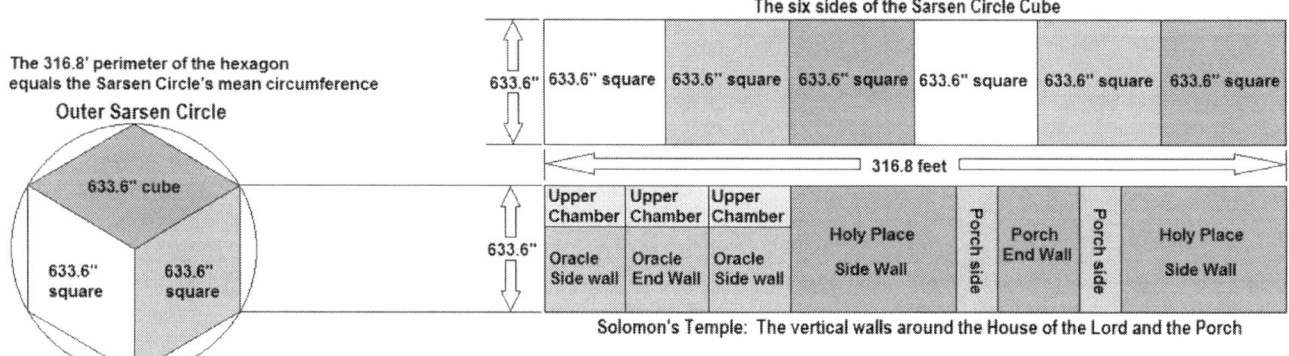

3.168 miles x 1 foot = the surface area of the 633.6" cube = the surface area of the Tower = the suface area of the walls around the House of the Lord and Porch, therefore the vertical faces = the surface area of two 633.6" cubes, interesting because the volume of the two cubes = the volume of the House of the Lord and the Tower. The two 633.6" cubes volume x 316.8 = the full design Pyramid's volume, two 633.6" cubes are symbolized by the designs of Stonehenge and Silbury Hill (i.e. 1 each).

Jeremiah 32:20 Which hast set signs and wonders in the land of Egypt, *even* unto this day, and in Israel, and among *other* men; and hast made thee a name, as at this day.

It says God set these signs and wonders which means he was the cause, they are the Great Pyramid (wonder in the land of Egypt), Solomon's Temple (wonder in Israel) and the wonders among other men refers to the Stonehenge group of monuments there are no other contenders and Stonehenge was already ancient before Jeremiah was born.

It has been thoroughly demonstrated that the designs of the English monuments and the Great Pyramid have the same Source and Solomon's Temple was also built to God's specifications. It has been demonstrated how the designs of the English monuments interact with the full design Great Pyramid so we would expect the same concerning the design of Solomon's Temple and our expectations are not in vain if the cubit referred to in the Bible's descriptions is the great cubit of 1.76 feet.

Ezekiel 40:5 And behold a wall on the outside of the house round about, and in the man's hand a measuring reed of six cubits *long* by the cubit and an hand breadth: so he measured the breadth of the building, one reed; and the height, one reed.

Above the Bible states six great cubits = 1 reed, we know they are great cubits because it says each is 1 handbreadth longer than the cubit they were using at the time

2 Ch 3:3 Now these *are the things wherein* Solomon was instructed for the building of the house of God. The length by cubits after the first measure *was* threescore cubits, and the breadth twenty cubits.

The first time a cubit is mentioned in the Bible is in the measurements of Noah's Ark predating the reference of any cubit in recorded history, therefore this must be the cubit after the first measure and if this cubit = 1.76 feet then the design of Noah's Ark also interacts with the design of Solomon's Temple, Stonehenge, Silbury Hill etc.

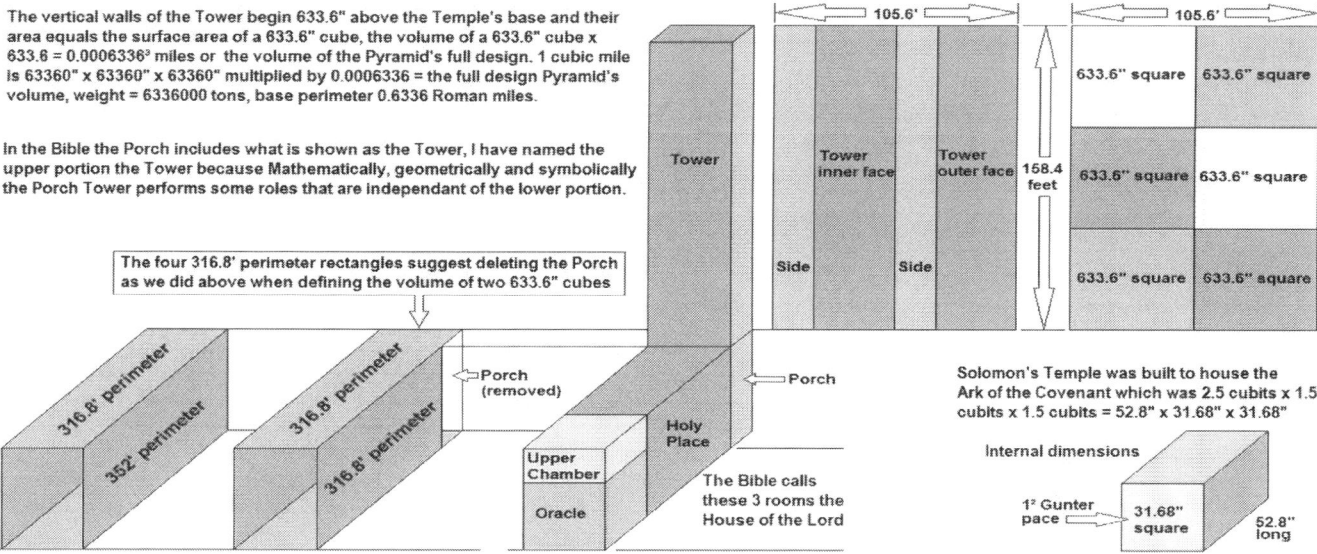

The Bible says Noah's Ark was 300 cubits long, 30 cubits high and 50 cubits wide, using the cubit of the first measure a cubit + a handbreadth i.e. the great cubit of 1.76 feet, the Ark was 528 feet long, 52.8 feet high and 88 feet wide.

I believe the Ark was a regular boat shape and the Biblical measurements refer to the symbolism of the usable storage space inside the Ark.

The Ark was probably 528' long and 52.8 feet high but the width would vary, author Jonathan Gray has written a good account of why the remains of Noah's Ark are a regular boat shape about 138 feet wide and how this relates to the Bible's 88 feet wide description.

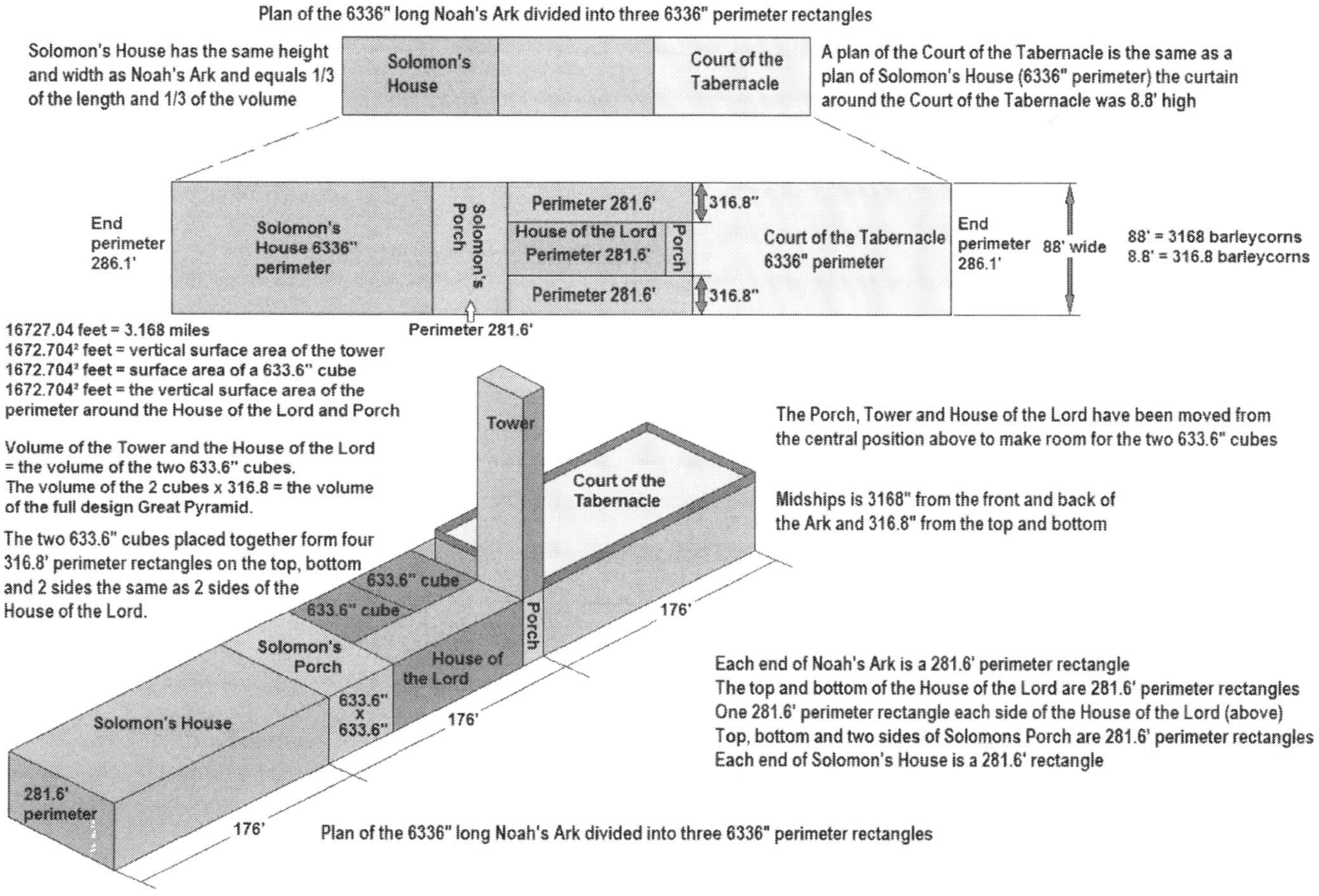

The numbers inside boxes are taken from the Musical Scale Chart, the numbers of music are more involved in the designs than is indicated below.

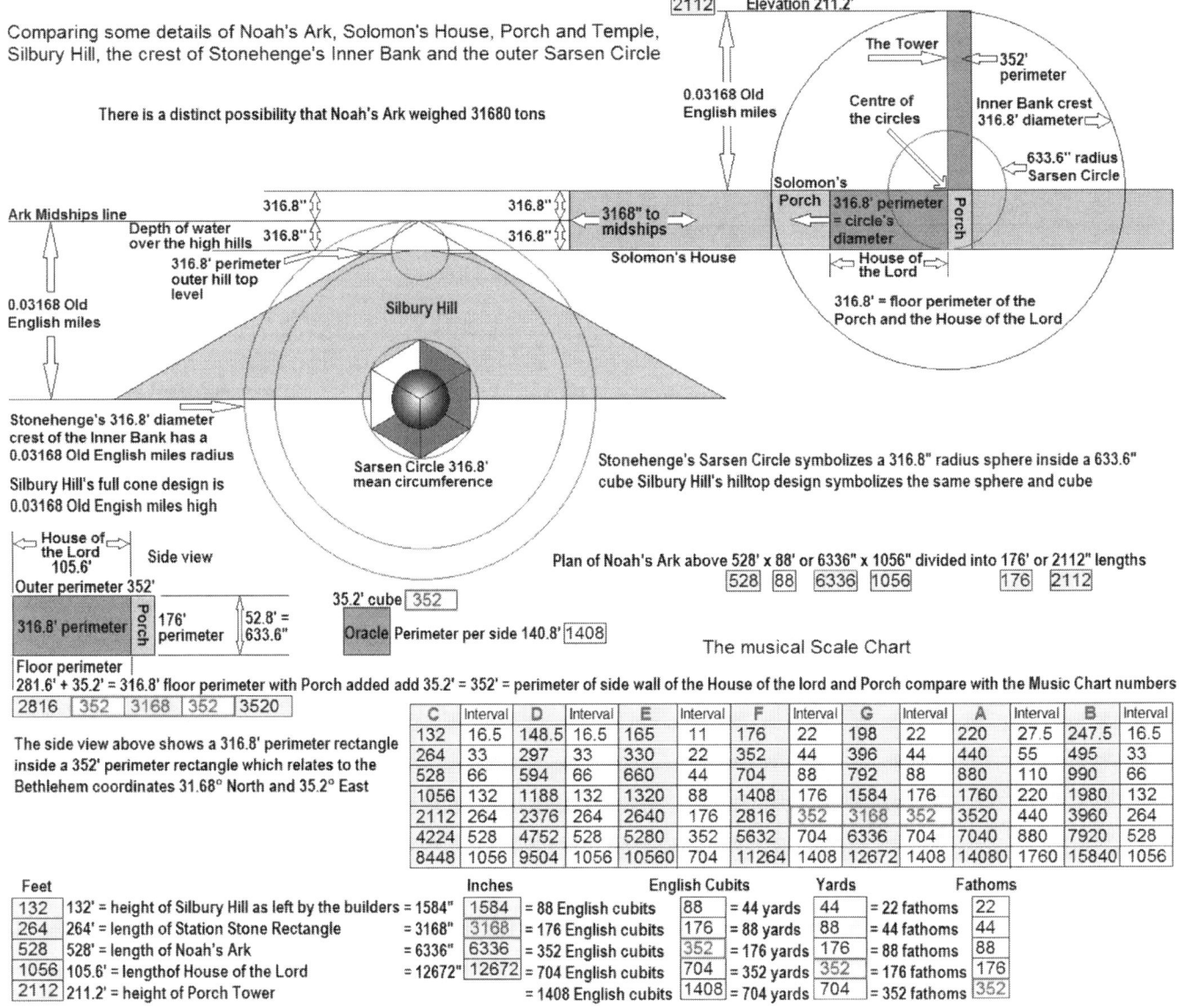

Jeremiah 32:20. Which hast set signs and wonders in the land of Egypt, *even* unto this day, and in Israel, and among *other* men; and hast made thee a name, as at this day. It is interesting that the prophet Jeremiah in the Bible verse above refers to the Great Pyramid as a wonder and we know it as one of the Seven Wonders of the World, the wonder in Israel refers to Solomon's Temple and the wonders among other men must refer to the Stonehenge group of monuments, there are no other contenders.

The Stonehenge group of monuments were already ancient in Jeremiah's time, Stonehenge, Silbury Hill, Avebury, Solomon's Temple and the Great Pyramid are referred to in the same verse of the Bible, and we have demonstrated how the designs of the English monuments profoundly interact with each other and the Great Pyramid therefore we would expect the design of Solomon's Temple to also interact with the other monuments on the list and it does if the cubit referred to in the Bible is the great cubit of 1.76 Imperial feet.

The mean circumference of the lintels on top of Stonehenge's Sarsen Circle is 316.8 feet and the mean length of each lintel is the Biblical reed of six great cubits = 10.56'.

The two Bible verses below refer to the Great Pyramid and have a Gematria of 5449 in the original Hebrew text which refers to the 5449 Pyramid inch height of the Pyramid from the socket level to the summit platform, and the 5449 Pyramid inches on the floor line from the entrance to the Dead End, the 5449 Pyramid inches also refers to the mean height of all the world's land above sea level.

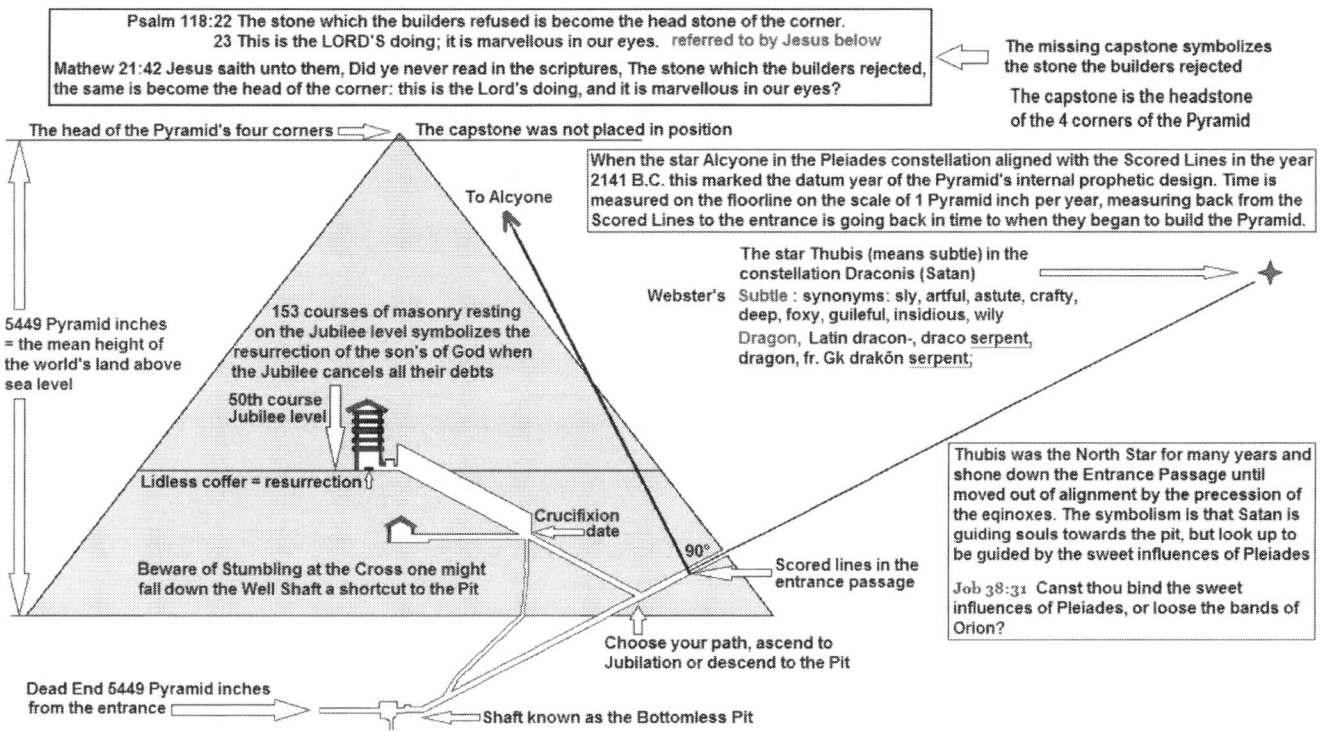

The Lord was born as an Israelite so that he could legally redeem his kindred people according to the law as a kinsman redeemer by paying the debt or the penalty they had incurred by their disobedience and breaking the vows they made at Mount Sinai. He also came as a descendant of the House of David which secured his right to the crown but he was rejected of men and like the Pyramid he was not crowned, but it's not over yet and one day the whole world will know he is the King of kings.

The Israelites were given the Ten Commandments at Mount Sinai and made a vow to obey which they broke this is why they needed to be saved from the penalty demanded by the law, Jesus purchased their freedom by paying the penalty himself this also paid everyone's penalty. The Bible says Israel was his treasure buried in the field and he purchased the whole field to gain the treasure, and the Bible says the field represents the world.

The lidless coffer symbolizes the resurrection and the coffer is resting on the 50th course of masonry which represents the Jubilee year, at the end of 7 sabbatical years (49 years) the Jobel (trumpet) was blown to announce the start of the Jubilee year. This was when all debts were cancelled, slaves were set free and people regained their land inheritance if they had sold it, the buyers of the land knew they only owned it until the Jubilee year. People could sell themselves into slavery and the courts could sell a law breaker into slavery if he had to pay compensation and did not have the means to do so. Slaves could buy their freedom at anytime if they had the money but if anyone else wanted to redeem them their master could refuse to release them unless the redeemer was a kinsman, hence the term kinsman redeemer.

When Jesus saved his people he saved the whole world, (he saved us from the second death) which is referred to by Paul in Romans 6:23 For the wages of sin is death; but the gift of God is eternal life through Jesus Christ our Lord. So eternal life is a gift through the Lord Jesus Christ and if you accept the gift it means you claim exemption from the second death (wages of sin = death penalty) because Jesus paid it for you anyone can refuse the free gift of salvation and pay the penalty personally, the law demands payment and it will be paid one way or another.

People who had no means of knowing the law are also judged with justice according to their deeds, this probably does not include the deliberately ignorant i.e. the ones who could have known and should have known but hid from the truth.

The lidless coffer and 153 courses of masonry are resting on the Jubilee level and the expression "Sons of God" (Beni Ha-Elohim) has a Gematria of 153 it occurs seven times in the Bible's Hebrew text. The symbolism is that the Sons of God (153) standing on the 50th course of masonry are celebrating (Jubilation) the Great Jubilee all debts have been cancelled they have been redeemed from sin (death) they have regained their inheritance and are resurrected which is symbolised by the lidless coffer.

The sign of the fish was a symbol used by followers of Jesus and there were 153 fishes caught in the net and not one was lost, this is also in harmony with the above, 8 x 153 = 1224 which is the Gematria of the Greek word for fishes, and 1224 is also the Gematria for the Greek words "the net", in the Biblical meaning of numbers 8 symbolizes a new beginning and relates to Jesus = 888 Gematria.

There are 7 days in a week and day 8 begins a new week. There are 7 notes per row on the Musical Scale Chart and the 8th note starts the next row, the chart's 7 rows of 7 notes remind me of the 7 sabbatical years from the end of the Jubilee to the beginning of the next heralded by the blowing of the Jobel (trumpet), also the 7 golden candlesticks in Solomon's Temple held 7 candles each.

Obviously no one was buried in the Great Pyramid which symbolizes resurrection and there is no evidence of a burial, Herodotus said Khufu was buried in an obscure place, orthodox experts disagree and dismiss this, the fact that there is no evidence of anyone being buried in the Great Pyramid has no effect on their theories, who needs evidence when we have the opinions of experts?

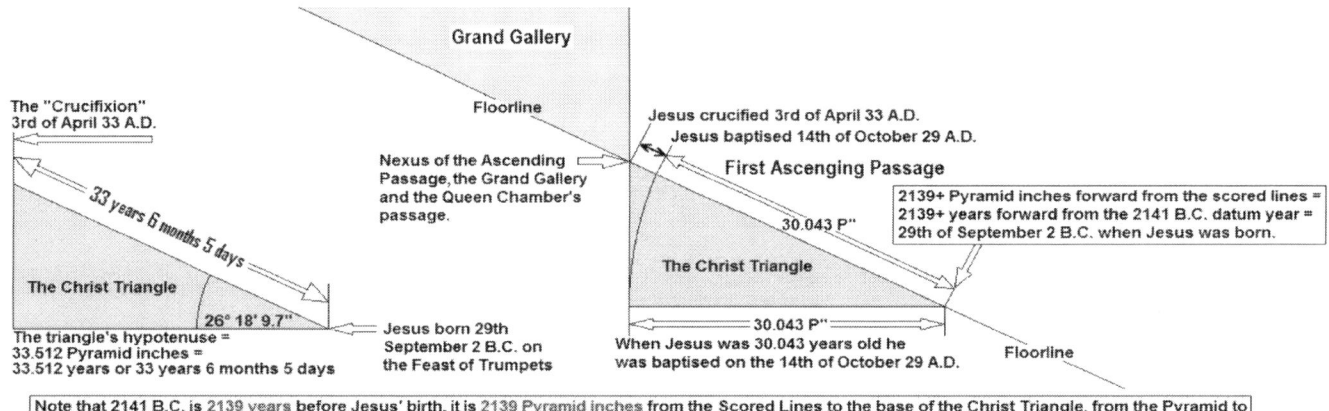

Jesus was born on the 29th of September 2 B.C. if our calendar had been formulated correctly Jesus would have been born in the year zero but our calendar skips the year zero going from 1 B.C. to 1 A.D. so the closest we can get to the correct date with our calendar is to pick the year before or the year after Jesus was born. Jesus was born very appropriately on the Feast of Trumpets according to the Pyramid's prophetic time scale and this is also borne out by careful analysis of relevant information found in the Bible along with all the relevant existent secular historical documents. Some estimations of the date of Jesus' birth are in error because not all the information available was taken into consideration, but if you think the Pyramid's Architect knew the real date you don't need opposing assumptions and opinions.

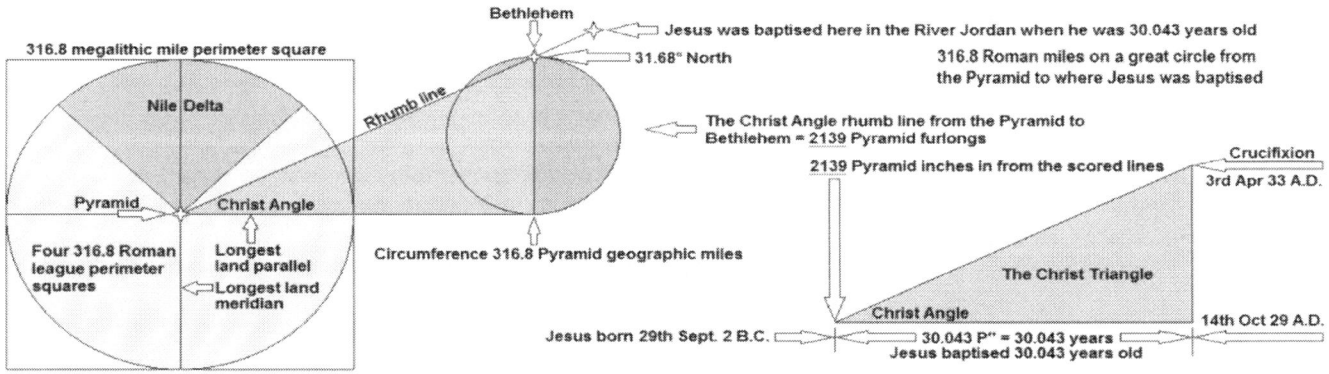

Below is one of many pictures found on Google Earth associated with the place where Jesus was baptised in the River Jordan, measuring 316.8 Roman miles from the Pyramid takes us to that position on the Jordan, whether this is exactly where Jesus was baptised or upstream or downstream I do not know and the Google Earth rule tool is not accurate enough to resolve this.

Have the locals selected the exact place where Jesus was baptised, there has always been a ford there which is where the Ark of the Covenant crossed and if there are steep banks either side this would restrict the area suitable for John the Baptist to do baptisms.

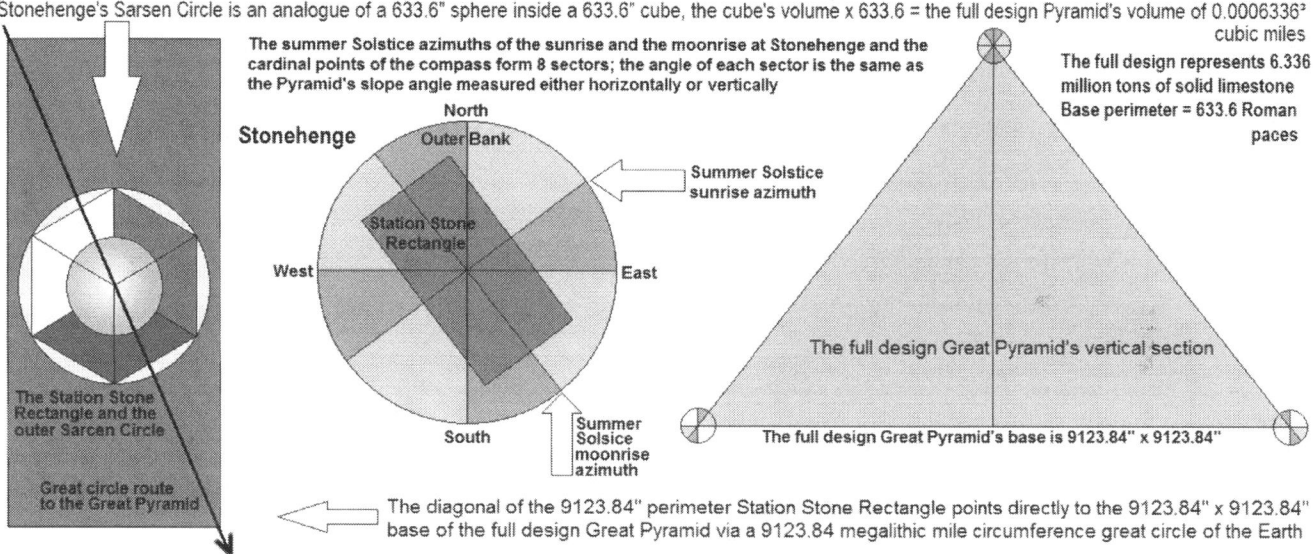

Supernatural Factors. The use of multiple units of measure before they were invented is supernatural, for example the Gunter Pace was invented about four hundred years ago by Edmund Gunter yet it was used in cosmic design before man existed and it has been demonstrated that all the units of measure used in this investigation were used by the Designer of the units of measure before any man ever used them.

This is not only prophetic but shows the Designer was in control from when he designed the units of measure until he caused man to use them.

Google Earth map

The earth's 7920 mile diameter x 864/275 = 24883.2 miles; divide by 100000 = 0.248832 miles which is the distance between the parallels or latitudes below. It is interesting that the ancient 864/275 method for π has been aplied here because this method for π is a major factor in the designs of Silbury Hill and the Avebury monument seen below.

0.248832 miles

864/275 x 0.3168 miles = 0.995328 miles

Parallel or line of latitude

I would like to explain something obvious that might be easily overlooked, Stonehenge and the other monuments are monuments to the Lord Jesus Christ who is the head of Christianity, therefore Christianity was not invented two and a half to three thousand years after Stonehenge was built as we are often told.

We are also told that Christianity is derived from Judaism but Judaism is an Anti Christ religion for example their Talmud says Jesus' father was a Roman soldier and his mother was a prostitute, the Bible says otherwise, and real Christians be they ever so few believe in the Bible no matter what the Talmud or the Koran or the New World Order's appointed experts say.

The 3168 numeric signature is revealed all over cosmic design when we use the units of measure that have been demonstrated to have the same Source as the cosmic design, this makes multiple references to the head of Christianity before man existed. So the message is that Christianity is the oldest religion even though it was known by other names i.e. it used to be called The Way and was first called Christianity in Antioch and this was a new label not a new religion because The Way and Christianity are exactly the same and Jesus said John 14:6 I am the way, the truth and the life: interesting that the first two words of the verse are I AM which is the name the Lord used when he sent Moses to talk to the Pharaoh and the next two words are The Way which is the old name for Christianity.

In the Bible's Hebrew Text His Name has a Gematria of 352, in the Greek text Lord Jesus Christ has a Gematria of 3168 and The Way has a Gematria of 352, the numbers 352 3168 352 shown side by side on the Musical Scale Chart translate via Gematria to His Name, Lord Jesus Christ, The Way and the Bethlehem coordinates 31.68° N. and 35.2° E. refer to His Name Lord Jesus Christ or alternatively 3168 = His Name.

The base perimeter and cross are formed by six 760.32' lines = 3041.28' total or 0.3168 MMi. which places the numeric signature of the Source on the megalithic mile. This is significant because on the 1 foot to 1 megalithic mile scale the measurements below relate to the measurements on the right that refer to the Pyramid's global position.

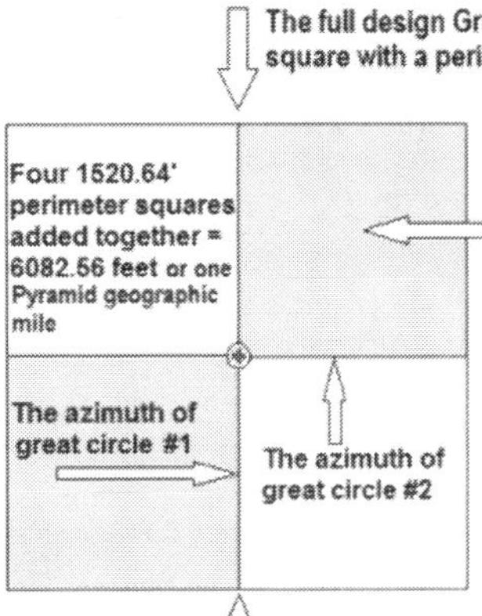

The full design Great Pyramid's base is 760.32' square with a perimeter of 3041.28'

Four 1520.64' perimeter squares added together = 6082.56 feet or one Pyramid geographic mile

1520.64' perimeter square 1520.64' per second rotation speed of the Equator

The azimuth of great circle #1
The azimuth of great circle #2

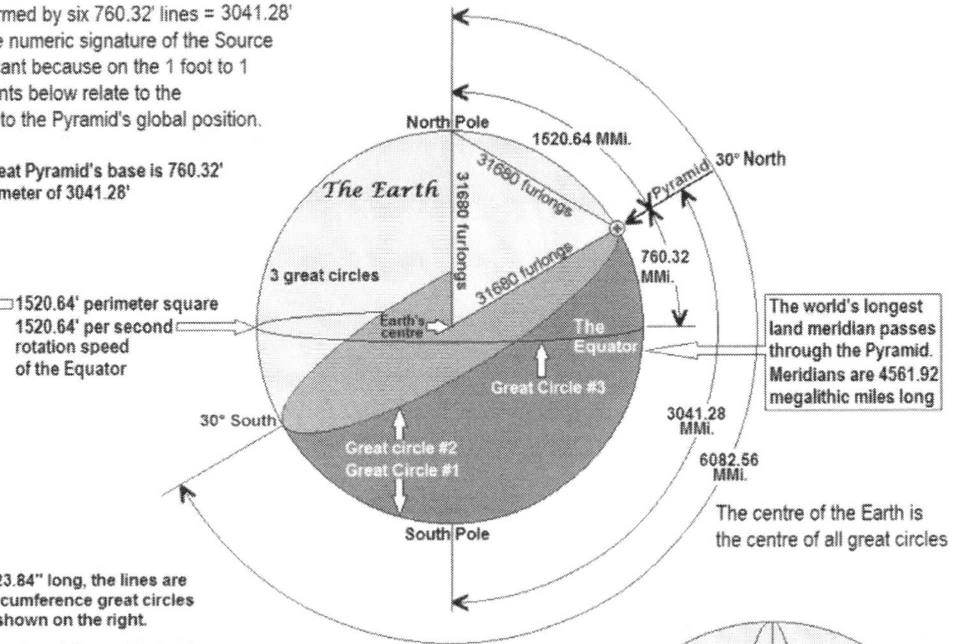

Each line of the cross in the base is 9123.84" long, the lines are on the azimuths of two 9123.84 MMi. circumference great circles crossing in the base, those circles are shown on the right.

The world's longest land meridian passes through the Pyramid. Meridians are 4561.92 megalithic miles long

The centre of the Earth is the centre of all great circles

The angle of the Pyramid's ascending and descending passages is called the Christ Angle

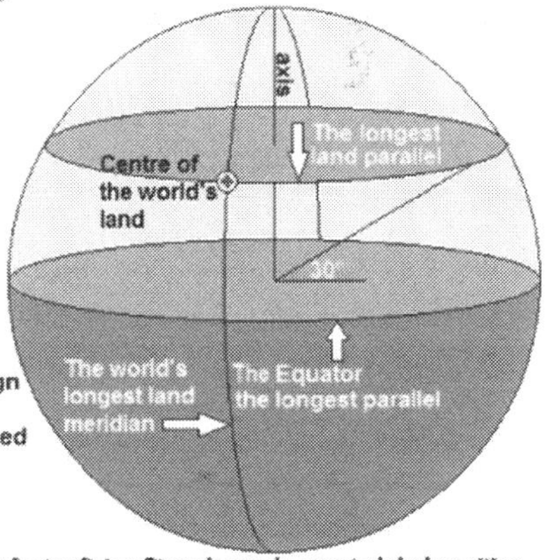

⊕ This symbol is used to represent the centre of the world's land
⊕ It is the site of the Great Pyramid
⊕ It is crossed by the world's longest land meridian
⊕ It is crossed by the world's longest land parallel
⊕ It is crossed by the world's longest river
⊕ It is crossed by the world's deepest and longest rift valley
⊕ It is the pivot point of the Nile Delta Quadrant
⊕ Where the world's longest river leaves the world;s biggest desert
⊕ Where the world's longest river enters the world's most fertile land
⊕ One of the few places on the earth's surface that can support the Pyramid's weight
⊕ Here is the junction of 4 right angle corners of 4 equal quarters of the world's land
⊕ Where the right angle corner of the Nile Delta Quadrant touches the centre of the 4 corners above
⊕ Here is the right angle junction of the stellar alignment marking the datum year of the Pyramid's prophetic design
⊕ From here a Christ Angle rhumb line to Bethlehem prophetically predicted the time and place of Jesus' birth
⊕ From here the Christ Angle rhumb line passes directly through Bethlehem to the place where Jesus was baptised
⊕ From here to where Jesus was baptised is 316.8 Roman miles
⊕ Half the world's land is north or south of this point
⊕ Half the world's land is east or west of this point
⊕ Half the area of the world's oceans is east or west of this point
⊕ A great circle aligned to one diagonal of Stonehenge's Station Stone Rectangle passes through here which is a factor fixing Stonehenge's exact global position

The photograph was taken in 1877, the building of the Aswan Dam caused the Nile to move away from the Pyramid a few miles to the East. The Pyramid could not have been closer to the Nile unless it was in the Nile. This is important concerning the designed unique global and regional features of the Pyramid's site.

The number 152064, the Great Pyramid's full design and its global position

The Great Pyramid's full design is an analogue of the 1520.64' circumference sphere shown below. The 1520.64' circumference = 18247.68" therefore the sphere is a model of the 18247.68 Roman league circumference Earth on the scale of 1" to 1 Roman league. Note that the 18247.68" circumference of the model's equator equals the distance the Earth's equator turned every second.

The Earth's 7920 mile diameter x 864/275 = 24883.2 miles circumference = 21600 Pyramid geographic miles. There are 21600 minutes in a circle therefore the Pyramid geographic mile is 1 minute of the Earth's circumference which is why it is called geographic. Note that this proves the 864/275 method for π is a factor in the formulation of the Pyramid geographic mile and this method for π is used throughout the Pyramid's full design and at Stonehenge

The 484' height of the Pyramid's full design x 1520.64' (Prime Measure) = the surface area of the sphere.

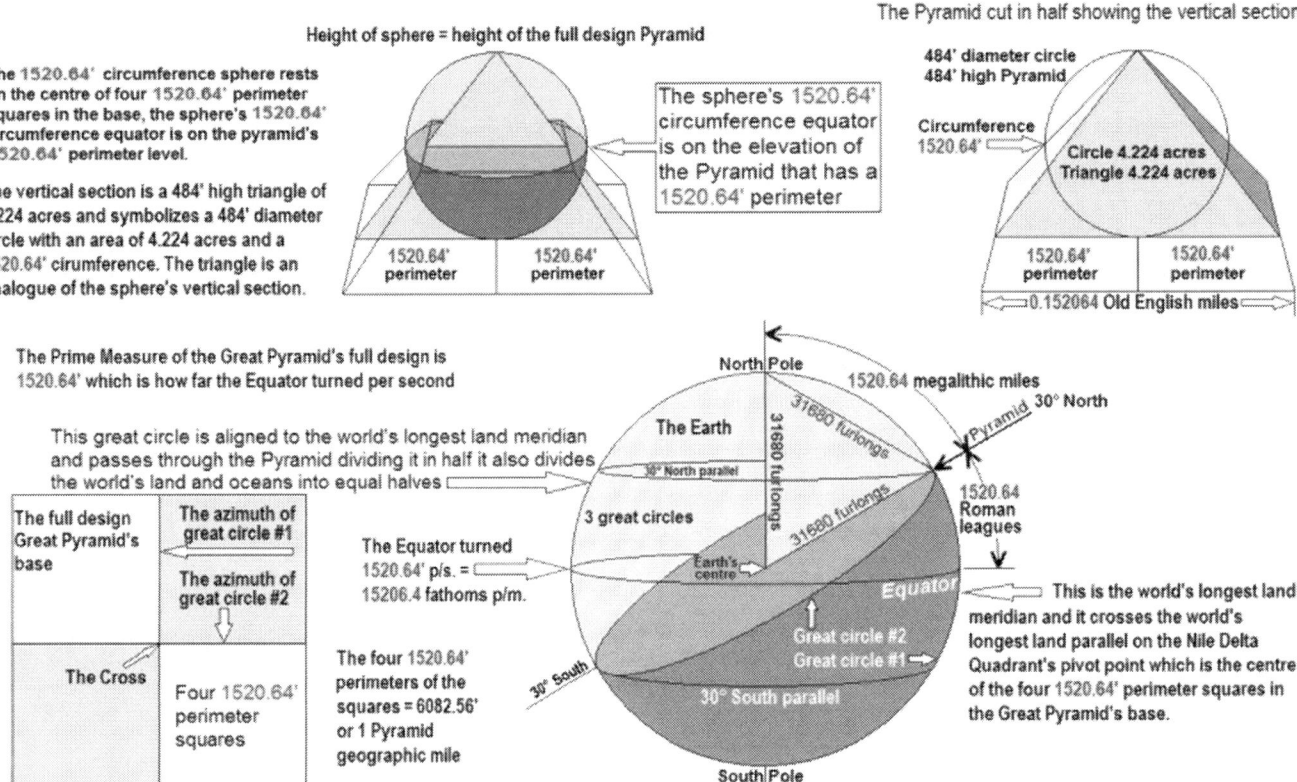

Each line of the cross is 9123.84" long, the lines are on the azimuths of the great circles #1 and #2 that are shown above

The world's longest land meridian and the world's longest land parallel form a cross through the centre of the Pyramid's base on the centre of the world's land on the pivot point of the Nile Delta Quadrant. The cross' 2 lines within the base total 1520.64' and they divide the base into four 1520.64' perimeter squares.

The cross' lines in the Pyramid's base total 1520.64' they divide the base into four 1520.64' perimeter squares. On the world's longest land meridian it is 1520.64 megalithic miles from the North Pole to the centre of the four 1520.64' perimeter squares. On the world's longest land meridian it is 1520.64 Roman leagues from the world's longest parallel (the equator) to the world's longest land parallel and the centre of the four 1520.64' perimeter squares.

The full design Great Pyramid's base

Each side of the full design Great Pyramid's base is 0.152064 Old English miles

The lines of the cross and the base perimeter equal 1520.64 yards

The Earth in its original perfect condition did not bulge on the equator it's circumference via 864/275 = 18247.68 Roman leagues, every second it turned 18247.68" = 1520.64' or 15206.4 fathoms per minute

It is 1520.64 megalithic miles to the North Pole from the centre of the four 1520.64' perimeter squares

The centre of the four 1520.64' perimeter squares is the pivot point of the Nile Delta Quadrant and the centre of the world's land. This is also where the world's longest land parallel crosses the world's longest land meridian. It is 1520.64 Roman leagues from that crossing to the world's longest parallel i.e. the equator, which turned 1520.64' per second.

From the centre of the four 1520.64' perimeter squares it is 1520.64 Roman leagues to the equator which turned 1520.64' per second

Chapter 19: Insights and Predictions about the Universe.

Particle Interactions of Key Numbers.

Particle	Number	Number	Anti-Particle
Neutron	1	8	Anti-Neutron
Proton	2	7	Anti-Proton
Electron	3	6	Positron
Neutrino	4	5	Anti-Neutrino

27 & 72 - Light.

The Integer Partition Theory predicts that Light should be produced by the interaction of the 2 and the 7, translated into particle physics as Proton and Anti-Proton.

We know the answer to this one already of course because when these two particles interact, they annihilate each other, producing a photon, the fundamental quantum of Light.

37 & 73 - Vibration.

The Integer Partition Theory predicts Sound or Vibration should be produced by the interaction of the 3 and 7, translated into particle physics as Electron and Anti-Proton. This gives the polarity to the Universe and is balanced in the middle by the number 55. I have not been able to find out about an interaction of these particles.

55 - The Central Axis – 'Tree of Life'.

The prediction is that at the centre of the Universe there is an interaction between 2 anti-neutrinos representing the number 55 and the answer to why 10 being the 10th Fibonacci number and the sum of the first ten numbers.

12 & 21.

The mirrored interaction between Neutron, 1 and Proton, 2.

36 & 63.

The mirrored interaction between Electron and Positron defines the polarity of the universe.

252.

Because the 5 is the inverted version of the number 2, I suspect that the multiplication 12 x 21 = 252 tells us something important where two protons are divided by an antineutrino.

(0)297 & 792(0).

Proton and Anti-Proton divided by energy.

3960.

The radius of the Earth exhibits the magnetic 3 6 9 (0) number group which is the containing infrastructure in which electricity may play out its dance.

137.

The number associated with the Fine Structure Constant looks to be a combination of the Neutron, Electron and Anti Proton.

3168.

The number for the Numerical World Soul - The interaction between two balanced number pairs, Electron and Positron, and Neutron and Anti-Neutron.

108(0).

The number for the mean radius of the Moon - The interaction between Energy and the Neutron Anti Neutron number pair.

19 & 91.

The interaction between a Neutron and Energy.

An Inner Sun?

10	42	19	23	1	1	2	3	5	7	9	8	5	1		12									
11	56	19	37	1	1	2	3	5	7	10	11	10	5	1		13								
12	77	30	47	1	1	2	3	5	7	11	13	15	12	6	1		14							
13	101	30	71	1	1	2	3	5	7	11	14	18	18	14	6	1		15						
14	135	45	90	1	1	2	3	5	7	11	15	20	23	23	16	7	1		16					
15	176	45	131	1	1	2	3	5	7	11	15	21	26	30	27	19	7	1		17				
16	231	67	164	1	1	2	3	5	7	11	15	22	28	35	37	34	21	8	1		18			
17	297	67	230	1	1	2	3	5	7	11	15	22	29	38	44	47	39	24	8	1		19		
18	385	97	288	1	1	2	3	5	7	11	15	22	30	40	49	58	57	47	27	9	1		20	
19	490	97	393	1	1	2	3	5	7	11	15	22	30	41	52	65	71	70	54	30	9	1		21
20	627	139	488	1	1	2	3	5	7	11	15	22	30	42	54	70	82	90	84	64	33	10	1	22
21	792	139	653	1	1	2	3	5	7	11	15	22	30	42	55	73	89	105	110	101	72	37	10	1
22	1002	195	807	1	1	2	3	5	7	11	15	22	30	42	56	75	94	116	131	136	119	84	40	11	1
23	1255	195	1060	1	1	2	3	5	7	11	15	22	30	42	56	76	97	123	146	164	163	141	94	44	11
24	1575	272	1303	1	1	2	3	5	7	11	15	22	30	42	56	77	99	128	157	186	201	199	164	108	48

If the Integer Partition table really is the algorithm of creation then the theory predicts the existence of an inner Sun located within the Holographic Plane and identified by the number 27 on the 16th Stage Value for the number 18.

Interestingly, 16 x 18 = 288 - an old friend we have seen throughout the book.

288 also appears on the number 18 as the total of the unlocked potentials for this number:-

$$40 + 49 + 58 + 57 + 47 + 27 + 9 + 1 = 288$$

The first scientific speculation that the Earth is in fact hollow originated in the late 17th century with the work of Edmund Halley, he of Halley's comet notoriety. In 1692, in order to explain anomalous compass readings, Halley theorised that the planet is in fact a series of nested, spherical shells, spinning in different directions, all surrounding a central core. Based on readings of the magnetic field and what he knew of the gravitational pull of the Sun and the Moon on the Earth, this new model could account for the inaccuracies in his readings of the magnetic fields of the planet.

Chapter 20: A Holographic Universe.

Of the theories out there in mainstream Science I have chosen the idea of the universe being a hologram that is projected by a machine based on Number, Geometry and Vibration or Music.

The main protagonists of the Hologrammic / Holographic Universe are David Bohm, Karl Pribram and Andrew Strominger. Bohm and Pribram's work is discussed in Michael Talbot's book - The Holographic Universe

I feel, while acknowledging that I am very far from qualified to be the best judge, that Andrew Strominger's idea that our universe is an image projected backwards in time from a hologram located at the boundary of the cosmos, in the infinite future, is the most likely scenario.

'As the image projects into the past, it fades away, becoming grainy and undefined, eventually fading to nothing. It's a bizarre notion, but over the past decade it has been gaining ever more credence, especially within the mathematical framework devised by string theorists. If correct, it could help explain how the universe, and time as we know it, came from nothing, as well as helping in the quest to unite quantum mechanics - the theory that governs particles on the small scale - and general relativity - which describes the large-scale cosmos - into one overarching theory of quantum gravity.'
http://www.fqxi.org/community/articles/display/176

English physicist, David Bohm, was the most avid early proponent of a holographic model of the universe. For Bohm, understanding the Cosmos as a whole was fundamental he called this wholeness the Holomovement, highlighting the constant motion in all things. My sense is that his theory is ostensibly correct and this fits nicely with my Integer Partition Theory.

Implicate & Explicate Orders.

This ever-emergent holomovement has two primary domains within which it orders and organizes itself - the pre-manifest, unseen, background field called the Implicate Order which are represented by the forming and containing 3 6 9 (0) Number Group and the manifest, experiential foreground field called the Explicate Order, represented by the 1 4 7 and 2 5 8 Number Groups, which interact to form all things which is the cosmos as we know it.

Bohm likened the Implicate as an immense underlying sea of energy and information, and the Explicate (specifically, the quantum activity) as an excitation of small waves rippling upon its surface.

From a holographic perspective, one could accurately equate this to a literal ocean of water with an immense unified ordering process (ocean currents) underlying a constantly moving surface of waves and interference patterns.

Reciprocal Relation.

Earlier I mentioned Walter Russell's work which spoke about a 9 octave waveform which repeats with 5.5 octaves containing the chemical elements we can perceive together with 3.5 octaves that we cannot. Looking at the relationship of these two numbers clearly showed the reciprocal nature of electricity and magnetism from a numerical perspective. This fits nicely with Bohm's two realms, the implicate and explicate, which are in what he called Reciprocal Relation, with each one informing and influencing the other in the on-going manifestation process.

At the quantum level, sub-atomic particles (electrons, for example) flicker in and out of existence at a very high rate, emerging into the explicate, and returning into the implicate.

As they do, the information gained by the particle while in the explicate manifest state is carried back (Enfolds) into the implicate state, influencing the underlying totality of background information in its ordering process.

This in turn influences the energy and information of the particle as it re-emerges (Unfolds) into the explicate manifest state, influencing and changing its ordering process... and on and on it goes.

'Rather than suggesting a continuous entity that moves 'through' time and space, the image of ordered enfoldment-unfoldment allows for a view of the electron as a perpetually emerging explicate structure, temporarily unfolding from an ordered implicate background, and then rapidly enfolding back into this background, in an on-going cycle.

By extension, the whole of experience can be understood as a flow of appearances resulting from such a cycle of enfoldment and unfoldment.

This enfolding-unfolding reciprocal information exchange occurs in nested scales of mutual influence, 'even between macroscopic processes and those at the atomic level, indicating the complexity of the pathways through which the qualitative infinity of nature may manifest.' David Bohm, Wholeness and the Implicate Order.

The dynamic flow process of the torus is an excellent model to depict this reciprocal relationship and nested mutual influence.

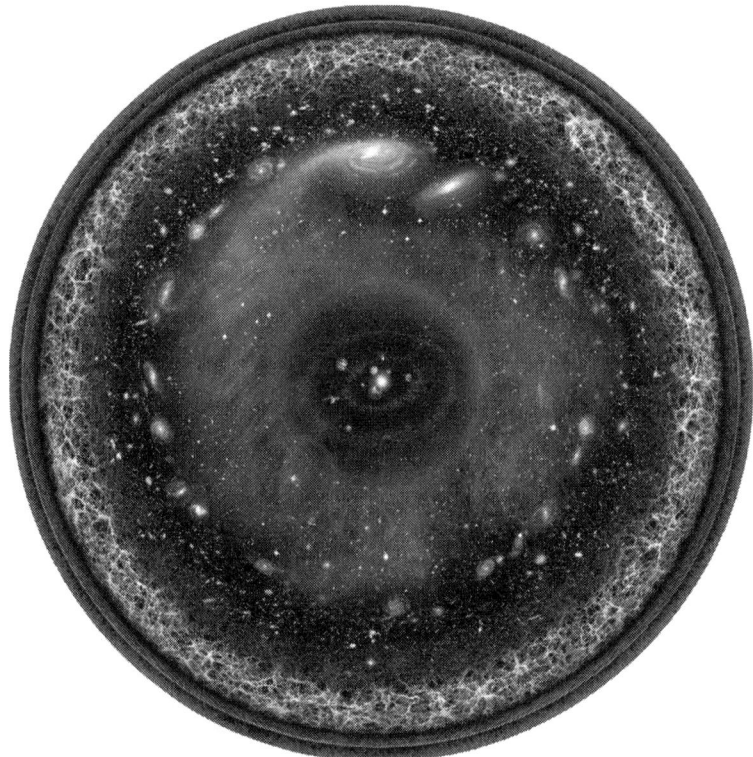

This is an illustrated logarithmic scale conception of the observable Universe with the Solar System at the centre, created by musician and artist Pablo Carlos Budassi.

Encircling the Solar System are the inner and outer planets, Kuiper belt, Oort cloud, Alpha Centauri star, Perseus Arm, Milky Way galaxy, Andromeda galaxy, other nearby galaxies, the cosmic web, cosmic microwave radiation, and invisible plasma produced by the Big Bang at the very edges.

The image is based on logarithmic maps of the Universe put together by Princeton University researchers, as well as images produced by NASA based on observations made by their telescopes and roving spacecraft.

The Princeton team, led by astronomers J Richard Gott and Mario Juric, based their logarithmic map of the Universe on data from the Sloan Digital Sky Survey.

Over the past 15 years it has been using a 2.5-metre, wide-angle optical telescope at Apache Point Observatory in New Mexico to create the most detailed three-dimensional maps of the Universe ever made, including spectra for more than 3 million astronomical objects.

Logarithmic maps are a really handy way of visualising something as inconceivably huge as the observable Universe, because each increment on the axes increases by a factor of 10 (or order of magnitude) rather than by equal increments.

Selected Articles.

How an Argument with Hawking suggested the Universe is a Hologram.

By John Timmer - https://arstechnica.com/science/2011/07/how-an-argument-with-hawking-suggested-the-universe-is-a-hologram/

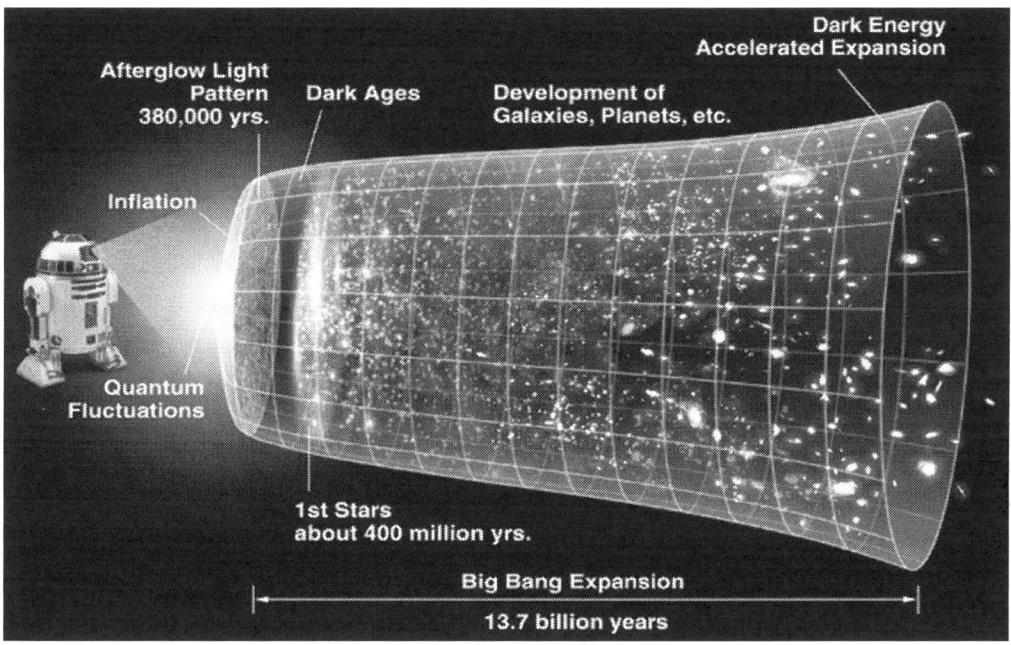

Illustration by NASA/WMAP Science Team, R2D2 © Lucasfilm

The proponents of string theory seem to think they can provide a more elegant description of the Universe by adding additional dimensions. But some other theoreticians think they've found a way to view the Universe as having one less dimension. The work sprung out of a long argument with Stephen Hawking about the nature of black holes, which was eventually solved by the realization that the event horizon could act as a hologram, preserving information about the material that's gotten sucked inside. The same sort of math, it turns out, can actually describe any point in the Universe, meaning that the entire content Universe can be viewed as a giant hologram, one that resides on the surface of whatever two-dimensional shape will enclose it.

That was the premise of panel at this summer's World Science Festival, which described how the idea developed, how it might apply to the Universe as a whole, and how they were involved in its development.

The whole argument started when Stephen Hawking attempted to describe what happens to matter during its lifetime in a black hole. He suggested that, from the perspective of quantum mechanics, the information about the quantum state of a particle that enters a black hole goes with it. This isn't a problem until the black hole starts to boil away through what's now called Hawking radiation, which creates a separate particle outside the event horizon while destroying one inside. This process ensures that the matter that escapes the black hole has no connection to the quantum state of the material that had gotten sucked in. As a result, information is destroyed. And that causes a problem, as the panel described.

As far as quantum mechanics is concerned, information about states is never destroyed. This isn't just an observation; according to panellist Leonard Susskind, destroying information creates paradoxes that, although apparently minor, will gradually propagate and eventually cause inconsistencies in just about everything we think we understand. As panellist Leonard Susskind put it, "all we know about physics would fall apart if information is lost."

Unfortunately, that's precisely what Hawking suggested was happening. "Hawking used quantum theory to derive a result that was at odds with quantum theory," as Nobel Laureate Gerard 't Hooft described the situation. Still, that wasn't all bad; it created a paradox and "Paradoxes make physicists happy."

"It was very hard to see what was wrong with what he was saying," Susskind said, "and even harder to get Hawking to see what was wrong."

The arguments apparently got very heated. Herman Verlinde, another physicist on the panel, described how there would often be silences when it was clear that Hawking had some thoughts on whatever was under discussion; these often ended when Hawking said 'rubbish.' 'When Hawking says 'rubbish,' he said, 'you've lost the argument.'

't Hooft described how the disagreement eventually got worked out. It's possible, he said, to figure out how much information has gotten drawn in to the black hole. Once you do that, you can see that the total amount can be related to the surface area of the event horizon, which suggested where the information could be stored. But since the event horizon is a two-dimensional surface, the information couldn't be stored in regular matter; instead, the event horizon forms a hologram that holds the information as matter passes through it. When that matter passes back out as Hawking radiation, the information is restored.

Susskind described just how counterintuitive this is. The holograms we're familiar with store an interference pattern that only becomes information we can interpret once light passes through them. On a micro-scale, related bits of information may be scattered far apart, and it's impossible to figure out what bit encodes what. And, when it comes to the event horizon, the bits are vanishingly small, on the level of the Planck scale (1.6×10^{-35} meters). These bits are so small, as 't Hooft noted, that you can store a staggering amount of information in a reasonable amount of space - enough to describe all the information that's been sucked into a black hole.

The price, as Susskind noted, was that the information is "hopelessly scrambled" when you do so.

From a Black Hole to the Universe.

Berkeley's Raphael Bousso was on hand to describe how these ideas were expanded out to encompass the Universe as a whole. As he put it, the math that describes how much information a surface can store works just as well if you get rid of the black hole and event horizon. (This shouldn't be a huge surprise, given that most of the Universe is far less dense than the area inside a black hole.) Any surface that encloses an area of space in this Universe has sufficient capacity to describe its contents. The math, he said, works so well that "it seems like a conspiracy." To him, at least.

Verlinde pointed out that things in the Universe scale with volume, so it's counterintuitive that we should expect its representation to them to scale with a surface area. That counter intuitiveness, he thinks, is one of the reasons that the idea has had a hard time being accepted by many.

When it comes to the basic idea - the Universe can be described using a hologram - the panel was pretty much uniform, and Susskind clearly felt there was a consensus in its favor. But, he noted, as soon as you stepped beyond the basics, everybody had their own ideas, and those started coming out as the panel went along. Bousso, for example, felt that the holographic principle was "your ticket to quantum gravity." Objects are all attracted via gravity in the same way, he said, and the holographic principle might provide an avenue for understanding why (if he had an idea about how, though, he didn't share it with the audience). Verlinde seemed to agree, suggesting that, when you get to objects that are close to the Planck scale, gravity is simply an emergent property.

But 't Hooft seemed to be hoping that the holographic principle could solve a lot more than the quantum nature of gravity - to him, it suggested there might be something underlying quantum mechanics. For him, the holographic principle was a bit of an enigma, since disturbances happen in three dimensions, but propagate to a scrambled two-dimensional representation, all while obeying the Universe's speed limit (that of light). For him, this suggests there's something underneath it all, and he'd like to see it be something that's a bit more causal than the probabilistic world of quantum mechanics; he's hoping that a deterministic world exists somewhere near the Planck scale. Nobody else on the panel seemed to be all that excited about the prospect, though.

What was missing from the discussion was an attempt to tackle one of the issues that plagues string theory: the math may all work out and it could provide a convenient way of looking at the world, but is it actually related to anything in the actual, physical Universe? Nobody even attempted to tackle that question. Still, the panel did a good job of describing how something that started as an attempt to handle a special case - the loss of matter into a black hole - could provide a new way of looking at the Universe. And, in the process, how people could eventually convince Stephen Hawking he got one wrong.

Juan Maldecena.

https://www.nature.com/news/simulations-back-up-theory-that-universe-is-a-hologram-1.14328

'A ten-dimensional theory of gravity makes the same predictions as standard quantum physics in fewer dimensions.

At a black hole, Albert Einstein's theory of gravity apparently clashes with quantum physics, but that conflict could be solved if the Universe were a holographic projection.

A team of physicists has provided some of the clearest evidence yet that our Universe could be just one big projection.

In 1997, theoretical physicist Juan Maldacena proposed that an audacious model of the Universe in which gravity arises from infinitesimally thin, vibrating strings could be reinterpreted in terms of well-established physics. The mathematically intricate world of strings, which exist in nine dimensions of space plus one of time, would be merely a hologram: the real action would play out in a simpler, flatter cosmos where there is no gravity.

Maldacena's idea thrilled physicists because it offered a way to put the popular but still unproven theory of strings on solid footing - and because it solved apparent inconsistencies between quantum physics and Einstein's theory of gravity. It provided physicists with a mathematical Rosetta stone, a 'duality', that allowed them to translate back and forth between the two languages, and solve problems in one model that seemed intractable in the other and vice versa (see 'Collaborative physics: String theory finds a bench mate'). But although the validity of Maldacena's ideas has pretty much been taken for granted ever since, a rigorous proof has been elusive.

In two papers posted on the arXiv repository, Yoshifumi Hyakutake of Ibaraki University in Japan and his colleagues now provide, if not an actual proof, at least compelling evidence that Maldacena's conjecture is true.

In one paper, Hyakutake computes the internal energy of a black hole, the position of its event horizon (the boundary between the black hole and the rest of the Universe), its entropy and other properties based on the predictions of string theory as well as the effects of so-called virtual particles that continuously pop into and out of existence (see 'Astrophysics: Fire in the Hole!'). In the other, he and his collaborators calculate the internal energy of the corresponding lower-dimensional cosmos with no gravity. The two computer calculations match.

"It seems to be a correct computation," says Maldacena, who is now at the Institute for Advanced Study in Princeton, New Jersey and who did not contribute to the team's work.

Regime Change.

The findings "are an interesting way to test many ideas in quantum gravity and string theory", Maldacena adds. The two papers, he notes, are the culmination of a series of articles contributed by the Japanese team over the past few years. "The whole sequence of papers is very nice because it tests the dual [nature of the universes] in regimes where there are no analytic tests."

"They have numerically confirmed, perhaps for the first time, something we were fairly sure had to be true, but was still a conjecture - namely that the thermodynamics of certain black holes can be reproduced from a lower-dimensional universe," says Leonard Susskind, a theoretical physicist at Stanford University in California who was among the first theoreticians to explore the idea of holographic universes.

Neither of the model universes explored by the Japanese team resembles our own, Maldacena notes. The cosmos with a black hole has ten dimensions, with eight of them forming an eight-dimensional sphere. The lower-dimensional, gravity-free one has but a single dimension, and its menagerie of quantum particles resembles a group of idealized springs, or harmonic oscillators, attached to one another.

Nevertheless, says Maldacena, the numerical proof that these two seemingly disparate worlds are actually identical gives hope that the gravitational properties of our Universe can one day be explained by a simpler cosmos purely in terms of quantum theory.

Andrew Strominger.

An Indian emperor was angry with a guru for teaching that everything is maya: an illusion. To prove him wrong, the emperor invited the guru to his palace and set a stampeding elephant on him. As the guru ran away, the emperor shouted after him, "Why do you run so fast, seeing that my elephant is only an illusion?" The guru replied, "Oh emperor, my running too is an illusion, everything in this world is an illusion."

Andrew Strominger, a string theorist at Harvard University, is one of a number of physicists who surprisingly agree with the guru in this ancient story - just swap the word "maya" for "holographic" and you're there. About ten years ago, Strominger had an outlandish idea. He mused that our universe is an image projected backwards in time from a hologram located at the boundary of the cosmos, in the infinite future. As the image projects into the past, it fades away, becoming grainy and undefined, eventually fading to nothing. It's a bizarre notion, but over the past decade it has been gaining ever more credence, especially within the mathematical framework devised by string theorists. If correct, it could help explain how the universe, and time as we know it, came from nothing, as well as helping in the quest to unite quantum mechanics - the theory that governs particles on the small scale - and general relativity - which describes the large-scale cosmos - into one overarching theory of quantum gravity. "It's one of the most speculative things I've ever worked on," admits Strominger. "But if it turns out to be right, then it's one of the most interesting things I've ever done."

The contrast with big bang cosmology is stark: there is no flick of a switch and everything is suddenly here. Running time forwards in this story, you have a pleasingly slow, continual process of creation as more and more of the hologram comes into view. In this picture, mind-bogglingly, even living beings would be projections from the future. "It wouldn't be the first time in physics when the unimaginable has been understood to be reality," retorts Strominger. He is sticking to his equations.

As crazy as this idea sounds, there's good reason why we should listen to this physics guru. Physicists acknowledge the strides Strominger - who has been awarded a $73,000 FQXi grant - has already made in our understanding of the basic structure of quantum gravity, black holes and string theory. "He has a unique perspective, which often allows him to solve apparently complex problems in a simple and clear way," says Alex Maloney, who is based at McGill University, Canada. (Read more about Maloney's FQXi-funded research in "The Holographic Universe") "His work is often quite prescient, and it is only in retrospect - sometimes years or decades later - that many of his papers have been fully appreciated," Maloney adds.

Early on Strominger knew that physics was for him: "The universe is such an amazing and wonderful place, there's nothing more enthralling than trying to understand it, and possibly make a small amount of progress," he says. But he is coy about his childhood and upbringing. The fact that he finished high school at 15, two years earlier than most, almost seems an embarrassment. Asked if he was a child prodigy, he says, "You'd better ask my parents!" He attributes his first appreciation of physics to his high school teacher, a Mexican immigrant to Boston, where he and his family lived at the time. "He didn't even have a PhD but he just loved physics and had a great ability to communicate it," says Strominger.

Black Holes and Holograms.

The roots of Strominger's latest ideas about a projected universe go back to the 1970s, when physicists Stephen Hawking and Jacob Bekenstein calculated that the information content of a black hole (described mathematically by its entropy) is proportional to its surface area. This was surprising because most people expected it would be related to its volume. Their discovery has been dubbed the "holographic principle" because it shows that information about a three-dimensional black hole is encoded on its two-dimensional surface, just like a hologram.

Then in 1996 Strominger and his Harvard colleague Cumrun Vafa derived a precise statistical description of a black hole's entropy in terms of microscopic energy states at the black hole's surface. In particular, they made a connection with the equations usually used to describe the behaviour of particles: quantum field theory. They realized that the equations they were using to describe the properties of the black hole were similar to those used to describe a system of particles using quantum field theory, but in a universe without gravity.

A year later Juan Maldecena of the Institute for Advanced Study (IAS) in Princeton, New Jersey, independently found a mathematical equivalence between two kinds of universes: the first universe contains particles that obey quantum field theory, but does not contain gravity; the second universe contains strings and gravity and has a special type of negatively curved space-time geometry (called an "anti de Sitter universe"), such as that thought to be found inside black holes. This may seem like a lot of obscure mathematical shuffling, but it's had a huge impact on theoretical physics. When physicists get stuck because their equations are too tough to solve, they can switch into the mirror universe where the math is often easier and finish their calculations there, before transferring their answers back again. (See "The Black Hole and the Babel Fish" for more about how the holographic principle is being used to predict the behavior of exotic materials in the lab.) "It is useful because problems that are hard in one formulation become easy in the other formulation," says Maldecena.

Maldecena's conjecture linking these two universes has proven a spectacular mathematical success, says Strominger. "We understand it in amazing detail, it's a very beautiful and precise statement and there is general agreement that there is overwhelming evidence that it is correct."

However, on the whole, the universe we see around us is not negatively curved, limiting the applications of Maldecena's correspondence. In fact, in the late 1990s, astronomers discovered the opposite: that the expansion of the universe is accelerating - something they attribute to some kind of "dark energy" pushing it outwards. They now believe that soon after the big bang, dark energy dominated the universe, and it may do so again in the far future, pulling the cosmos apart so ferociously that matter is ripped apart and gravity cannot hold stars and planets together. A cosmos in which dark energy plays the biggest role is known as a "de Sitter universe" (in contrast to the anti de Sitter universe that makes up one half of Maldecena's relationship). So the hunt has been on to find a similar correspondence principle to Maldecena's, but for a de Sitter, rather than an anti de Sitter universe.

That has not been easy, explains Strominger. "It's been very frustrating trying to understand de Sitter space within string theory, and I spent many years trying. It always ends up being complicated and not especially pretty," he says, adding modestly, "I just didn't succeed - but then I don't succeed in most things I do. But when you're trying and not succeeding you're always learning something."

Strange Gravity.

Indeed, through this endeavor, Strominger learned of a little-known theory describing interacting particles by a Russian physicist Misha Vasiliev. Although it's related to string theory, it includes a raft of new massless particles with unusual properties that have never been observed. "Everybody agrees that Vasiliev gravity is a strange theory. But nevertheless it seems to be mathematically self-consistent and in many ways much simpler to analyse than string theory," says Strominger. Crucially, Strominger and colleagues Tom Hartman also from IAS and Dionysius Anninos at Stanford University, California, found that the Vasiliev's bizarre model for the universe easily accommodated de Sitter space.

Using Vasiliev's model for gravity, Strominger and his colleagues have taken Maldacena's correspondence for anti de Sitter space, and applied it to de Sitter space. It represents the first example of how a universe that is more physically similar to the one in which we live can be understood in terms of information at a boundary surface, holographically projected back in time. "This is a very radical thing, it amounts to a total revolution in the way that we view the universe," says Strominger. "We're very far from seeing that this is how it works in our own universe, but nevertheless, it is a small step in the right direction. So people are interested in it and they want to take the next step," he says.

'It amounts to a total revolution in the way we view the universe' - Andrew Strominger

Maldecena agrees that the work is "very interesting" because it is the first example of the correspondence with positive curvature, which is more directly related to the universe we live in. Although it holds for a very special and somewhat strange theory of gravity, it gives a strong hint that our actual universe could one day be described using a similar correspondence, he adds. In turn, this will be a powerful tool for understanding quantum gravity and even how our universe emerged from nothing in the first place.

Strominger hopes to move his correspondence onto firmer ground, mathematically speaking, by finding more examples of how it can work. He has a lot of calculating planned for the next part of the journey, up to what he calls "the next bend in the road," including trying to find a solution that looks even more like the universe that we live in, starting with a big bang. "It's not a vague question, we can translate it into some real mathematics and I expect we'll be able to solve it in the next year or two," says Strominger.

Whether or not our universe is really a holographic projection from the future is still far from any sort of observational test because it has only been shown to work using this strange Vasiliev model for the universe, rather than our real universe. "We are asking the most difficult questions we can conceive of," says Strominger. "It takes a long time to understand what the question is, it takes a long time to understand the answer, and it takes an even longer time to test it. You have to be a little crazy to work on these things because they go on for years." We just have to hope that our holographic projections last long enough for physicists to test the answer—and perhaps upset a few emperors along the way. - https://fqxi.org/community/articles/display/176

A Strange New Theory of How Space-Time is Emerging.

"A metaphorical chip holding all the programming for our universe stores information like a quantum computer." This is the radical insight to the foundation of our Universe developed by Mark Van Raamsdonk, a professor of theoretical physics at the University of British Columbia, that says that the world we see around us is a projection from a set of rules written in simpler, lower-dimensional physics - just as the 2D code in a computer's memory chip creates an entire virtual 3D world. "What Mark has done is put his finger on a key ingredient of how space-time is emerging: entanglement," says Gary Horowitz, who studies quantum gravity at the University of California Santa Barbara. Horowitz says this idea has changed how people think about quantum gravity, though it hasn't yet been universally accepted. "You don't come across this idea by following other ideas. It requires a strange insight," Horowitz adds. "He is one of the stars of the younger generation."

"We're trying to construct a dictionary," says Van Raamsdonk, that allows physicists to translate descriptions of our complex universe into simpler terms. If they succeed, they will have found the biggest jigsaw piece in the puzzle of a Grand Unified Theory - something that can describe all of the forces of our universe, at all scales from the atomic to the galactic. That puzzle piece is, specifically, something that can describe gravity within the framework of quantum mechanics, which governs physics on small scales. Such a unified theory is needed to explain the extreme scenarios of a black hole or the first moments of the universe." The catalyst for Van Raamsdonk's theory was a 1998 paper by Juan Maldacena a theoretical astrophysicist at Princeton's Institute for Advanced studies that proposed that to understand quantum gravity through string theory, you can look instead to the much more ordinary, well-described system of quantum mechanics called quantum field theory that concluded that it seems that all the information about our complex multi-dimensional world can be described using a simpler, lower-dimensional language - just as a 3D image is projected from the 2D screen of a hologram, or a 3D computer gaming world created from a 2D memory chip. "After that, people wrote thousands of papers just testing whether that could be true," says Van Raamsdonk. "No one has actually proven it, but we're as certain about it as about anything in physics," he added. - http://www.physics-astronomy.com/2014/09/a-strange-new-theory-of-how-space-time.html#.VP4NIFWUctM

Is the Universe a Computer Simulation?

Physicists have devised a new experiment to test if the universe is a computer. A philosophical thought experiment has long held that it is more likely than not that we're living inside a machine.

The theory basically goes that any civilisation which could evolve to a 'post-human' stage would almost certainly learn to run simulations on the scale of a universe. And that given the size of reality - billions of worlds, around billions of suns - it is fairly likely that if this is possible, it has already happened.

And if it has? Well, then the statistical likelihood is that we're located somewhere in that chain of simulations within simulations. The alternative - that we're the first civilisation, in the first universe - is virtually (no pun intended) absurd. And it's not just theory. We previously reported that researchers at the University of Bonn in Germany had found evidence the Matrix was less than fiction.

Professor Martin Savage at the University of Washington says while our own computer simulations can only model a universe on the scale of an atom's nucleus, there are already "signatures of resource constraints" which could tell us if larger models are possible. This is where it gets complex.

Essentially, Savage said that computers used to build simulations perform "lattice quantum chromodynamics calculations" - dividing space into a four-dimensional grid. Doing so allows researchers to examine the force which binds subatomic particles together into neutrons and protons - but it also allows things to happen in the simulation, including the development of complex physical "signatures", that researchers don't program directly into the computer. In looking for these signatures, such as limitations on the energy held by cosmic rays, they hope to find similarities within our own universe.

And if such signatures do appear in both? Boot up, baby. We're inside a computer. (Maybe). "If you make the simulations big enough, something like our universe should emerge," Savage told the University of Washington news service.

Zohreh Davoudi, one of Savage's students, goes further: "The question is, 'Can you communicate with those other universes if they are running on the same platform?'", she said. Now *that* would be a long-distance phone call. - http://www.huffingtonpost.co.uk/2012/12/12/physicists-universe-simulation-test-university-of-washington-matrix_n_2282745.html

Are We Living in a Computer Simulation?

Physicists say they may have evidence that the universe is a computer simulation.

How? They made a computer simulation of the universe. And it looks sort of like us.

A long-proposed thought experiment, put forward by both philosophers and popular culture, points out that any civilisation of sufficient size and intelligence would eventually create a simulation universe if such a thing were possible.

And since there would therefore be many more simulations (within simulations, within simulations) than real universes, it is therefore more likely than not that our world is artificial.

Now a team of researchers at the University of Bonn in Germany led by Silas Beane say they have evidence this may be true.

In a paper named 'Constraints on the Universe as a Numerical Simulation', they point out that current simulations of the universe - which do exist, but which are extremely weak and small - naturally put limits on physical laws.

Technology Review explains that "the problem with all simulations is that the laws of physics, which appear continuous, have to be superimposed onto a discrete three dimensional lattice which advances in steps of time."

What that basically means is that by just *being* a simulation, the computer would put limits on, for instance, the energy that particles can have within the program.

These limits would be experienced by those living within the sim - and as it turns out, something which looks just like these limits do in fact exist.

For instance, something known as the Greisen-Zatsepin-Kuzmin, or GZK cut off, is an apparent boundary of the energy that cosmic ray particles can have. This is caused by interaction with cosmic background radiation. But Beane and co.'s paper argues that the pattern of this rule mirrors what you might expect from a computer simulation.

Naturally, at this point the science becomes pretty tricky to wade through - and we would advise you read the paper itself to try and get the full detail of the idea.

But the basic impression is an intriguing one.

Like a prisoner in a pitch-black cell, we may never be able to see the 'walls' of our prison -- but through physics we may be able to reach out and touch them. - http://arxiv.org/pdf/1210.1847v2.pdf

Chapter 21: Conclusion.

As a non-academic and with no experience of having ever written or produced a book, this has been a bit of a tricky book to compile. What I hope I have achieved in doing so is to show how Numbers, Geometry and Music may just be at the fundament of everything we call physical reality. I certainly don't pretend to have all the answers, far from it, but what I feel I may have been somewhat successful in doing is simply illustrating an enormous number of 'coincidences' which, when taken together, may have opened a small crack in a doorway to understanding that has been shut for a very long time.

The Mayans looked upon the people alive today in the modern era with awe. They were the masters of timing. They knew and understood the wheel of time, recognising that everything that has once been, will be again. I believe they knew of a time in great antiquity when we had a perfect understanding of the Universe and the reality of our predicament in this mortal coil. They knew this time would come again in this modern era.

It is simply not good enough, that as we hurtle through space at mind boggling speeds on spaceship Earth, that in some parts of the spaceship people are drowning in champagne and cocaine, where in others, people are unable to access even the most basic necessities for survival. There are no problems that cannot be solved by 7+ billion of us if only there was the Will to succeed in producing the same chance for everyone. It is our Science that is just not good enough. A full understanding of Physics, Chemistry, Biology and Mathematics, combined with Logic and Reason could allow us to produce unlimited energy and resources of any nature and remove the illusion of scarcity that drives Man's inhumanity to Man, for good.

The strength of the Integer Partition Theory lies in its potential to be applied ubiquitously. It is tethered to so many different areas of knowledge and understanding, so much physical evidence that, when understood properly, is shouting the numerical structure of the universe, that it may never be forgotten. I believe it is the glue that underpins the work of so many great men and women that alludes to the Truth.

I hope to have conveyed the simplicity and elegance of my theory throughout this work by showing how Number underpins all Geometry which in turn underpins Music. I have shown how every single numerical sequence essential to Nature exhibits numbers working in specific symmetrical pairings and groups. I have shown how the decimal system of Numbers can be said to have been discovered and not invented in the same way that Music has, through their innate musicality. I have clearly demystified Prime Numbers using the simplest of means in modular arithmetic and discerned their structure when we consider the square pyramidal nature when we consider a three dimensional version of the image shown below, and in doing so, we can perhaps better understand both the Great Pyramid's connection to Light and the enormous importance of its dimensions.

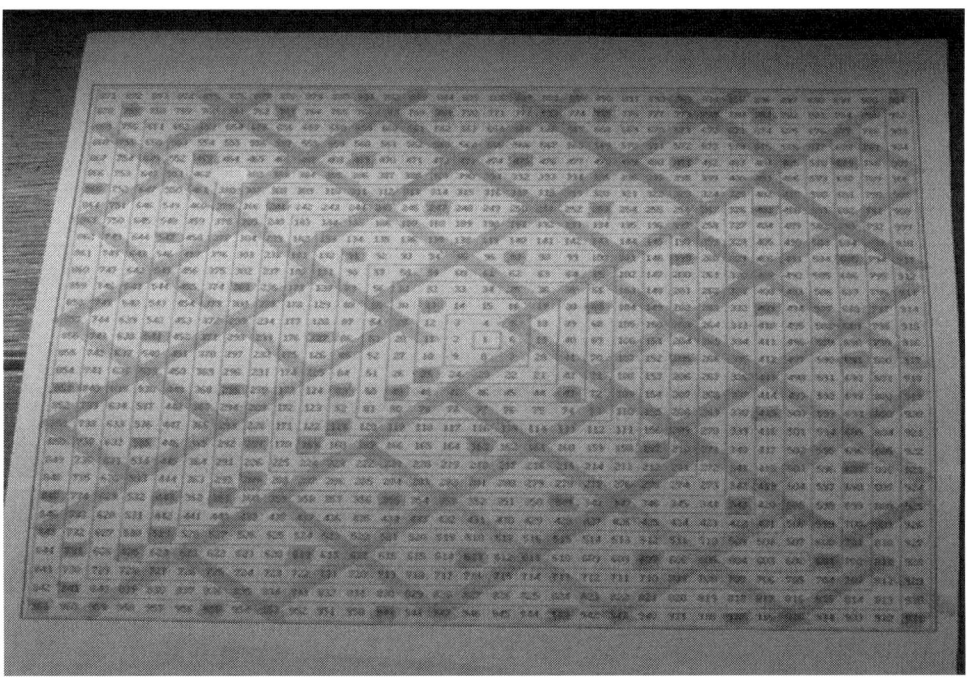

I have elucidated on how the decimal system of Number can be represented with universal ubiquity in physical reality. The magnetic containing infrastructure of the 3 6 and 9 (0) which allows electricity to play out its merry dance and represented by the 1 4 7 and 2 5 8 number groups that are exclusive to Prime Numbers.

By examining the way in which Hydrogen is distributed among the Bases which code for the group of Amino Acids common to all biology we can see the musicality of our DNA. It is very clearly a resonance and not something that could have progressed incrementally from some inanimate primordial soup. Not one Proton or Neutron can be displaced or removed. Yes, our DNA can adjust to prevailing environments to produce different effects to ensure survival but no to the idea of Darwinism and the theory of evolution within DNA. The group of Amino Acids that must be present to sustain all biology has never been any different than it is now. I have also shown how DNA is deeply connected to the Earth and how the group of Amino Acids may very well be linked to Group Theory via the Sporadic Groups and which may open up a complete understanding of all the higher dimensions of DNA.

Aside from obvious biological connections, the work on DNA has deep ramifications also in the area of Chemistry and shines a light on the work of the incredible Dr Walter Russell and in particular his octave wave theory of the chemical elements which I believe to be absolutely correct. The work I have done shows conclusively that there exists an octave relationship between Hydrogen and Carbon.

With Talal Ghannam's insights in his brilliant and brave book, The Mystery of Numbers, to hand, we can begin to understand and apply the Integer Partition Theory and make predictions about how the universal machine actually works. I am guessing at which mainstream scientific theories and frameworks may be applicable, for sure, but collaboration with the right people can advance this theory and explain everything. We live in a numerical construct, a machine, the algorithm for which, when fully interpreted, can liberate Man entirely. Because of the importance of this particular numerical sequence the discovery in 2011 by Dr Ken Ono of the generating function for this sequence takes on even more significance than solving a problem that mathematicians had previously thought to be impossible, even by the great Ramanujan who discovered patterns within it.

The Integer Partition Theory is the only theory to date which provides a definitive answer to the central mystery of natural philosophy, the fine structure constant and its value as the reciprocal of the number 137. Statistician I. J. Good argued that 'a numerological explanation would only be acceptable if it came from a more fundamental theory that also provided a Platonic explanation of the value.' I believe that I have done that, at least to some extent. That the measured numbers are very slightly different from those identified through scientific observation, is because it is my belief that this theory shows the ideal which can never be measured exactly from within the construct.

Science has been politicised and splintered to advance and protect vested interests and so that power through combined knowledge is reserved for the very few who have access to all of it. I am not naïve enough to believe that the scientific community will embrace my theory with open arms but hopefully it will catch the attention of a few who can see its elegance and potential.

There is so much more research that I could have included in this book but to do so would have meant it becoming too unwieldy and turgid for the average reader. For those of you with an appetite, my website at www.newunderstandings.com serves as a repository for all of it. Access to it is via a one-time subscription and has until now largely been used by collaborators. In due course, all of the research will be up there and will include any new work that I wish to make public.

I do hope you have enjoyed this book and have drawn your own conclusions from it. Far be it from me to tell you what to believe is true about the world we live in or if you must believe in one God or another. Religion is what Man has done to create self-aggrandising institutions for the purpose of control through hierarchy.

Upon reading this, if you have done work or know of work which you think is relevant to the overarching theory of order in the universe, and how it is evidenced here on Earth, then please do get in touch with me via anthony@marketharmony.com - I am sure there will be future editions in which your work may be included, with your permission.

Yours, most sincerely.

Anthony Morris 03/03/2018.

ABOUT THE AUTHOR

Anthony Morris was born on January 14th 1969, the elder of the two sons of the late Christopher, and Isabel Morris. He is married to Mary and they have three children, Anthony 19, Maria 16, and Samantha 10.

Anthony balances a portfolio career, analysing global financial markets since 1995 and working with individuals to improve their personal performance since 2003.

Performance Coach

Anthony is trained as a Clinical Hypnotherapist, trained in Ericksonian Hypnosis, Psychotherapy and Neuro Linguistics. He operates from 96 Harley Street, London and specialises in helping people all over the world to recover quickly from depression, stress and anxiety through the teaching of his own systematic approach to create effective mental resilience. These same strategies may also be utilized by anyone wishing to access more of their latent potential, their genius. In addition he loves to work with anyone that has a really big ambitious goal, particularly those in the sports and investment industries where Anthony's passions and direct experience lie. Further information is available at www.shineagentsofchange.com

Further; Anthony has produced an online course which is designed to replicate what he teaches his clients who are suffering from depression and anxiety which has become a pandemic in the modern era. To access this please visit www.shinecourses.com where you will see a video interview with Anthony talking about the course and where you can sign up for a free introductory course before deciding if you would like to take the main course which is called 'Surviving Your Self.'

2018 will see the introduction of a further course, 'Being Your Best You' which will employ largely the same strategies but focused on achieving personal performance excellence and in tandem with the 'Surviving Your Self' course will be offered as a complete solution in the corporate world, not only for dealing with acute problems presenting in the workplace but also as a proactive and pre-emptive strike against mental health issues that people suffer from. In addition, the 'Being Your Best You' course will be framed as an investment in the organisations employees, the only sustainable source of competitive advantage. Some of the more progressive companies understand that this will work its way into the bottom line through improved culture, communication, focus, creativity, productivity and ultimately cost savings. For more information please contact him at anthony@marketharmony.com

Financial Markets

Anthony has studied financial markets for over 20 years through the lens of a Harmonic Grid which has allowed him to time financial markets across all asset classes with great success. He came to the view that this apparent order cannot only be confined to financial markets and must in fact be ubiquitous to any oscillating data set, in the behavior of people, societies, the world and indeed the universe. The Numerical Universe is an attempt at revealing the underlying mathematical logic that supports this idea of order in the markets and order in the universe. Anthony has worked with some of the biggest names in investment management and currently writes 2 monthly newsletters; The Global Macro and Commodity Markets Digest. For more information, please contact him at anthony@marketharmony.com

Made in the USA
Lexington, KY
02 December 2019